Cosmology in Scalar-Tensor Gravity

Fundamental Theories of Physics

*An International Book Series on The Fundamental Theories of Physics:
Their Clarification, Development and Application*

Editor:
ALWYN VAN DER MERWE, *University of Denver, U.S.A.*

Editorial Advisory Board:
JAMES T. CUSHING, *University of Notre Dame, U.S.A.*
GIANCARLO GHIRARDI, *University of Trieste, Italy*
LAWRENCE P. HORWITZ, *Tel-Aviv University, Israel*
BRIAN D. JOSEPHSON, *University of Cambridge, U.K.*
CLIVE KILMISTER, *University of London, U.K.*
PEKKA J. LAHTI, *University of Turku, Finland*
ASHER PERES, *Israel Institute of Technology, Israel*
EDUARD PRUGOVECKI, *University of Toronto, Canada*
TONY SUDBURY, *University of York, U.K.*
HANS-JÜRGEN TREDER, *Zentralinstitut für Astrophysik der Akademie der
  Wissenschaften, Germany*

Cosmology in Scalar-Tensor Gravity

by

Valerio Faraoni
*University of Northern British Columbia,
Prince George, Canada*

KLUWER ACADEMIC PUBLISHERS
DORDRECHT / BOSTON / LONDON

A C.I.P. Catalogue record for this book is available from the Library of Congress.

ISBN 1-4020-1988-2

Published by Kluwer Academic Publishers,
P.O. Box 17, 3300 AA Dordrecht, The Netherlands.

Sold and distributed in North, Central and South America
by Kluwer Academic Publishers,
101 Philip Drive, Norwell, MA 02061, U.S.A.

In all other countries, sold and distributed
by Kluwer Academic Publishers,
P.O. Box 322, 3300 AH Dordrecht, The Netherlands.

Printed on acid-free paper

All Rights Reserved
© 2004 Kluwer Academic Publishers
No part of this work may be reproduced, stored in a retrieval system, or transmitted
in any form or by any means, electronic, mechanical, photocopying, microfilming, recording
or otherwise, without written permission from the Publisher, with the exception
of any material supplied specifically for the purpose of being entered
and executed on a computer system, for exclusive use by the purchaser of the work.

Printed in the Netherlands.

To Louine, Donovan,
and my parents

Contents

Preface		ix
Acknowledgments		xi
1. SCALAR-TENSOR GRAVITY		1
1	Introduction	1
2	Brans-Dicke theory	8
3	Brans-Dicke cosmology in the Jordan frame	10
4	The limit to general relativity	13
5	Relation to Kaluza-Klein theory	17
6	Brans-Dicke theory from Lyra's geometry	20
7	Scalar-tensor theories	22
7.1	Effective Lagrangians and Hamiltonians	27
8	Motivations for scalar-tensor theories	29
9	Induced gravity	31
10	Generalized scalar-tensor theories	32
11	Conformal transformation techniques	32
11.1	Conformal transformations	33
11.2	Brans-Dicke theory	38
11.3	Kaluza-Klein cosmology	41
11.4	Scalar-tensor theories	45
11.5	Generalized scalar-tensor theories	46
12	Singularities of the gravitational coupling	48
2. EFFECTIVE ENERGY-MOMENTUM TENSORS AND CONFORMAL FRAMES		55
1	The issue of the conformal frame	55
1.1	The first viewpoint	56

		1.2	The second viewpoint	61
		1.3	The third viewpoint	64
		1.4	Other viewpoints	69
		1.5	Einstein frame or Jordan frame?	70
		1.6	Energy conditions in relativistic theories	70
		1.7	Singularity theorems and energy conditions	72
	2	Effective energy-momentum tensors	74
		2.1	Time-dependence of the gravitational coupling	76
		2.2	Conservation equations for the various $T_{ab}^{(J)}[\phi]$	78
3.	GRAVITATIONAL WAVES			83
	1	Introduction		83
	2	Einstein frame scalar-tensor waves		84
		2.1	Gravitational waves in the Einstein frame	84
		2.2	Corrections to the geodesic deviation equation	85
	3	Gravitational lensing by scalar-tensor gravitational waves		87
		3.1	Jordan frame analysis	87
		3.2	Einstein frame analysis	90
		3.3	Propagation of light through a gravitational wave background	91
4.	EXACT SOLUTIONS OF SCALAR-TENSOR COSMOLOGY			93
	1	Introduction		93
	2	Exact solutions of Brans-Dicke cosmology		95
		2.1	$K = 0$ FLRW solutions	96
		2.1.1	The O'Hanlon and Tupper solution	96
		2.1.2	The Brans-Dicke dust solution	97
		2.1.3	The Nariai solution	98
		2.1.4	Other solutions with cosmological constant	100
		2.1.5	Generalizing Nariai's solution	101
		2.1.6	Phase space analysis for $K = 0$ and $V(\phi) = 0$	102
		2.1.7	Phase plane analysis for $K = 0$ and $V(\phi) = \Lambda\phi$	103
		2.2	$K = \pm 1$ solutions and phase space for $V = 0$	104
		2.3	Phase space for any K and $V = m^2\phi^2/2$	108
		2.3.1	The Dehnen-Obregon solution	108
		2.4	Bianchi models	109
		2.4.1	Bianchi V universes	109
	3	Exact solutions of scalar-tensor theories		110

5.	THE EARLY UNIVERSE	115
	1 Introduction	115
	2 Extended inflation	116
	2.1 The original extended inflationary scenario	117
	2.2 Alternatives	118
	3 Hyperextended inflation	120
	4 Real inflation?	121
	5 Constraints from primordial nucleosynthesis	122
6.	PERTURBATIONS	127
	1 Introduction	127
	2 Scalar perturbations	128
	3 Tensor perturbations	136
7.	NONMINIMAL COUPLING	143
	1 Introduction	143
	1.1 Generalized inflation	145
	1.2 Motivations for nonminimal coupling	146
	1.3 Which value of ξ?	147
	2 Effective energy-momentum tensors	150
	2.1 Approach à la Callan-Coleman-Jackiw	151
	2.2 Effective coupling	152
	2.3 A mixed approach	154
	2.4 Discussion	156
	2.5 Energy conditions in FLRW cosmology	157
	2.6 Nonminimal coupling and gravitational waves	159
	3 Conformal transformations	160
	4 Inflation and $\xi \neq 0$: the unperturbed universe	164
	4.1 Necessary conditions for generalized inflation	165
	4.1.1 Specific potentials	166
	4.2 The effective equation of state with nonminimal coupling	169
	4.3 Critical values of the scalar field	171
	5 The slow-roll regime of generalized inflation	173
	5.1 Derivation of the stability conditions	178
	5.2 Slow-roll parameters	182
	6 Inflation and $\xi \neq 0$: perturbations	183
	6.1 Density perturbations	184

		6.2	Tensor perturbations	191
	7	Conclusion		192
8.	THE PRESENT UNIVERSE			197
	1	Present acceleration of the universe and quintessence		197
		1.1	Coupled quintessence	205
		1.2	Multiple field quintessence	207
		1.3	Falsifying quintessence models	207
	2	Quintessence with nonminimal coupling		208
		2.1	Models using the Ratra-Peebles potential	209
		2.2	Necessary conditions for accelerated expansion	210
		2.3	Doppler peaks with nonminimal coupling	212
	3	Superquintessence		213
		3.1	An exact superaccelerating solution	217
		3.2	Big Smash singularities	218
	4	Quintessence in scalar-tensor gravity		221
	5	Conclusion		226

References	229
Index	265

Preface

Scalar-tensor gravity was introduced as a definite theory in the early 1960's owing to cosmological reasons and, since its inception, has been consistently applied to cosmology. Since then, string theorists using a different approach have discovered many similarities between string and scalar-tensor theory, prompting further investigation of the latter. Such independent motivations and approaches have spurred the interest of many researchers in scalar-tensor gravity resulting in rapid growth in this area. As a consequence it has become increasingly difficult for the researcher to find his or her own way through the vast literature. This book constitutes the first attempt to review the subject. At present, there is no such overview in the literature on cosmology.

The book is primarily of interest to researchers and postgraduate students in theoretical cosmology, general relativity, alternative theories of gravity, and the phenomenology of string theories; and secondarily valuable to all readers interested in theoretical physics and astrophysics at the graduate level.

Although the main purpose of this book is to provide a map of the field, another primary purpose is to address methodological issues in scalar-tensor gravity and to focus attention on problems that have appeared relatively recently, and are most likely to occupy the researchers for a long time. An extensive bibliography guides the reader to more detailed literature on particular subjects and provides a useful reference for future research in scalar-tensor gravity.

While discussing the applications of scalar-tensor gravity to cosmology, the first three chapters of the book retain a fairly general point of view. The remaining chapters adopt the specific viewpoint of the cosmologist and foray into more specialized issues. Among the subjects treated are the distinguishing features of scalar-tensor gravity, effective energy-momentum tensors, gravitational waves in scalar-tensor cosmology, specific scalar-tensor theories, exact cosmological solutions and cosmological perturbations, scenarios of inflation in the early universe and quintessence in the present universe. Topics such as the limit to general relativity and the issue of the conformal frame — found only in research articles and somewhat controversial — are discussed critically. The last chapter on scalar-tensor models of quintessence is a snapshot of current research in a field that is evolving very quickly and brings the reader in contact with recent ideas on quintessence and the acceleration of the universe at the time of writing. The issue of which of the current models (if any) of quintessence correctly describes our cosmos has far-reaching consequences for the ultimate fate of the universe.

Prince George

November 2003

VALERIO FARAONI

Acknowledgments

I am grateful to many people with whom I had discussions — systematic or occasional — but always useful, and to all the colleagues who brought up interesting points. Among others, I wish to thank S. Sonego, the late D.W. Sciama, G.F.R. Ellis, M. Bruni, B. Bertotti, A. Labeyrie, S. Bellucci, Y.M. Cho, G. Esposito-Farèse, F. Occhionero, C. Baccigalupi, S. Matarrese, S. Odintsov, V. Sahni, L.H. Ford, E. Calzetta, S. Liberati, V. Mukhanov, W. Israel, S. Mukohyama, J. Cervantes-Cota, R. Fakir, M. Castagnino, R. Casadio, F.I. Cooperstock, C. Bracco, R. Easther, L. Brenig, and P. Teyssandier.

Support during the preparation of this book came from my wife Louine, and inspiration from my son Donovan. At age eighteen months, he was curious about the world and physics, always looking for particles on the floor and sticking his finger in every black hole he could find. He also showed an early interest in all kinds of strings.

Finally, I wish to thank the editor of the Kluwer Academic series *Fundamental Theories of Physics*, Professor Alwyn van der Merwe, for his invitation and encouragement to write this work, and Doctor Sabine Freisem and the staff from Kluwer Academic Publishers for their friendly advice and assistance through the writing.

Chapter 1

SCALAR-TENSOR GRAVITY

1. Introduction

Nature's simplest field, the scalar field, has been the protagonist of physical theories for a long time. Even before the publication of the theory of general relativity G. Nordström formulated a conformally flat scalar theory of gravity in 1912 [673, 674, 675, 676] which was regarded by Einstein as a serious competitor to general relativity for some time [312]. Today, a distinguishing feature of modern scalar-tensor cosmology is that the gravitational coupling is time-dependent: this idea dates back to the work of Dirac in 1937 related to the large number hypothesis ([293, 294, 295], see also Refs. [201, 549, 843, 85]). Dirac noticed that dimensionless combinations of cosmological constants and fundamental physical constants are connected by a relation that arises naturally, provided that one of the "constants" is allowed to change over cosmological timescales. Dirac's choice was to let the gravitational coupling G become time-dependent, while keeping the other fundamental constants fixed [295]. Next, the development of this idea by P. Jordan in the following decade took the form of a complete gravitational theory in which G was promoted to the role of a gravitational scalar field [519, 521]. From an effective point of view, the latter somehow behaves as a form of matter and therefore satisfies a generalized conservation equation built into the theory [519, 521].

Finally, the idea of a scalar-tensor theory reached full maturity with the work of Brans and Dicke [152]. In 1961 they published a new theory that was to become the prototype of the alternative theory to Einstein's general relativity. The theory contains a scalar field which describes gravity together with the metric tensor familiar from general relativity.

The primary motivation for searching for a new gravitational theory comes from cosmology, with the new theory explicitly incorporating Mach's principle, which is seen as left out or marginalized in Einstein's theory. The latter contains some — but not all — aspects of Mach's principle, and even has solutions (plane-fronted waves with parallel rays) that are regarded as explicitly anti-Machian [688]. The gravitational scalar field is the inverse of the gravitational coupling which is constant in general relativity. In Brans-Dicke theory, however, the gravitational coupling is variable and is determined by all matter in the universe. A distinguishing feature of the theory is that the cosmological distribution of matter affects local gravitational experiments.

The new theory of Brans and Dicke has been followed by many attempts (some successful and some not) to formulate other theories of gravity alternative to general relativity [806, 909]. The deviations of these theories from general relativity are forced to be small by present-day experiments performed in the Solar System [909, 910]. Among the many alternative theories of gravity proposed in the literature, we are interested in scalar-tensor theories, which generalize Brans and Dicke's theory. Scalar-tensor theories include Brans-Dicke theory, the theory of a scalar field nonminimally coupled to the Ricci curvature of spacetime and induced gravity.

In no other area of gravitational physics are scalar-tensor theories as important as they are in cosmology because, although the deviations from general relativity are extremely small today, the situation may change going into the future or extrapolating back to the past. Since the natural time scale for the evolution of the scalar field is cosmological, deviations from general relativity manifest themselves in the context of cosmology. For this reason, the discussion of scalar-tensor theories in this book is restricted to the aspects relevant for cosmology.

Interest in Brans-Dicke and scalar-tensor gravity declined in the 1970s, probably due to the more and more stringent constraints imposed on these theories by Solar System experiments. The observational limits can only be satisfied by assuming large values of $|\omega|$ where ω is a free parameter contained in the theory of Brans and Dicke which is naturally expected to be of order unity. The lower bound on $|\omega|$ kept becoming larger and larger as the experiments became more and more accurate, which corresponds to fine-tuning ω to satisfy the observational limits. This period of abandonment of scalar-tensor gravity was followed by a surge of interest owing to the new importance of scalar fields in modern unified theories, in particular the dilaton of string theories. The low-energy limit of the bosonic string theory yields a Brans-Dicke theory with parameter $\omega = -1$ and scalar-tensor gravity exhibits certain similarities to supergravity and string theory (apart from pointing out these similarities, string theories are not discussed in this book). At the classical level, Brans-Dicke theory can be derived from classical Kaluza-Klein theory, after compactification of the extra spatial dimensions, albeit only for a discrete set of values of the coupling parameter ω. Dimensional reduction is a common feature of modern unified theories, and the simple case of Kaluza-Klein theory is treated in this book, in order to provide the reader with an idea of the physical reasons why compactification of the extra dimensions generates scalar-tensor gravity. A gravitational scalar field is an essential feature of supergravity, superstring, and M-theories.

Another reason for renewed interest in Brans-Dicke and scalar-tensor theories is the fact that plausible mechanisms have been discovered at the classical level that allow the parameter ω, which is variable in scalar-tensor gravity, to assume values of order unity in the early universe and diverge later, thus reducing gravity to general relativity at later epochs in the history of the universe. Finally, renewed interest in scalar-tensor gravity is also due to the use of these theories in inflationary scenarios of the universe, following the larger and larger consensus obtained by the idea of inflation after Guth's and Linde's seminal papers in the early 1980s. While the attractor solution of general relativity in phase space — the de Sitter exponentially expanding universe — fails to provide a graceful exit from an inflationary epoch, its analogue in Brans-Dicke theory — a power-law attractor solution — does the job. More recently, the astonishing discovery that the universe is undergoing an accelerated expansion today, and that approximately 70% of the energy of the universe is in a form with negative pressure (*dark energy*), has spurred interest in scalar-tensor theories which account for such features with a long-range gravitational scalar field. The last chapter of this book is devoted to a discussion of the present universe in scalar-tensor models.

In the weak field limit when viable alternative theories of gravity approach general relativity, the differences between these theories become small, and one can deal with most alternative theories in a unified way by using the Parametrized Post-Newtonian (PPN) formalism [909]. No analogue of this formalism exists in cosmology, where the strong field regime makes these theories differ greatly from one another at early times in the history of the universe and possibly also in the far future. The detection of cosmological effects peculiar to scalar-tensor gravity would deepen our understanding of the properties of space and time in strong gravitational fields.

The purpose of this book is to introduce the reader who already has some familiarity with general relativity and cosmology to the many facets of scalar-tensor cosmology and its methods. Rather than presenting a catalogue of exact solutions and mathematical techniques, the focus is on the physics and the reader is referred to the original papers for details. Different points of view are presented on issues that are still controversial and on topics rarely covered in the literature at the time of writing.

Notations and conventions

The following notations and conventions are used in this book. The metric signature is $-+++$ (or $-++...+$ in $n > 4$ spacetime dimensions). For ease of comparison with the other literature, units are used in which the speed of light c and the reduced Planck constant $\hbar$ assume the value unity. G is Newton's constant, the Planck mass is $m_{pl} = G^{-1/2}$ in these units, and $\kappa \equiv 8\pi G$. Small Latin indices assume the values 0, 1, 2, 3, while Greek indices assume the values 1, 2, 3. While allowing for $(4+d)$ spacetime dimensions (only one of which is timelike) in the discussion of Kaluza-Klein and string theories prior to compactification, the value $d = 0$ is assumed in the rest of this book. Capital Latin indices assume the values 0, 1, 2, 3, ..., (3+d) in higher-dimensional gravity.

A comma denotes ordinary differentiation, ∇_a is the covariant derivative operator, and g denotes the determinant of the metric tensor g_{ab}, while η_{ab} is the Minkowski metric, which takes the form $\mathrm{diag}(-1,1,1,1)$ in Cartesian coordinates in four dimensions. ϵ^{abcd} is the totally antisymmetric Levi-Civita tensor. Round and square brackets around indices denote symmetrization and antisymmetrization, respectively, which include division by the number of permutations of the indices, for example:

$$A_{(ab)} \equiv \frac{A_{ab} + A_{ba}}{2}, \qquad A_{[ab]} \equiv \frac{A_{ab} - A_{ba}}{2}. \tag{1.1}$$

The Riemann and Ricci tensors are given in terms of the Christoffel symbols Γ^d_{ab} by

$$R_{abc}{}^d = \Gamma^d_{ac,b} - \Gamma^d_{bc,a} + \Gamma^e_{ac}\Gamma^d_{eb} - \Gamma^e_{bc}\Gamma^d_{ea}, \tag{1.2}$$

$$R_{ac} \equiv R_{abc}{}^b = \Gamma^b_{ac,b} - \Gamma^b_{bc,a} + \Gamma^d_{ac}\Gamma^e_{de} - \Gamma^d_{bc}\Gamma^b_{da}, \tag{1.3}$$

and $R \equiv g^{ab} R_{ab}$ is the Ricci curvature. $\Box \equiv g^{cd}\nabla_c\nabla_d$ is d'Alembert's operator. A tilde denotes quantities defined in the Einstein frame, while a caret denotes quantities defined in a higher-dimensional space prior to the compactification of the extra dimensions. Apart from Section 1.6, torsion is not considered in this book. A subscript 0 usually denotes quantities evaluated at the present instant of time in the dynamics of the universe.

Equations of big bang and inflationary cosmology

The standard equations of big bang and inflationary cosmology in general relativity are recalled for comparison with the corresponding equations in scalar-tensor gravity. The reader is referred to standard textbooks (e.g., [898, 892, 541, 569]) for a comprehensive treatment.

The Friedmann-Lemaitre-Robertson-Walker metric (hereafter "FLRW") is

$$ds^2 = -dt^2 + a^2(t) \left[\frac{dr^2}{1 - Kr^2} + r^2 \left(d\theta^2 + \sin^2\theta\, d\varphi^2 \right) \right] \quad (1.4)$$

in comoving polar coordinates (t, r, θ, φ), where the normalized curvature index K can assume the values 0 or ± 1. The Einstein field equations in this metric reduce to the ordinary differential equations for the scale factor $a(t)$

$$H^2 = \frac{\kappa}{3}\rho - \frac{K}{a^2}, \quad (1.5)$$

$$\frac{\ddot{a}}{a} = \dot{H} + H^2 = -\frac{\kappa}{6}(\rho + 3P), \quad (1.6)$$

where an overdot denotes differentiation with respect to the comoving time t, $H \equiv \dot{a}/a$ is the Hubble parameter, ρ and P are the energy density and pressure of the cosmic fluid, respectively. A cosmological constant Λ, if present, can be described as a perfect fluid with density and pressure

$$P^{(\Lambda)} = -\rho^{(\Lambda)} - \frac{\Lambda}{\kappa}. \quad (1.7)$$

The energy density and pressure satisfy the conservation equation

$$\dot{\rho} + 3H(P + \rho) = 0, \quad (1.8)$$

which follows from the covariant conservation equation $\nabla^b T_{ab} = 0$ for the energy-momentum tensor of the cosmic fluid. A critically open ($K = 0$) universe has energy density (*critical density*)

$$\rho_c = \frac{3H^2}{\kappa} \quad (1.9)$$

and one can express the energy density $\rho^{(i)}$ of the i-th component of a multi-component cosmic fluid in a FLRW universe in units of the critical density by introducing the corresponding Ω-parameter

$$\Omega^{(i)} \equiv \frac{\rho^{(i)}}{\rho_c}. \quad (1.10)$$

Since $\rho = \sum_i \rho^{(i)}$, the Hamiltonian constraint (1.5) in a $K=0$ universe can be written as

$$1 = \sum_i \Omega^{(i)} \,. \tag{1.11}$$

In standard inflationary theories, the universe is dominated by a scalar field ϕ minimally coupled to the Ricci curvature and with potential energy density $V(\phi)$; standard inflation corresponds to acceleration of the universe $\ddot{a} > 0$ and is described by the action

$$S = \int d^4x \sqrt{-g} \left[\frac{R}{2\kappa} - \frac{1}{2} g^{cd} \nabla_c \phi \nabla_d \phi - V(\phi) \right] \,, \tag{1.12}$$

where no matter action is included other than ϕ, since the inflaton ϕ dominates the dynamics during inflation. The stress-energy tensor of ϕ assumes the canonical form

$$T_{ab} = \nabla_a \phi \nabla_b \phi - \frac{1}{2} g_{ab} \nabla^c \phi \nabla_c \phi - g_{ab} V(\phi) \,, \tag{1.13}$$

which can be put in the form of the energy-momentum tensor of a perfect fluid with energy density and pressure (with respect to a comoving observer with four-velocity u^a)

$$\rho = T_{ab} u^a u^b = \frac{\left(\dot{\phi}\right)^2}{2} + V(\phi) \,, \tag{1.14}$$

$$P = T_{ab} h^{ab} = \frac{T_{ii}}{g_{ii}} = \frac{\left(\dot{\phi}\right)^2}{2} - V(\phi) \,, \tag{1.15}$$

where $h_a{}^b$ is the projection operator on the three-dimensional spatial sections defined by $h_{ab} \equiv g_{ab} + u_a u_b$. The scalar field potential $V(\phi)$ is usually taken to be positive because during inflation, in which $\dot{\phi} \simeq 0$, $\rho \simeq V(\phi)$ is dominated by the potential of the scalar and is therefore required to be non-negative. This requirement follows from the known properties of matter and is needed in order to produce a regime with $\ddot{a} > 0$. The Friedmann equations (1.5) and (1.6) assume the form

$$H^2 = \frac{\kappa}{3} \left[\frac{\left(\dot{\phi}\right)^2}{2} + V \right] - \frac{K}{a^2} \,, \tag{1.16}$$

$$\dot{H} = -H^2 - \frac{\kappa}{3} \left[\left(\dot{\phi}\right)^2 - V \right] \,, \tag{1.17}$$

while the scalar ϕ satisfies the Klein-Gordon equation

$$\ddot{\phi} + 3H\dot{\phi} + \frac{dV}{d\phi} = 0 \,. \tag{1.18}$$

In practice the curvature term can be omitted from eq. (1.16) since, even starting from anisotropic initial conditions, the universe is taken extremely closely to a spatially flat FLRW universe in most models of inflation with a single scalar field and without fine-tuning (*cosmic no-hair*). With a few exceptions, $K = 0$ is a robust prediction of inflation [569].

Only two of the three equations (1.16)-(1.18) are independent. In fact, if $\dot{\phi} \neq 0$, the Klein-Gordon equation (1.18) can be derived from the conservation equation (1.8) or, equivalently, from eqs. (1.16) and (1.17).

The equations of inflation (1.16)-(1.18) and those satisfied by cosmological perturbations are solved in the *slow-roll approximation*

$$\frac{\left(\dot{\phi}\right)^2}{2} << V \,, \qquad \ddot{\phi} << H\dot{\phi} \,, \tag{1.19}$$

in which eqs. (1.16)-(1.18) reduce to

$$H^2 \simeq \frac{\kappa V}{3} \,, \tag{1.20}$$

$$3H\dot{\phi} \simeq -\frac{dV}{d\phi} \,. \tag{1.21}$$

The slow-roll approximation assumes that the solution $\phi(t)$ of the field equations rolls slowly (i.e., with negligible $\ddot{\phi}$) over a shallow section of the scalar field potential $V(\phi)$, which then mimics the effect of a cosmological constant $\Lambda \simeq V(\phi)$. As a consequence, the expansion of the universe is almost de Sitter-like,

$$a(t) = a_0 \exp\left[H(t)\, t\right] \,, \tag{1.22}$$

where

$$H(t) = H_0 + H_1\, t + \ldots \tag{1.23}$$

and the constant term H_0 dominates in the expression of $H(t)$. There are de Sitter solutions $H =$const., $\phi =$const., that are attractors in the phase space of the dynamical system (1.16)-(1.18), which justifies the slow-roll approximation. Inflation stops when the potential $V(\phi)$ ends its plateau and quickly decreases to a zero minimum. Then ϕ quickly accelerates toward this minimum, overshoots it, and oscillates around it. These oscillations are damped by particle creation due to the explicit coupling of ϕ to other fields or to the Ricci curvature (*reheating*), dissipating the kinetic energy of ϕ and raising the temperature of the universe after the inflationary expansion. The superluminal expansion — which lasts long enough to allow for approximately 60 e-folds of cosmic expansion — solves the flatness, horizon and monopole problems of the standard big bang model and provides a mechanism for the generation of density perturbations. These are generated by quantum fluctuations of the scalar field (accompanied by gravitational waves consisting of fluctuations of the metric tensor) and seed the formation of large scale structures. The latter begin to grow after the end of the inflationary and radiation eras. Inflation ends when the slow-roll parameters

$$\epsilon \equiv \frac{1}{2\kappa}\left(\frac{V'}{V}\right)^2 , \qquad \eta \equiv \frac{1}{\kappa}\frac{V''}{V} \qquad (1.24)$$

(where a prime denotes differentiation with respect to ϕ) become of order unity.

2. Brans-Dicke theory

The Jordan-Fierz-Brans-Dicke theory of gravity [519, 521, 375, 152] commonly referred to as Brans-Dicke (hereafter "BD") theory is the prototype of gravitational theories alternative to general relativity. The action in the so-called Jordan frame is

$$S^{(BD)} = \frac{1}{16\pi}\int d^4x\,\sqrt{-g}\left[\phi R - \frac{\omega}{\phi}g^{cd}\nabla_c\phi\nabla_d\phi - V(\phi)\right] + S^{(m)} , \qquad (1.25)$$

where

$$S^{(m)} = \int d^4x\,\sqrt{-g}\,\mathcal{L}^{(m)} \qquad (1.26)$$

is the action describing ordinary matter (i.e., any form of matter different from the scalar field ϕ) and ω is a dimensionless parameter. The factor ϕ in the denominator of the second term in brackets in the action (1.25) is introduced to make ω dimensionless. Matter is not directly coupled to ϕ, in the sense that the Lagrangian density $\mathcal{L}^{(m)}$ does not depend on ϕ — "minimal coupling" for matter — but ϕ is directly coupled to the Ricci scalar. The gravitational field is described by the metric tensor g_{ab} and by the BD scalar field ϕ, which together with the matter variables describe the dynamics. The scalar field potential $V(\phi)$ constitutes a natural generalization of the cosmological constant and may reduce to a constant, or to a mass term; it is sometimes included when BD theory is used in theories of the early universe or in quintessential scenarios of the present universe. Although free scalar fields are legitimate in classical gravitational theories, they are uncommon in high energy physics because usually quantum corrections produce interactions leading to potentials $V(\phi)$.

The original motivation for the introduction of BD theory was the search for a theory containing Mach's principle, which is not completely or explicitly embodied in general relativity. A local problem such as the study of the BD equivalent of the Schwarzschild solution of general relativity takes into account the cosmological matter distributed in the universe, and that the gravitational coupling is generated by this matter through the cosmological field ϕ. By contrast, in general relativity the gravitational coupling is constant and one can consider the Schwarzschild solution by neglecting the rest of the universe.

By varying the action (1.25) with respect to g_{ab} and using the well known properties [555]

$$\delta\left(\sqrt{-g}\right) = -\frac{1}{2}\sqrt{-g}\, g_{ab}\, \delta g^{ab}, \tag{1.27}$$

$$\delta\left(\sqrt{-g}\, R\right) = \sqrt{-g}\left(R_{ab} - \frac{1}{2}g_{ab}R\right)\delta g^{ab} \equiv \sqrt{-g}\, G_{ab}\, \delta g^{ab}, \tag{1.28}$$

where G_{ab} is the Einstein tensor, one obtains the field equation

$$\begin{aligned}G_{ab} =\ & \frac{8\pi}{\phi}T_{ab}^{(m)} + \frac{\omega}{\phi^2}\left(\nabla_a\phi\nabla_b\phi - \frac{1}{2}g_{ab}\nabla^c\phi\nabla_c\phi\right) \\ & + \frac{1}{\phi}\left(\nabla_a\nabla_b\phi - g_{ab}\,\Box\,\phi\right) - \frac{V}{2\phi}g_{ab},\end{aligned} \tag{1.29}$$

where

$$T_{ab}^{(m)} \equiv \frac{-2}{\sqrt{-g}}\frac{\delta}{\delta g^{ab}}\left(\sqrt{-g}\,\mathcal{L}^{(m)}\right) \tag{1.30}$$

is the stress-energy tensor of ordinary matter. Variation of the action with respect to ϕ yields

$$\frac{2\omega}{\phi}\Box\phi + R - \frac{\omega}{\phi^2}\nabla^c\phi\nabla_c\phi - \frac{dV}{d\phi} = 0. \tag{1.31}$$

By taking the trace of eq. (1.29),

$$R = \frac{-8\pi T^{(m)}}{\phi} + \frac{\omega}{\phi^2}\nabla^c\phi\nabla_c\phi + \frac{3\,\Box\,\phi}{\phi} + \frac{2V}{\phi}, \tag{1.32}$$

and using the resulting eq. (1.32) to eliminate R from eq. (1.31), one obtains

$$\Box\phi = \frac{1}{2\omega + 3}\left[8\pi T^{(m)} + \phi\frac{dV}{d\phi} - 2V\right]. \tag{1.33}$$

As made clear by eq. (1.33), the scalar ϕ has non-conformal matter (i.e., matter with trace $T^{(m)} \equiv T^{(m)a}{}_a \neq 0$) as its source, but the scalar is not directly coupled to $T^{(m)}_{ab}$ or $\mathcal{L}^{(m)}$. The field ϕ acts back on ordinary matter only through the metric tensor g_{ab} in the manner described by eq. (1.29). The term proportional to $\phi \, dV/d\phi - 2V$ in the right hand side of eq. (1.33) vanishes if the potential $V(\phi)$ consists of a pure mass term.

The form of the action (1.25) or of the field equation (1.29) suggests that the BD field ϕ plays the role of the inverse of the gravitational coupling,

$$G_{eff}(\phi) = \frac{1}{\phi}, \qquad (1.34)$$

which becomes a function of the spacetime point. For this reason, the range of values $\phi > 0$ corresponding to attractive gravity is usually chosen. The BD parameter ω in the action (1.25) is a free parameter of the theory: from the theoretical point of view, a value of ω of order unity (*gravitational coupling strength*) would be natural, and it does appear in the low-energy limit of string theories. However, values of ω of this order are excluded by the available tests of gravitational theories in the weak field limit, for a massless and for a light scalar field ϕ. A light scalar is one that has a range larger than the size of the Solar System or of the laboratory used to test gravity.

Apart from the exceptions discussed in Section 1.4, the larger the value of ω, the closer BD gravity is to general relativity. Time delay experiments in the Solar System [731, 909] place the often quoted constraint $\omega \geq 500$ on this parameter. A more recent lower bound on ω is $\omega > 3300$ [910].

Although the theory is certainly viable in the limit of large ω — in which it reduces to general relativity — many physicists feel that the large value of ω required to satisfy the experimental constraints amounts to a fine-tuning that makes BD theory unattractive from the physical point of view. Renewed interest in BD theory and its generalizations, collectively known as scalar-tensor theories, arose with the discovery that BD theory is closely related to the Kaluza-Klein compactification of extra spatial dimensions, a feature common to all modern unified theories. Furthermore, the low energy limit of the gravitational sector of bosonic string theory yields a BD theory with $\omega = -1$ (see Section 1.8).

An alternative formulation of BD theory in the so-called Einstein conformal frame [290] is often discussed in the literature, and is presented in Section 1.11.2.

3. The equations of Brans-Dicke cosmology in the Jordan frame

In this section we derive the BD field equations for a FLRW universe described by the metric (1.4). The BD scalar ϕ only depends on the cosmic time t; in this metric one has

$$\nabla^c \phi \nabla_c \phi = -\left(\dot{\phi}\right)^2 , \qquad (1.35)$$

$$\Box \phi = -\left(\ddot{\phi} + 3H\dot{\phi}\right) = -\frac{1}{a^3}\frac{d}{dt}\left(a^3 \dot{\phi}\right) . \qquad (1.36)$$

Under the assumption that the stress-energy tensor of ordinary matter (1.30) assumes the form corresponding to a perfect fluid with energy density $\rho^{(m)}$ and pressure $P^{(m)}$,

$$T_{ab}^{(m)} = \left(P^{(m)} + \rho^{(m)}\right) u_a u_b + P^{(m)} g_{ab} , \qquad (1.37)$$

the time-time component of the BD field equation (1.29) yields the constraint equation

$$H^2 = \frac{8\pi}{3\phi}\rho^{(m)} + \frac{\omega}{6}\left(\frac{\dot{\phi}}{\phi}\right)^2 - H\frac{\dot{\phi}}{\phi} - \frac{K}{a^2} + \frac{V}{6\phi} , \qquad (1.38)$$

which provides a first integral. The trace equation (1.32) yields, using $R = 6\left(\dot{H} + 2H^2 + K/a^2\right)$

$$\dot{H} + 2H^2 + \frac{K}{a^2} = -\frac{4\pi T^{(m)}}{3\phi} - \frac{\omega}{6}\left(\frac{\dot{\phi}}{\phi}\right)^2 + \frac{1}{2}\frac{\Box \phi}{\phi} + \frac{V}{3\phi} . \qquad (1.39)$$

By using eqs. (1.38), (1.33) and the expression of the trace

$$T^{(m)} = 3P^{(m)} - \rho^{(m)} , \qquad (1.40)$$

one obtains

$$\begin{aligned}\dot{H} =\ & \frac{-8\pi}{(2\omega + 3)\phi}\left[(\omega + 2)\rho^{(m)} + \omega P^{(m)}\right] - \frac{\omega}{2}\left(\frac{\dot{\phi}}{\phi}\right)^2 \\ & + 2H\frac{\dot{\phi}}{\phi} + \frac{K}{a^2} + \frac{1}{2(2\omega + 3)\phi}\left(\phi\frac{dV}{d\phi} - 2V\right) .\end{aligned} \qquad (1.41)$$

In addition, eq. (1.33) for the scalar field reduces to

$$\ddot{\phi} + 3H\dot{\phi} = \frac{1}{2\omega + 3}\left[8\pi\left(\rho^{(m)} - 3P^{(m)}\right) - \phi\frac{dV}{d\phi} + 2V\right] . \qquad (1.42)$$

The combination $2V - \phi \, dV/d\phi$ in eqs. (1.41) and (1.42) vanishes when $V(\phi) = m^2 \phi^2/2$ and in this case the scalar field mass does not directly affect the dynamics of ϕ, but is only felt through the scale factor which depends on V through eq. (1.38). When $V = 0$ or $m^2 \phi^2/2$ and in the absence of ordinary matter, or in the presence of a radiative fluid with equation of state $P^{(m)} = \rho^{(m)}/3$, eq. (1.42) can be integrated, yielding the solution $\phi = $const. or the equation

$$\dot{\phi} = \frac{C}{a^3}, \tag{1.43}$$

where C is an integration constant.

Equations (1.38) and (1.42) can be regarded as the two independent equations regulating the dynamics of spatially homogeneous and isotropic BD cosmology. When ordinary matter is described by a perfect fluid stress-energy tensor $T^{(m)}_{ab}$, the conservation equation $\nabla^b T^{(m)}_{ab} = 0$ becomes

$$\left(\rho^{(m)}\right)^{\cdot} + 3H \left(\rho^{(m)} + P^{(m)}\right) = 0. \tag{1.44}$$

Eq. (1.42) is of second order in $\phi(t)$, while eqs. (1.38) and (1.44) are of first order in a, ϕ and ρ, and one must specify an equation of state to solve them. Usually, the equation of state

$$P^{(m)} = (\gamma - 1) \rho^{(m)} \tag{1.45}$$

with $\gamma =$constant is assigned, and the equations solved. Then, one immediately integrates eq. (1.44) obtaining

$$\rho^{(m)} = \frac{C'}{a^{3\gamma}}, \tag{1.46}$$

where C' is an integration constant.

The contribution of the scalar field ϕ to the right hand side of eq. (1.29) is often described as that of another cosmic fluid. One cannot assign *a priori* a constant equation of state for this ϕ-fluid since ϕ is a dynamical variable and its effective equation of state is time-dependent. In practice, one prescribes a form of the potential $V(\phi)$ instead of specifying an equation of state for the ϕ-fluid. From the conceptual point of view, ϕ should be regarded as part of the gravitational field and not as a form of matter. The interpretation of ϕ as an extra matter field disguises a forced interpretation of the BD field equations as effective Einstein equations, while they actually are different field equations of a theory that is not general relativity.

When H, ϕ and $\rho^{(m)}$ are chosen as dynamical variables, the initial data set $\left(H_0, \phi_0, \dot{\phi}_0, \rho_0^{(m)}\right)$ must be specified in addition to the parameters

ω, K and the function $V(\phi)$. By contrast, in general relativity one only needs the initial data $\left(H_0, \rho_0^{(m)} \right)$.

The equations for spatially flat FLRW models in vacuum ($T_{ab}^{(m)} = 0$) BD cosmology enjoy a duality symmetry [576, 577]. By redefining the scale factor and the BD field as follows,

$$\alpha \equiv \ln a , \tag{1.47}$$

$$\Phi \equiv -\ln(G\phi) , \tag{1.48}$$

the duality transformation takes the form [576, 577]

$$\alpha \longrightarrow \left(\frac{3\omega + 2}{3\omega + 4} \right) \alpha - 2 \left(\frac{\omega + 1}{3\omega + 4} \right) \Phi , \tag{1.49}$$

$$\Phi \longrightarrow -\left(\frac{6}{3\omega + 4} \right) \alpha - \left(\frac{3\omega + 2}{3\omega + 4} \right) \Phi . \tag{1.50}$$

This transformation generalizes the scale factor duality present in the effective action of string theories [878, 420, 859, 435]

$$\alpha \longrightarrow -\alpha , \tag{1.51}$$

$$\Phi \longrightarrow \Phi - 6\alpha . \tag{1.52}$$

This duality is reproduced by eqs. (1.49) and (1.50) for $\omega = -1$: the value of the BD parameter that yields low-energy string theory.

4. The limit to general relativity

As stated in the previous section, in the $\omega \longrightarrow \infty$ limit of the BD parameter BD theory becomes indistinguishable from general relativity. Although this is the standard belief [898, 909] and indeed it is true in most situations, the statement is not valid in general. Several exact solutions of BD theory fail to yield the corresponding general relativistic solution in the $\omega \to \infty$ limit [626, 743, 746, 745, 691, 775, 39]. The "anomalous" asymptotic behavior seems to be associated with solutions corresponding to vanishing trace of the matter energy-momentum tensor, $T^{(m)} = 0$ [58]. This issue needs to be understood because the belief that BD gravity reduces to general relativity in the $\omega \to \infty$ limit[1] is the basis

---

[1]The $\omega \to \infty$ limit of BD theory is also used as a criterion to determine the analogue of general relativity in two dimensions [560].

for constraining BD theory using its weak field limit in the analysis of Solar System experiments [909].

For the exact BD solutions exhibiting the "wrong" asymptotics, the BD field has the asymptotic behaviour

$$\phi = \phi_0 + \mathrm{O}\left(\frac{1}{\sqrt{\omega}}\right) \qquad (1.53)$$

as $\omega \to \infty$, instead of the usual one

$$\phi = \phi_0 + \mathrm{O}\left(\frac{1}{\omega}\right) \qquad (1.54)$$

reported in standard textbooks (e.g., [898]).

There is now a better understanding of BD solutions associated with vanishing trace $T^{(m)}$ that do not reduce to their counterparts in general relativity. One way of understanding this phenomenon is by relating it to an invariance property enjoyed by BD theory under special conformal transformations [346, 349].

Let us consider BD theory with a free scalar, $V = 0$. When $T^{(m)} = 0$, the entire BD action (1.25) is invariant under a class of transformations $\{\mathcal{F}_\alpha\}$ (defined explicitly below). The change of the BD parameter $\omega \to \tilde{\omega}$ is equivalent to a mapping of BD theory into an equivalence class. The equivalence relation is defined as follows: if two actions of the type (1.25) are related by a transformation $\mathcal{F}_\alpha$, they are said to be equivalent (it is straightforward to check that this defines an equivalence relation). The $\omega \to \infty$ limit can also be regarded as a parameter change that moves the theory within the same equivalence class. Hence, this limit cannot yield general relativity, which lies outside of the equivalence class. On the contrary, when $T^{(m)} \neq 0$, the (restricted) conformal invariance is broken and the $\omega \to \infty$ limit, similarly to any other parameter change $\omega \to \tilde{\omega}$, does not move BD theory within a restricted equivalence class. Hence general relativity can be obtained in this limit.

To see how this argument works in detail, consider first the purely gravitational sector of BD theory (1.25) (without $S^{(m)}$) on a BD spacetime (M, g_{ab}, ϕ), where M is a four-dimensional smooth manifold. The conformal transformation[2]

$$g_{ab} \longrightarrow \tilde{g}_{ab} = (G\phi)^{2\alpha} g_{ab} \qquad \left(\alpha \neq \frac{1}{2}\right), \qquad (1.55)$$

---

[2]We consider a region of spacetime in which $\phi \neq 0$; a zero of ϕ would cause trouble with the effective gravitational coupling (1.34).

Scalar-tensor gravity

and the scalar field redefinition

$$G\phi \longrightarrow G\tilde{\phi} = (G\phi)^{1-2\alpha} \,, \tag{1.56}$$

transform the BD Lagrangian density into

$$\mathcal{L}^{(BD)}\sqrt{-g} = \sqrt{-\tilde{g}}\left[\tilde{\phi}\tilde{R} - \frac{\tilde{\omega}}{\tilde{\phi}}\,\tilde{g}^{cd}\,\tilde{\nabla}_c\tilde{\phi}\,\tilde{\nabla}_d\tilde{\phi}\right] \,, \tag{1.57}$$

where $\tilde{g}$ and $\tilde{R}$ are the metric determinant and the Ricci curvature associated with the new metric $\tilde{g}_{ab}$, respectively, and

$$\tilde{\omega} = \frac{\omega - 6\alpha(\alpha-1)}{(1-2\alpha)^2} \tag{1.58}$$

is the new BD parameter. BD gravity is then invariant in form under the transformation defined by eqs. (1.55) and (1.56). This invariance property is often regarded as an analogue of the conformal invariance of string theories [221, 222, 224, 862, 545].

When the parameter α varies, the transformations $\mathcal{F}_\alpha$

$$(M, g_{ab}, \phi) \xrightarrow{\mathcal{F}_\alpha} \left(M, \tilde{g}_{ab}, \tilde{\phi}\right) \tag{1.59}$$

defined by eqs. (1.55) and (1.56) span the entire range of real values of the parameter $\tilde{\omega}$. Two BD spacetimes are said to be equivalent if they can be related by a transformation $\mathcal{F}_\alpha$. It is straightforward to check that this defines an equivalence relation, and that the set of BD spacetimes $\left(M, g_{ab}^{(\omega)}, \phi^{(\omega)}\right)$ is divided into equivalence classes $\mathcal{E}$.

In general, when matter represented by a (symmetric) tensor $T_{ab}^{(m)}$ is added to the gravitational sector of the theory, the conformal invariance described above is broken. However, if the trace $T^{(m)}$ vanishes, the conservation equation

$$\nabla^b T_{ab}^{(m)} = 0 \tag{1.60}$$

regulating the dynamics of this matter is conformally invariant [892]. Since the $T_{ab}^{(m)}$ of conformal matter is not affected by conformal transformations, the full BD action (1.25) is again conformally invariant. The parameter change $\omega \to \tilde{\omega}$ is again equivalent to a transformation $\mathcal{F}_\alpha$ moving BD theory within an equivalence class $\mathcal{E}$ that does not contain general relativity. Similarly, the parameter change $\omega \to \tilde{\omega}$ with increasingly large values of $\tilde{\omega}$ is equivalent to the limit $\omega \to \infty$ and cannot reproduce Einstein's theory. The failure to obtain general relativity as the $\omega \to \infty$ limit of BD solutions when $T^{(m)}$ vanishes is thus explained.

In addition, the asymptotic behaviour of the BD field can be easily obtained using this approach [346, 349].

There is no loss of generality beginning from the value $\omega = 0$ of the BD parameter. In fact, one can obtain any value of the parameter ω from $\omega = 0$ by using a transformation $\mathcal{F}_\alpha$. The solution of eq. (1.58) with respect to α yields

$$\alpha = \frac{1}{2}\left(1 \pm \frac{\sqrt{3}}{\sqrt{2\tilde{\omega}+3}}\right) \tag{1.61}$$

for $\tilde{\omega} > -3/2$. In addition, the situation is symmetric for $\tilde{\omega} < -3/2$: BD theory with $\omega = -3/2$ — which is conformally invariant with respect to a different class of conformal transformations — is sometimes studied [681, 280, 60]. As $\tilde{\omega} \to \infty$ and $\alpha \to 1/2$, one obtains

$$G\tilde{\phi} \approx 1 \mp \left(\frac{3}{2\tilde{\omega}}\right)^{1/2} \ln(G\phi) , \tag{1.62}$$

the "anomalous" asymptotic behaviour expressed by eq. (1.53). The value of the "old" BD scalar ϕ corresponding to $\omega = 0$ is not affected by the limit $\tilde{\omega} \to \infty$ and the only dependence of $\tilde{\phi}$ on $\tilde{\omega}$ is the one made explicit in eq. (1.62). In terms of the field equations when $\tilde{\omega} \to +\infty$, the second term on the right hand side of eq. (1.29)

$$\frac{\tilde{\omega}}{\tilde{\phi}^2}\left(\tilde{\nabla}_a\tilde{\phi}\tilde{\nabla}_b\tilde{\phi} - \frac{1}{2}\tilde{g}_{ab}\tilde{\nabla}^c\tilde{\phi}\tilde{\nabla}_c\tilde{\phi}\right) \tag{1.63}$$

does not tend to zero, but rather to

$$\frac{3}{2}\left(\nabla_a\phi\nabla_b\phi - \frac{1}{2}g_{ab}\nabla^c\phi\nabla_c\phi\right) , \tag{1.64}$$

where eq. (1.62) was used, and the field equations (1.29) do not reduce to the Einstein equations with the same $T_{ab}^{(m)}$. The asymptotic behaviour of the BD scalar then determines whether a certain solution of the BD field equations converges to the corresponding solution of the Einstein equations or not. It can be proven [349] that the expression (1.63) cannot vanish identically, and is always present in the field equations (1.29), with the only exceptions being Einstein spaces with $\phi =$constant [642], and the first order field equations in the weak field approximation [70, 206].

The fact that the $\omega \to \infty$ limit of BD theory does not always produce general relativity signals a conceptual difference between BD and Einstein theories. In BD theory the gravitational coupling is determined by all matter in the universe, according to Mach's principle. Hence, it is rather pointless (although technically possible) to consider vacuum solutions, which trivially have $T^{(m)} = 0$, in BD theory. An exception is the situation in which the scalar ϕ dominates all forms of matter. This is the case near the big bang singularity of FLRW spaces at which ϕ behaves as a stiff fluid [646]. Conformal matter, which also has $T^{(m)} = 0$, does not act as a source of ϕ in eq. (1.33) and behaves as vacuum in this respect. On the contrary, Einstein's theory, with its fixed gravitational coupling, allows one to consider meaningful vacuum[3] solutions with $R_{ab} = 0$. The $\omega \to \infty$ limit then will not reduce BD solutions with $T^{(m)} = 0$ to general-relativistic solutions which correspond to a conceptually different physical situation [642].

From a more general point of view, the reader should keep in mind that the limit of a spacetime as a parameter varies may not be well defined even in general relativity, or it may depend on the coordinate system employed. For example, the limit for diverging mass parameter M of the Schwarzschild solution

$$ds^2 = -\left(1 - \frac{2M}{r}\right) dt^2 + \left(1 - \frac{2M}{r}\right)^{-1} dr^2 + r^2 \left(d\theta^2 + \sin^2\theta \, d\varphi^2\right) \tag{1.65}$$

can be either Minkowski space or a Kasner space [427]. It is desirable to study the limiting process in a coordinate-independent approach [691, 692].

5. Relation to Kaluza-Klein theory

Added interest in BD theory comes from the fact that it can be derived from Kaluza-Klein theory [530, 537, 234, 52, 44, 687] with d extra spatial dimensions, obtaining the BD parameter

$$\omega = -\frac{(d-1)}{d}. \tag{1.66}$$

In this derivation, the BD scalar field has a geometrical origin in the determinant of the metric defined on the submanifold of the extra dimensions. This feature constitutes an essential ingredient of modern unified theories. In the simplest version of classical Kaluza-Klein theory [44, 52, 687] with a single dilaton field, one begins with the spacetime $(M \otimes K, \hat{g}_{AB})$, where M is a 4-dimensional manifold with one timelike dimension and K is a submanifold with d spatial dimensions ($d \geq 1$). A

---

[3] Note that in this case the gravitational coupling G disappears from the field equations, while this does not happen for the BD field $\phi = G_{eff}^{-1}$ in the BD vacuum field equations.

caret denotes $(4+d)$-dimensional quantities. The $(4+d)$-dimensional metric is written as

$$(\hat{g}_{AB}) = \begin{pmatrix} \hat{g}_{ab} & 0 \\ 0 & \hat{\phi}_{\alpha\beta} \end{pmatrix}, \qquad (1.67)$$

where $A, B, \ldots = 0, 1, \ldots, (3+d)$, $a, b, \ldots = 0, 1, 2, 3$, and $\alpha, \beta, \ldots = 4, 5, \ldots, (3+d)$.

In the context of Kaluza-Klein cosmology, it is assumed that there are no off-diagonal terms in the metric (1.67). These were present in the original Kaluza-Klein [530, 537] theory introduced with the purpose of unifying electromagnetism (described by the electromagnetic potential A_a corresponding to off-diagonal terms in eq. (1.67)) and gravity (described by the 4-dimensional metric). A gauge field corresponding to off-diagonal terms in (1.67) would introduce a preferred direction and spoil the spatial homogeneity and isotropy of the FLRW metric. It is also assumed that $\hat{g}_{ab}$ is a FLRW metric on M and $\hat{\phi}_{ab}$ is a diagonal Riemannian metric on K.

The starting point of Kaluza-Klein cosmology is vacuum general relativity in $(4+d)$ dimensions, described by the action

$$S = \int d^{(4+d)}x \sqrt{-\hat{g}} \, \mathcal{L}^{(4+d)} = \frac{1}{16\pi\hat{G}} \int d^{(4+d)}x \sqrt{-\hat{g}} \left(\hat{R} + \hat{\Lambda} \right), \qquad (1.68)$$

where $\hat{g} = \det(\hat{g}_{AB})$, $\hat{R}$ is the Ricci curvature associated with the metric $\hat{g}_{AB}$, $\hat{\Lambda}$ and $\hat{G}$ are the $(4+d)$-dimensional cosmological constant and gravitational coupling, and $g_{AB} = g_{AB}(x^a)$ only depends on the position on M. The higher-dimensional stress-energy tensor describing higher-dimensional matter different from $\hat{\Lambda}$ is assumed to vanish in order to consider a regime (e.g., inflation) in which the dynamics of the universe are dominated by a single scalar field (the Kaluza-Klein scalar defined below and obtained from pure geometry), neglecting all other forms of matter.

By introducing the determinant of the metric of the extra dimensions,

$$\varphi \equiv \left| \det(\hat{\phi}_{\alpha\beta}) \right|, \qquad (1.69)$$

and the symmetric tensor

$$\rho_{\alpha\beta} \equiv \varphi^{-1/d} \hat{\phi}_{\alpha\beta}, \qquad (1.70)$$

it is, by definition, $|\det(\rho_{\alpha\beta})| = 1$.

It is assumed that the extra spatial dimensions are curled up on a microscopic scale of size l. Then, the integral over the $(4+d)$ dimensions in the action (1.68) is split into the product of an integral over the four spacetime dimensions $\int d^4x$ and of an integral over the remaining d dimensions $\int d^d x$. Performing this last integration, the action reduces to

$$S = \frac{V^{(l)}}{16\pi \hat{G}} \int d^4x \sqrt{-g} \sqrt{\varphi} \left[\left(R + R_K + \hat{\Lambda} \right) + \frac{(d-1)}{4d} \frac{g^{ab} \nabla_a \varphi \nabla_b \varphi}{\varphi^2} \right], \quad (1.71)$$

where $V^{(l)}$ is the volume of the compact manifold K of the extra dimensions and R_K is the Ricci curvature of K.

In terms of the scalar field

$$\phi \equiv \sqrt{\varphi}, \quad (1.72)$$

and setting $G = \hat{G}/V^{(l)}$, the action is

$$S = \frac{1}{16\pi G} \int d^4x \sqrt{-g} \left[\phi \left(R + R_K + \hat{\Lambda} \right) + \frac{(d-1)}{d} \frac{g^{ab} \nabla_a \phi \nabla_b \phi}{\phi} \right]. \quad (1.73)$$

The action (1.73) describes a BD theory with parameter ω given by eq. (1.66) in which the scalar ϕ does not carry dimensions.

Kaluza-Klein theory provides an aesthetically appealing geometrical origin of the cosmological scalar field due to the presence of extra spatial dimensions in vacuum, higher dimensional, general relativity. This attractive geometrical feature spurred motivation for studies of Kaluza-Klein cosmology (see Ref. [687] for a review). By contrast, in most studies of inflation or quintessence, a cosmological scalar field is introduced by hand. From a practical point of view, the derivation of BD theory from Kaluza-Klein gravity provides a technique for generating exact solutions in one theory starting from familiar solutions in the other [130].

In recent literature, Kaluza-Klein theory is reformulated in the Einstein conformal frame by means of a conformal transformation and a redefinition of the scalar field (see Section 1.11.3). In fact, the Jordan frame Kaluza-Klein theory, reduced to BD theory after compactification, is incompatible with the experimental limit $|\omega| \geq 500$.

Furthermore, doubts on the possibility of Kaluza-Klein dimensional reduction are cast by the following argument [606]. In the $(4 + d)$-dimensional point of view, during dimensional reduction, spacetime is highly anisotropic since three spatial dimensions expand while the other

d contract. A quantized matter field on this $(4+d)$-dimensional spacetime may isotropize it due to quantum particle creation (which is associated with negative pressure), a phenomenon known to occur in four-dimensional spacetime as well [916, 917, 494, 135, 493]. This isotropization process forbids the contraction of the extra dimensions accompanied by the simultaneous expansion of the ordinary three dimensions, which is instead a highly anisotropic phenomenon. From the classical point of view, the presence of a cosmological constant $\hat{\Lambda}$ in the higher-dimensional spacetime would tend to generate inflation and isotropize the $(4+d)$-dimensional spacetime, as described by the cosmic no-hair theorems of four spacetime dimensions [891, 650, 608, 637, 516, 825, 610]. Starting with an anisotropic Bianchi model, inflation generically proceeds and leads to a FLRW metric. This, or the argument of Ref. [606], seems to indicate that the $\hat{\Lambda}$-term should be removed from the $(4+d)$-dimensional action (1.68). However, the Casimir effect on the space of compact dimensions may have an opposing effect [43, 42, 606].

Kaluza-Klein theory reformulated in the Einstein frame is discussed in Section 1.11.3.

6. Brans-Dicke theory from Lyra's geometry

An objection was raised that the BD field appearing in the BD action (1.25) — describing the gravitational field together with the metric tensor — has no geometric origin, while in Einstein's theory the gravitational sector of the action is purely geometric, adding to the aesthetic appeal of the latter. One can reply to this objection by pointing out the geometric origin of the BD scalar in the determinant of the extra spatial dimensions in Kaluza-Klein theory. Another response is to generalize the concept of Lorentzian manifold of general relativity to encompass Lyra manifolds and, in this approach, the gravitational sector of the BD action (1.25) only contains geometrical terms [815]. The Lyra geometry [598] is occasionally used in alternative cosmologies [793, 794, 460, 795, 796, 126, 805, 515, 722]. A n-dimensional Lyra spacetime (M, ψ, g_{ab}) consists of a smooth manifold M (at least $\mathcal{C}^2$) equipped with a smooth scalar field ψ (called the *gauge function*) with the dimensions of the inverse of a length, and a connection (*Lyra connection*). The connection coefficients are given by

$$\Gamma^c_{ab} = \frac{1}{\psi} \{^c_{ab}\} + \frac{s+1}{\psi^2} g^{cd} \left(g_{bd}\, \partial_a \psi - g_{ab}\, \partial_d \psi \right), \qquad (1.74)$$

where s is a constant, $\{^c_{ab}\}$ are the usual Christoffel symbols, and the second term on the right hand side of eq. (1.74) describes torsion. The

Lyra connection is metric-preserving, i.e.,

$$\nabla_a g_{bc} = 0 \,. \tag{1.75}$$

The torsion tensor is given by

$$T_{ab}{}^c = \frac{s}{\psi^2} \left(g_b^c \nabla_a \psi - g_a^c \nabla_b \psi \right) \,. \tag{1.76}$$

Note that the connection is torsion-free if $s = 0$ and the gauge function ψ plays the role of a torsion potential.

The *Lyra curvature tensor* is defined as

$$K_{dab}{}^c = \frac{1}{\psi^2} \left[\partial_a (\psi \Gamma^c_{db}) - \partial_d (\psi \Gamma^c_{ab}) + \Gamma^e_{db} \Gamma^c_{ea} - \Gamma^e_{ab} \Gamma^c_{ed} \right] \tag{1.77}$$

and its contractions $K_{ac} \equiv K_{abc}{}^b$ and $K \equiv g^{ab} K_{ab}$ can be formed in a similar manner to the Riemann tensor in general relativity. Note however that K_{ab} and K do not coincide with R_{ab} and R because the definition (1.77) of K_{abcd} yields

$$K_{dab}{}^c = \frac{R_{dab}{}^c}{\psi^2} + \frac{2(s+1)}{\psi} H_{dab}{}^c + \frac{2(s+1)^2}{\psi^2} L_{dab}{}^c \,, \tag{1.78}$$

where $R_{dab}{}^c$ is the usual Riemann tensor and $H_{dab}{}^c$ and $L_{dab}{}^c$ are two extra tensors obtained as combinations of g_{ab}, ψ, and $\nabla_c \psi$ [598]. Note that $K_{dab}{}^c$ is dimensionless. The covariant volume element of Lyra's geometry is $\psi^4 \sqrt{-g} \, d^n x$ and is also dimensionless. In Lyra's geometry in four dimensions, the analogue of the gravitational sector of the Einstein-Hilbert action is the dimensionless action

$$S = \int d^4 x \, \psi^4 \sqrt{-g} \, K \,. \tag{1.79}$$

Using the expression of the Lyra scalar curvature [598, 815]

$$K = \frac{R}{\psi^2} + \frac{2(s+1)}{\psi^3} (1-n) \,\Box\, \psi \tag{1.80}$$
$$+ \frac{1}{\psi^4} \left[(s+1)^2 (3n - n^2 - 2) - 2(s+1)(2-n) \right] \nabla^c \psi \nabla_c \psi$$

with $n = 4$, dropping the integral of a total divergence — which can be transformed to a boundary term giving zero contribution to the variation of the Lyra action (1.79)) — one obtains [815]

$$S = \int d^4 x \sqrt{-g} \left(\psi^2 R - 4\omega \, g^{cd} \nabla_c \psi \nabla_d \psi \right) \,, \tag{1.81}$$

where
$$\omega = \frac{3\left(s^2 - 1\right)}{2}. \tag{1.82}$$

Finally, setting $\phi \equiv \psi^2$ yields the canonical BD action

$$S = \int d^4x \sqrt{-g} \left(\phi R - \frac{\omega}{\phi} g^{cd} \nabla_c \phi \nabla_d \phi \right). \tag{1.83}$$

Lyra's action is appealing because it contains no dimensional coupling. This accords to the idea that a fundamental theory should have no dimensional coupling built into it and be scale-invariant and that effective couplings arise as vacuum expectation values of certain fields, as in Sakharov's induced gravity. When matter is introduced in the Lyra action, scale-invariance is broken.

However, the absence of dimensional couplings that makes Lyra's action attractive is at odds with the philosophy that leads to the introduction of BD theory, namely that Mach's principle requires the effective coupling ϕ^{-1} to be determined by all matter in the universe. Therefore, the derivation of the gravitational sector of BD theory from pure geometry may seem purposeless. Moreover, a mass term or a more general potential for the BD field does not arise naturally in the Lyra geometry — no more than a graviton mass would fit in Einstein's theory. On the other hand, exponential potentials for the scalar field do arise naturally in Kaluza-Klein theory. Notwithstanding these difficulties, the geometric derivation of BD theory from pure Lyra geometry constitutes an interesting alternative to the derivation from Kaluza-Klein theory. Another possibility for "geometrizing" the BD field in vacuum is given in the context of the geometrodynamics program in Ref. [710].

7. Scalar-tensor theories

Scalar-tensor theories of gravity ([121, 887, 678], see also [909]) generalize BD theory and are described in the Jordan frame by the action

$$S^{(ST)} = \int d^4x \sqrt{-g} \left[\frac{f(\phi)}{2} R - \frac{\omega(\phi)}{2} \nabla^c \phi \nabla_c \phi - V(\phi) \right] + S^{(m)}, \tag{1.84}$$

where $S^{(m)} = \int d^4x \sqrt{-g}\, \mathcal{L}^{(m)}$ does not explicitly depend on the scalar field ϕ. The distinguishing feature of the action (1.84) is that the BD parameter $\omega(\phi)$ is now a function of the BD-like scalar and therefore varies with the spacetime point. In the spatially homogeneous and isotropic universe relevant for cosmology, ω only varies with the cosmic time. The potential $V(\phi)$ is often included in the action when studying the early universe or the present-day accelerated universe, as is customary for most scalar fields in particle physics and inflationary theories. The action (1.84) contains the BD action (1.25) as the special case

$$f(\phi) = \frac{\phi}{8\pi}, \qquad \omega(\phi) = \frac{\omega_0}{8\pi\phi}, \tag{1.85}$$

where ω_0 is a constant and the potential is rescaled by a factor 16π. Also, scalar-tensor theories with multiple scalar fields are systematically investigated [254, 120, 727].

The field equations derived by varying the action (1.84) and adapted to the spatially flat FLRW metric (1.4) when matter is described by a fluid of density $\rho^{(m)}$ and pressure $P^{(m)}$ are:

$$3fH^2 = \rho^{(m)} + \frac{\omega}{2}\left(\dot{\phi}\right)^2 + V - 3H\dot{f}, \tag{1.86}$$

$$-2f\dot{H} = \rho^{(m)} + P^{(m)} + \omega\left(\dot{\phi}\right)^2 + \ddot{f} - H\dot{f}, \tag{1.87}$$

$$\ddot{\phi} + 3H\dot{\phi} + \frac{1}{2\omega}\left[\omega'\left(\dot{\phi}\right)^2 - f'R + 2V'\right] = 0, \tag{1.88}$$

where a prime denotes differentiation with respect to ϕ and $\dot{f} = f'\dot{\phi}$; while [501]

$$\nabla^b T^{(m)}_{ab} = 0. \tag{1.89}$$

There are several ways to present scalar-tensor theories. One form of the action alternative to (1.84) often encountered in the literature is

$$S = \frac{1}{16\pi}\int d^4x \sqrt{-g}\left[\phi R - \frac{\omega(\phi)}{\phi}g^{ab}\nabla_a\phi\nabla_b\phi - V(\phi)\right] + S^{(m)}. \tag{1.90}$$

This action can be recast as one with canonical kinetic energy term for a different scalar field φ,

$$S = \frac{1}{16\pi}\int d^4x \sqrt{-g}\left[f(\varphi)R - \frac{1}{2}g^{ab}\nabla_a\varphi\nabla_b\varphi - U(\varphi)\right] + S^{(m)} \tag{1.91}$$

by setting [678]

$$\phi = f(\varphi), \tag{1.92}$$

where the function f is defined by the equation

$$\omega(\phi) = \frac{f(\varphi)}{2\left(df/d\varphi\right)^2} \tag{1.93}$$

and
$$U(\varphi) = V[f(\varphi)] . \tag{1.94}$$

The alternative form (1.91) of the action can be obtained whenever the function $f(\varphi)$ admits regular inverse f^{-1}. This is not the case, for example, if $f(\phi)$ is represented by a series with even powers of ϕ [573, 856].

The equations of motion obtained by varying the action (1.90) are

$$\begin{aligned} R_{ab} - \frac{1}{2}g_{ab}R &= \frac{8\pi}{\phi} T_{ab}^{(m)} + \frac{\omega(\phi)}{\phi^2} \left(\nabla_a \phi \nabla_b \phi - \frac{1}{2} g_{ab} \nabla^c \phi \nabla_c \phi \right) \\ &+ \frac{1}{\phi} \left(\nabla_a \nabla_b \phi - g_{ab} \Box \phi \right) - \frac{V}{2\phi} g_{ab} , \end{aligned} \tag{1.95}$$

$$\Box \phi = -\frac{\phi}{2\omega}\left(R - \frac{dV}{d\phi}\right) - \frac{1}{2}\left(\nabla^c \phi \nabla_c \phi\right)\left(\frac{1}{\omega}\frac{d\omega}{d\phi} - \frac{1}{\phi}\right). \tag{1.96}$$

The last equation can be rearranged by taking the trace of eq. (1.95),

$$R = -\frac{8\pi}{\phi} T^{(m)} + \frac{\omega}{\phi^2} \nabla^c \phi \nabla_c \phi + 3\frac{\Box \phi}{\phi} + \frac{2V}{\phi}, \tag{1.97}$$

and substituting it into eq. (1.96) to obtain

$$\Box \phi = \frac{1}{2\omega + 3}\left(8\pi T^{(m)} - \frac{d\omega}{d\phi}\nabla^c \phi \nabla_c \phi + \phi \frac{dV}{d\phi} - 2V\right). \tag{1.98}$$

Ordinary matter satisfies the conservation equation

$$\nabla^b T_{ab}^{(m)} = 0 . \tag{1.99}$$

In the following we make use of the form that the field equations assume in the FLRW metric

$$ds^2 = -dt^2 + a^2(t) \left[\frac{dr^2}{1 - Kr^2} + r^2 \left(d\theta^2 + \sin^2 \theta \, d\varphi^2 \right) \right] , \tag{1.100}$$

under the assumption that ordinary matter is described by a perfect fluid stress-energy tensor

$$T_{ab}^{(m)} = \left(P^{(m)} + \rho^{(m)}\right) u_a u_b + P^{(m)} g_{ab} . \tag{1.101}$$

The time-time component of eq. (1.95) (called *Hamiltonian constraint* in general relativity) yields

$$H^2 = \frac{8\pi}{3\phi}\rho^{(m)} - H\left(\frac{\dot\phi}{\phi}\right) + \frac{\omega(\phi)}{6}\left(\frac{\dot\phi}{\phi}\right)^2 - \frac{K}{a^2} + \frac{V}{6\phi}, \tag{1.102}$$

while the equation for the scalar field becomes

$$\ddot{\phi} + \left(3H + \frac{\dot{\omega}}{2\omega + 3}\right)\dot{\phi}$$
$$= \frac{1}{2\omega + 3}\left[8\pi\left(\rho^{(m)} - 3P^{(m)}\right) - \phi\frac{dV}{d\phi} + 2V\right]. \quad (1.103)$$

The trace equation (1.97) yields

$$6\left(\dot{H} + 2H^2 + \frac{K}{a^2}\right) = \frac{8\pi}{\phi}\left(\rho^{(m)} - 3P^{(m)}\right) - \omega\left(\frac{\dot{\phi}}{\phi}\right)^2$$
$$+ 3\frac{\Box\phi}{\phi} + \frac{2V}{\phi}. \quad (1.104)$$

By substituting into the last equation the values of H^2 and of $\Box\phi$ obtained, respectively, from eq. (1.102) and from the scalar field equation (1.103), one obtains

$$\dot{H} = \frac{-8\pi}{(2\omega + 3)\phi}\left[(\omega + 2)\rho^{(m)} + \omega P^{(m)}\right] - \frac{\omega}{2}\left(\frac{\dot{\phi}}{\phi}\right)^2 + \frac{K}{a^2}$$
$$+ 2H\frac{\dot{\phi}}{\phi} + \frac{1}{2(2\omega + 3)\phi}\left[\phi\frac{dV}{d\phi} - 2V + \frac{d\omega}{d\phi}\left(\dot{\phi}\right)^2\right] \quad (1.105)$$

while

$$\left(\rho^{(m)}\right)^{\cdot} + 3H\left(\rho^{(m)} + P^{(m)}\right) = 0. \quad (1.106)$$

These equations differ from the equations of BD cosmology by the terms in $\dot{\omega} = \dot{\phi}\, d\omega/d\phi$, and by the fact that now ω is a function of ϕ. Only two of the equations (1.102), (1.103), and (1.105) are independent.

From the cosmological point of view, a time-dependent parameter $\omega[\phi(t)]$ is more attractive than the constant parameter of BD theory. In a FLRW universe, ϕ and $\omega(\phi)$ only depend on the cosmic time t. This opens the door to the possibility that ω be of order unity in the early universe. The latter would then be allowed to behave significantly differently than in general relativity. Later, during the radiation- or matter-dominated era, ω could tend to large values so that the present day constraint $\omega > 500$ is satisfied [678, 107, 830, 81, 256, 257, 426, 413, 258]. In this way, the present-day limits on ω are circumvented, while significant deviations from general relativity are still permitted early in the cosmic history. For example, the extended inflationary scenario of the universe [552, 897, 553, 572, 557] illustrates this philosophy; this scenario only works for low values of ω ($\omega \leq 20$), which are forbidden by present-day constraints. To rescue its cosmologically interesting features, the extended inflationary scenario was modified into hyperextended inflation, based on a scalar-tensor theory with a time-dependent parameter ω. The latter assumes large values after inflation, thanks to the introduction of a potential $V(\phi)$ with a minimum in which ϕ settles after the end of the inflationary epoch [484, 81, 543, 830, 573, 249, 567, 568, 442]. This way of evading the constraints on the deviations from general relativity is only acceptable if there is no fine-tuning. In other words, a natural mechanism must exist that induces convergence of the scalar-tensor theory to general relativity. Many authors have investigated the possibility that general relativity behaves as an attractor for scalar-tensor theories [107, 830, 81, 411, 256, 257, 82, 639, 895, 413, 426, 769, 836, 83, 770, 258, 132, 798], either during the radiation- or matter-dominated eras of the standard big bang model, or even during inflation. While there are scalar-tensor theories with the desired attractor behaviour, this feature is not shared by all scalar-tensor theories.

The conditions for a scalar-tensor theory to converge to general relativity are [678]

$$\omega \to \infty \quad \text{and} \quad \frac{1}{\omega^3} \frac{d\omega}{d\phi} \to 0 \,. \tag{1.107}$$

Both of these conditions should be kept in mind since there are theories in which the dynamical solutions satisfy the first limit but not the second one. Examples are theories with

$$\omega(\phi) = \frac{\omega_0}{(1 - \phi/\phi_0)^\alpha} \tag{1.108}$$

with ω_0 and ϕ_0 constants ($\phi_0^{-1} = G$ is the present value of the gravitational coupling) and $0 < \alpha < 1/2$, as $\phi \to \phi_0$ from below [83].

When the scalar-tensor action is written in the form (1.90), most expressions of the coupling function encountered in the literature — although by all means not the only meaningful ones — seem to fall into a few categories [83]:

$$2\omega(\phi) + 3 = \frac{1}{B\,|G\phi - 1|^\delta} \quad \left(\delta > \frac{1}{2}\right), \tag{1.109}$$

$$2\omega(\phi) + 3 = \frac{1}{B\,|\ln(G\phi)|^\delta} \quad \left(\delta > \frac{1}{2}\right), \tag{1.110}$$

and

$$2\omega(\phi) + 3 = \frac{1}{B\,\left|(G\phi)^\delta - 1\right|} \quad (\delta > 0) \,, \tag{1.111}$$

where $B > 0$ is a constant. These three classes of theories exhibit different behaviours at early times, but they all coincide asymptotically in the limit to general relativity $G\phi \to 1$, in which

$$2\omega(\phi) + 3 \approx \frac{1}{C \left|\ln(G\phi)\right|^\alpha} \qquad (1.112)$$

since $\ln(G\phi) \approx G\phi - 1$ and $(G\phi)^\delta - 1 \approx \delta \ln(G\phi)$ in this limit.

Other forms of $\omega(\phi)$ suggested in the literature are (1.109) with $\delta < 0$ [78],

$$2\omega(\phi) + 3 = \frac{B}{G\phi}, \qquad (1.113)$$

where B is again a positive constant [786, 82], or the polynomial

$$2\omega(\phi) + 3 = \sum_{k=1}^{n} A_k \, (G\phi)^{n_k}, \qquad (1.114)$$

where $n_k \geq 0$ [57].

All these functional dependences are motivated by the fact that they lend themselves to the exact solution of the relevant field equations. If gravity is described by a scalar-tensor theory, the form of the coupling function is unknown and attention should not be restricted to the forms (1.109)-(1.114) of $\omega(\phi)$.

The existence of Noether symmetries in the equations describing FLRW scalar-tensor cosmologies is studied in Refs. [270, 178, 179, 182, 577, 183, 184, 773, 772]. These symmetries give rise to conserved quantities, and are only present when suitable restrictions are imposed on the form of the functions $f(\phi)$ and $\omega(\phi)$ in the action (1.84).

7.1 Effective Lagrangians and Hamiltonians

If the function $f(\phi)$ in the scalar-tensor action (1.84) admits regular inverse, one can consider this action with the BD parameter ω set to unity without loss of generality. The corresponding field equations (1.86)-(1.88) (with $\omega = 1$, $\dot{\omega} = 0$) in a $K = 0$ FLRW metric can be derived from an effective Lagrangian, or from the corresponding effective Hamiltonian. The Lagrangian is computed from the action (1.84) expressed in this particular metric. The result is [270, 178, 179, 182, 183, 184, 181, 773]

$$L(a, \phi) = 3a\dot{a}^2 f(\phi) + 3a^2 \dot{a}\dot{\phi}\frac{df}{d\phi} + a^3 \left[V(\phi) - \frac{(\dot{\phi})^2}{2} - P^{(m)}\right], \qquad (1.115)$$

in terms of the generalized coordinates a and ϕ. It is straightforward to verify that the Euler-Lagrange equations

$$\frac{d}{dt}\left(\frac{\partial L}{\partial \dot{a}}\right) - \frac{\partial L}{\partial a} = 0, \qquad \frac{d}{dt}\left(\frac{\partial L}{\partial \dot{\phi}}\right) - \frac{\partial L}{\partial \phi} = 0, \qquad (1.116)$$

reproduce the ordinary differential equations (1.87) and (1.88) with $\omega = 1$. The former are the only two independent equations of motion, while the Hamiltonian constraint (1.86) can be regarded as a constraint on the initial data, as usual. The Hamiltonian associated with the Lagrangian (1.115) is

$$E = p_a \dot{a} + p_\phi \dot{\phi} - L = 3a^2 \dot{a} \dot{f} + 3a \dot{a}^2 f - \frac{a^3}{2}\dot{\phi}^2 - a^3 V + a^3 P^{(m)}. \quad (1.117)$$

Another approach to effective Lagrangian and Hamiltonian descriptions includes the special case of a massive conformally coupled scalar field described by

$$f(\phi) = \kappa^{-1} - \frac{1}{6}\phi^2, \qquad (1.118)$$

with potential

$$V(\phi) = \frac{m^2}{2}\phi^2 + \lambda \phi^4 + V_0, \qquad (1.119)$$

where V_0 is a constant, which is studied in relation to the dynamical properties of a spatially flat FLRW universe [174, 474, 140, 742, 197, 651, 357]. By defining the new variables

$$\psi \equiv \sqrt{\frac{\kappa}{6}}\, a, \qquad \varphi \equiv a\phi, \qquad (1.120)$$

and using conformal time η, the field equations are transformed to

$$\frac{d^2\psi}{d\eta^2} - \frac{\kappa\, m^2}{6}\varphi^2\psi - 4V_0\psi^3 = 0, \qquad (1.121)$$

$$\frac{d^2\varphi}{d\eta^2} + \frac{6\, m^2}{\kappa}\varphi\psi^2 + 4\lambda\varphi^3 = 0. \qquad (1.122)$$

These can be derived from the effective Lagrangian

$$L(\varphi, \psi) = \frac{1}{2}\left(\frac{d\varphi}{d\eta}\right)^2 - \frac{18}{\kappa^2}\left(\frac{d\psi}{d\eta}\right)^2 - \frac{3\, m^2}{\kappa}\varphi^2\psi^2 - \lambda\varphi^4 - \frac{36 V_0}{\kappa^2}\psi^4, \qquad (1.123)$$

or from the associated Hamiltonian

$$E(\varphi, \psi) = \frac{1}{2}\left(\frac{d\varphi}{d\eta}\right)^2 - \frac{18}{\kappa^2}\left(\frac{d\psi}{d\eta}\right)^2 + \frac{3m^2}{\kappa}\varphi^2\psi^2 + \lambda\varphi^4 + \frac{36V_0}{\kappa^2}\psi^4 \,. \tag{1.124}$$

Effective Lagrangians and Hamiltonians are also found in the context of Bianchi models [663, 664, 577, 367].

8. Motivations for scalar-tensor theories

BD and scalar-tensor theories have repeatedly generated interest since their inception for several reasons. First, a gravitational scalar field appears in scalar-tensor theories together with the metric tensor g_{ab}, and a fundamental scalar coupled to gravity is an unavoidable feature of superstring, supergravity, and M-theories [387, 173, 593, 443, 717]. Scalar fields are ubiquitous in particle physics and in cosmology. Examples are the Higgs boson of the Standard Model, the superpartner of spin 1/2 particles in supergravity, the string dilaton appearing in the supermultiplet of the higher-dimensional graviton, or non-fundamental fields like composite bosons and fermion condensates.

Second, the gravitational sector of BD theory exhibits a restricted conformal invariance under the transformations $\mathcal{F}_\alpha$ (described in Section 1.4) that mimics the conformal invariance of string theories at high energy [221, 222, 254, 862, 545, 151]. The duality symmetry of string theory (1.51) and (1.52) has an analogue in the symmetry (1.49) and (1.50) of BD theory.

Perhaps the strongest motivation for the study of scalar-tensor theories comes from the fact that the low energy limit of the bosonic string theory corresponds to a BD theory with $\omega = -1$ [173, 387][4]. At tree level, the low energy action of the bosonic string theory in the string frame is

$$S = \frac{1}{2k_D^2}\int d^D x \sqrt{-g}\; e^{-2\Phi}\left[R[g] - 4\,g^{ab}\partial_a\Phi\,\partial_b\Phi\right.$$
$$\left. - \frac{1}{12}H_{abc}H^{abc} - \frac{D-26}{l_S^2}\right], \tag{1.125}$$

where Φ is the dimensionless string dilaton, the 3-form $H_{abc} = \partial_{[a}b_{bc]}$ is the strength of the Kalb-Ramond field b_{ab}, k_D is the D-dimensional gravitational constant, and l_S is the string scale. It is customary to set $H_{abc}H^{abc} = 0$ and to retain only the string dilaton. This simplifying choice is justified by the fact that in a spatially homogeneous and

---

[4]A BD theory with $\omega = -3$ is also obtained from a less conventional string theory [324].

isotropic cosmology the 3-form field reduces to $H_{0bc} = 0$, $H_{123} = h(t)$, and it can be modeled by a perfect fluid [545]. Therefore, inclusion of ordinary matter models the $H_{abc}H^{abc}$ term in FLRW cosmology.

In four dimensions, the redefinition of the dilaton

$$\varphi = e^{-2\Phi} \qquad (1.126)$$

then yields the BD action with $\omega = -1$

$$S = \frac{1}{2k^2} \int d^4x \sqrt{-g} \left[\varphi R + \frac{1}{\varphi} g^{cd} \partial_c \varphi \, \partial_d \varphi \right] . \qquad (1.127)$$

When matter is added, the dilaton of string theory formulated in the Einstein frame couples directly to ordinary matter, while the Jordan frame BD scalar seen so far does not (see Section 1.11.1 for an explanation of this terminology).

In general, a p-brane in d dimensions leads, after compactification, to a BD theory with parameter [303]

$$\omega = -\frac{(d-1)(p-1) - (p+1)^2}{(d-2)(p-1) - (p+1)^2} , \qquad (1.128)$$

which reduces to -1 for a string (a 1-brane). The conformal transformation

$$g_{ab} \to \tilde{g}_{ab} = \exp\left[\frac{4(\phi_0 - <\phi>)}{D-2}\right] g_{ab} \qquad (1.129)$$

brings the action (1.125) into the Einstein frame form

$$\begin{aligned} S &= \int d^D x \sqrt{-\tilde{g}} \left[\tilde{R} - \frac{4}{D-2} \tilde{g}^{cd} \tilde{\nabla}_c \tilde{\phi} \tilde{\nabla}_d \tilde{\phi} - \frac{1}{12} e^{-\frac{8\tilde{\phi}}{D-2}} H_{abc} H^{abc} \right. \\ &\quad \left. - \frac{D-26}{3l_S^2} e^{\frac{4\tilde{\phi}}{D-2}} \right] , \end{aligned} \qquad (1.130)$$

where $\tilde{\phi} \equiv \phi - <\phi>$. Further similarities between scalar-tensor and string theories include the coupling of the dilaton to ordinary matter in the Einstein frame, which is similar to the coupling of the BD scalar in the conformally rescaled version of the theory. Also, there are similar corrections to the geodesic equation, violations of the equivalence principle described as non-metricity of the theory, and corrections to the geodesic deviation equation. Further, the issue of the conformal frame (see Chapter 2) is similar for scalar-tensor theories [363] and for string theories [287, 27, 672].

Additional interest in BD and scalar-tensor gravity originates from the extended and hyperextended inflationary scenarios of the early universe discussed in Chapter 5. Other motivations arise from the realization that in the Randall-Sundrum [729, 728] brane-world, and in similar models (see Ref. [288] for a review), BD theory is recovered in the 4-dimensional brane (a domain wall) moving in a higher-dimensional space [416, 62]. In the Randall-Sundrum model a 4-brane of positive tension is embedded in a 5-dimensional anti-de Sitter space and non-gravitational physics is effectively confined to this brane, without compactification of the extra (fifth) spatial dimension. A configuration with two parallel branes, one with positive and one with negative tension, is also studied [729, 728]. In the case in which the negative tension brane — where we do not live — is empty, one computes the BD parameter on this negative tension brane and on the positive tension brane (denoted by $+$, and which we inhabit) as [416]

$$\omega_{\pm} = \frac{3}{2}\left[\exp\left(\pm\frac{2\,d}{l}\right) - 1\right], \qquad (1.131)$$

where l is a length scale related to the tension of the wall and d is the value of the fifth coordinate at the position of the brane with negative tension.

9. Induced gravity

Induced gravity is formally a scalar-tensor theory described by the action

$$S = \int d^4x\,\sqrt{-g}\left[\frac{\epsilon\phi^2}{2}R - \frac{1}{2}g^{cd}\nabla_c\phi\nabla_d\phi - V(\phi)\right] + S^{(m)}, \qquad (1.132)$$

where

$$G_{eff}(\phi) = \frac{1}{8\pi\epsilon\,\phi^2} \qquad (1.133)$$

is regarded as the effective gravitational coupling. The theory was originally introduced by Sakharov [761] (see also [915, 15]) and is motivated by the idea that gravity may not be fundamental. Starting from quantum field theory in flat spacetime, the gravitational action is generated by first loop effects and the value of the gravitational coupling G_{eff} is computed from particle physics as the vacuum expectation value of $\xi\phi^2$ (see Ref. [883] for a modern perspective and for a summary of the variations on the main theme occurring in the literature).

The gravitational sector of the theory can be recast in the BD form

$$S = \int d^4x\,\sqrt{-g}\left[\frac{\varphi}{2}R - \frac{\omega}{2\varphi}g^{cd}\nabla_c\varphi\nabla_d\varphi - U(\varphi)\right], \qquad (1.134)$$

where
$$\phi = \sqrt{\varphi}, \qquad \omega = \frac{1}{4}, \qquad (1.135)$$
and
$$U(\varphi) = V\left(\sqrt{\varphi}\right). \qquad (1.136)$$

Induced gravity is studied in cosmology to implement inflation in the early universe [820, 12, 719, 595, 721, 553, 335, 498, 500, 524, 525] or present-day quintessence [709].

10. Generalized scalar-tensor theories

The distinction between scalar-tensor and generalized scalar-tensor theories is a matter of nomenclature and different terminology is used by different authors for the same class of theories. In practice, one should inspect the relevant action or field equations to understand what kind of theory is under consideration, keeping in mind that sometimes different mathematical expressions of the action describe the same theory using different variables.

To fix the terminology here, we say that a *generalized scalar-tensor* (or *non-linear*) theory of gravity is one described by the action

$$S = \int d^4x \, \frac{\sqrt{-g}}{16\pi} \left[f(\phi, R) - \frac{\eta}{2} \nabla^c \phi \nabla_c \phi \right] + S^{(m)}, \qquad (1.137)$$

where $f(\phi, R)$ is an arbitrary regular function of the scalar field ϕ and of the Ricci curvature. f is possibly non-linear (hence the alternative name "non-linear" theory of gravity) and η is a parameter of the theory. In particular, quadratic Lagrangians describing higher derivative theories of gravity [832] and used in the context of quantum gravity are included in the action (1.137). Omitting ordinary matter for simplicity, the field equations are (e.g., [609])

$$\frac{\partial f}{\partial R}\left(R_{ab} - \frac{1}{2}g_{ab}R\right) = \frac{\eta}{2}\left(\nabla_a \phi \nabla_b \phi - \frac{1}{2}g_{ab}\nabla^c \phi \nabla_c \phi\right)$$
$$+ \frac{1}{2}g_{ab}\left(f - \frac{\partial f}{\partial R}R\right) + \nabla_a \nabla_b \left(\frac{\partial f}{\partial R}\right) - g_{ab} \Box \left(\frac{\partial f}{\partial R}\right), \qquad (1.138)$$

$$\eta \Box \phi = -\frac{\partial f}{\partial \phi}. \qquad (1.139)$$

The particular case in which $f(\phi, R)$ is a linear function of the Ricci scalar,

$$f(\phi, R) = f(\phi) R - V(\phi), \qquad (1.140)$$

recovers scalar-tensor theories. Generalized scalar-tensor theories are only occasionally used in cosmology and, in the rest of this book, we focus on the "ordinary" scalar-tensor theories defined in Section 1.7.

11. Conformal transformation techniques

The use of conformal transformations as a mathematical tool has become widespread in works on scalar-tensor theories and other alternative theories of gravity, especially in the cosmological context (see Refs. [613] and [363] for reviews). The idea is to perform a conformal transformation of the spacetime metric and to redefine the scalar field in order to obtain new dynamical variables $\tilde{g}_{ab}$ and $\tilde{\phi}$, in terms of which the gravitational sector of the theory assumes the form familiar from general relativity and $\tilde{\phi}$ has canonical kinetic energy density. The new set of variables $\left(\tilde{g}_{ab}, \tilde{\phi}\right)$ is called the *Einstein conformal frame*, as opposed to (g_{ab}, ϕ), which constitutes the *Jordan frame*. The transformation to the Einstein frame is generated by the scalar field ϕ present in the theory. While the conformal transformation technique has proven itself to be a useful mathematical tool, its use has opened a debate on whether the two conformal frames are physically equivalent or, alternatively, the issue of which conformal frame is physical. As we shall see, the disagreement originates from misunderstandings of the conformal transformation, an on-going source of confusion among cosmologists.

The two conformal frames are actually physically equivalent provided that one takes into account the fact that, in the Einstein frame, the units of length, mass and time employed scale with the scalar field in an appropriate way. Many authors forget this point, attributing physical meaning to the Einstein frame with *fixed* units, and discarding the Jordan frame. This procedure results in a theory that is physically different from the original one. Other authors discard the Einstein frame as unphysical and only accept the Jordan frame as physical. Finally, a third class of authors regard the Jordan frame and the Einstein frame with fixed units as physically equivalent, which is incorrect. This issue is discussed in detail in the next chapter: first we expose the mathematical technique of conformal transformations.

11.1 Conformal transformations

Given a spacetime (M, g_{ab}) where M is a smooth manifold of dimension $n \geq 2$ and g_{ab} a Lorentzian or Riemannian metric on it, the point-dependent rescaling of the metric tensor

$$g_{ab} \longrightarrow \tilde{g}_{ab} = \Omega^2 g_{ab} \,, \qquad (1.141)$$

where the *conformal factor* $\Omega(x)$ is a nowhere vanishing, regular[5] function, is called a *Weyl* or *conformal* transformation. By rescaling the metric, the lengths of spacelike and timelike intervals and the norms of spacelike and timelike vectors are changed, while null vectors and null intervals of the metric g_{ab} remain null in the rescaled metric $\tilde{g}_{ab}$. The light cones are not changed by the transformation (1.141) and the spacetimes (M, g_{ab}) and $(M, \tilde{g}_{ab})$ have the same causal structure; the converse is also true [892]. A vector that is timelike, spacelike, or null with respect to the metric g_{ab} has the same character with respect to $\tilde{g}_{ab}$, and vice-versa.

If one performs the Arnowitt-Deser-Misner [47] decomposition of the metric given by

$$g_{ab}\,dx^a\,dx^b = \left(-N^2 + N_i N^i\right)dt^2 + 2N_j dt\,dx^j + h_{ij}dx^i\,dx^j\,, \qquad (1.142)$$

where N and N^i are the lapse function and the shift vector respectively, one immediately derives the transformation properties of the metric decomposition from eq. (1.141):

$$\tilde{N} = \Omega N\,, \qquad (1.143)$$
$$\tilde{N}^i = N^i\,, \qquad (1.144)$$
$$\tilde{h}_{ij} = \Omega^2 h_{ij}\,. \qquad (1.145)$$

The Arnowitt-Deser-Misner [47] mass of an asymptotically flat spacetime does not change under the conformal transformation and scalar field redefinition [193].

The following transformation properties of geometric quantities [839, 892] are useful:

$$\tilde{g}^{ab} = \Omega^{-2} g^{ab}\,, \qquad \tilde{g} = \Omega^{2n}\,g\,, \qquad (1.146)$$

where n is the dimension of the spacetime manifold M;

$$\tilde{\Gamma}^a_{bc} = \Gamma^a_{bc} + \Omega^{-1}\left(\delta^a_b \nabla_c \Omega + \delta^a_c \nabla_b \Omega - g_{bc} \nabla^a \Omega\right)\,, \qquad (1.147)$$

$$\begin{aligned}\widetilde{R_{abc}{}^d} &= R_{abc}{}^d + 2\delta^d_{[a} \nabla_{b]} \nabla_c(\ln\Omega) - 2g^{de}g_{c[a}\nabla_{b]}\nabla_e(\ln\Omega) \\ &+ 2\nabla_{[a}(\ln\Omega)\delta^d_{b]}\nabla_c(\ln\Omega) - 2\nabla_{[a}(\ln\Omega)g_{b]c}g^{de}\nabla_e(\ln\Omega) \\ &- 2g_{c[a}\delta^d_{b]}g^{ef}\nabla_e(\ln\Omega)\nabla_f(\ln\Omega)\,,\end{aligned}$$

$$(1.148)$$

---

[5] A conformal continuation beyond points where the conformal factor becomes singular is sometimes possible [157, 158, 159].

$$\tilde{R}_{ab} = R_{ab} - (n-2)\nabla_a\nabla_b(\ln\Omega) - g_{ab}g^{ef}\nabla_f\nabla_e(\ln\Omega)$$
$$+ (n-2)\nabla_a(\ln\Omega)\nabla_b(\ln\Omega)$$
$$- (n-2)g_{ab}g^{fe}\nabla_f(\ln\Omega)\nabla_e(\ln\Omega), \qquad (1.149)$$

$$\tilde{R} \equiv \tilde{g}^{ab}\tilde{R}_{ab} = \Omega^{-2}\left[R - 2(n-1)\,\Box\,(\ln\Omega)\right.$$
$$\left. - (n-1)(n-2)\frac{g^{ab}\nabla_a\Omega\nabla_b\Omega}{\Omega^2}\right], \qquad (1.150)$$

for $n \geq 2$. In the case of $n = 4$ spacetime dimensions, the transformation property of the Ricci scalar can be written as

$$\tilde{R} = \Omega^{-2}\left[R - \frac{6\,\Box\,\Omega}{\Omega}\right]$$
$$= \Omega^{-2}\left[R - \frac{12\,\Box\,(\sqrt{\Omega})}{\sqrt{\Omega}} + \frac{3g^{ab}\nabla_a\Omega\nabla_b\Omega}{\Omega^2}\right]. \qquad (1.151)$$

The Weyl tensor $C_{abc}{}^d$ with the last index contravariant is conformally invariant,

$$\widetilde{C_{abc}{}^d} = C_{abc}{}^d, \qquad (1.152)$$

but the same tensor with indices raised or lowered with respect to $C_{abc}{}^d$ is not. This property justifies the name of *conformal tensor* often used for $C_{abc}{}^d$ [587]. If the original metric g_{ab} is Ricci-flat ($R_{ab} = 0$), the conformally transformed metric $\tilde{g}_{ab}$ is not (cf. eq. (1.149)): in the conformally transformed world the conformal factor Ω plays the role of a form of matter. The consequences of this fact are evident in Section 3.2 in the discussion of Einstein frame and Jordan frame for gravitational waves in scalar-tensor gravity. However, if the Weyl tensor vanishes in one frame, it also vanishes in the conformally related frame: conformally flat metrics are mapped into conformally flat metrics, a property used in cosmology when mapping FLRW universes. In particular, de Sitter spaces with scale factor $a(t) = a_0 \exp(Ht)$ and a constant scalar field as a material source are mapped into de Sitter spaces, a feature related to the scale invariance of the exponential function and used in Section 7.3.

The source-free Klein-Gordon equation $\Box\phi = 0$ satisfied by the scalar field of BD theory in the absence of ordinary matter and self-interactions is not conformally invariant. However, its generalization

$$\Box\phi - \frac{n-2}{4(n-1)}R\phi = 0 \qquad (1.153)$$

for $n \geq 2$ is conformally invariant [701, 210]. In studies of the early universe, a potential energy density $V(\phi)$ for the scalar field is often introduced, yielding

$$\Box \phi - \xi\, R\, \phi - \frac{dV}{d\phi} = 0 \qquad (1.154)$$

instead of eq. (1.153), where ξ is a dimensionless coupling constant. It is reasonable to allow for the possibility that the scalar ϕ acquire a mass or other potential at high energies. If the evolution of ϕ is sufficiently fast, the potential term can be neglected with respect to the kinetic terms. Eq. (1.154) is conformally invariant in four spacetime dimensions if $\xi = 1/6$ and $V = 0$ or $V = \lambda \phi^4$ [701, 172, 892]. Even a constant potential V, equivalent to a cosmological constant in the action, corresponds to an effective mass for the scalar (not to be identified with a real mass [356]) which breaks conformal invariance [602].

In four spacetime dimensions, Maxwell's equations are conformally invariant, while the equation satisfied by the electromagnetic four potential A^a,

$$\Box A_a - R^b{}_a A_b = -4\pi j_a \qquad (1.155)$$

(where j^a is the four-current) is not conformally invariant [250, 95]. Quantum corrections to classical electrodynamics, including the generation of mass terms and the conformal anomaly, break the conformal invariance. When a field of arbitrary spin is considered in general spacetime dimensions, the conditions for conformal invariance are varied; they are studied in Ref. [510].

In general, the covariant conservation equation for a (symmetric) stress-energy tensor $T^{(m)}_{ab}$ representing ordinary matter,

$$\nabla^b T^{(m)}_{ab} = 0 \qquad (1.156)$$

is not conformally invariant [892]. In scalar-tensor theories in which the conformal factor Ω of eq. (1.141) depends on the BD-like scalar ϕ, the conformally transformed $\tilde{T}^{(m)}_{ab}$ satisfies the equation

$$\tilde{\nabla}^b \tilde{T}^{(m)}_{ab} = -\frac{d}{d\phi}\left[\ln \Omega(\phi)\right] \tilde{T}^{(m)} \tilde{\nabla}_a \phi \,. \qquad (1.157)$$

The conservation equation (1.156) then appears to be conformally invariant for a matter component that has vanishing trace $T^{(m)}$ of the energy-momentum tensor (this feature is associated with light-like behaviour and conformal invariance). Examples are the Maxwell field and a radiative fluid with equation of state $P^{(m)} = \rho^{(m)}/3$. Unless $T^{(m)} = 0$, eq. (1.157) describes an exchange of energy and momentum between matter and scalar field.

Scalar-tensor gravity

Since the geodesic equation ruling the motion of free particles in general relativity can be derived from the conservation equation (1.156), it immediately follows that timelike geodesics of the original metric g_{ab} are not geodesics of the rescaled metric $\tilde{g}_{ab}$. Particles that were in free fall in the universe (M, g_{ab}) are now subject to a force proportional to $\tilde{\nabla}^a \phi$ in the rescaled world $(M, \tilde{g}_{ab})$, a sort of fifth force. The definition of the stress-energy tensor

$$\tilde{T}_{ab}^{(m)} = \frac{-2}{\sqrt{-\tilde{g}}} \frac{\delta \left(\sqrt{-\tilde{g}}\, \mathcal{L}^{(m)}\right)}{\delta \tilde{g}^{ab}}, \qquad (1.158)$$

together with the rescaling (1.141) of the metric, yields

$$\tilde{T}_{ab}^{(m)} = \Omega^{-2} T_{ab}^{(m)}, \qquad \widetilde{T_a{}^{b\,(m)}} = \Omega^{-4} T_a{}^{b\,(m)}, \qquad \tilde{T}^{ab} = \Omega^{-6} T^{ab\,(m)}, \qquad (1.159)$$

and

$$\tilde{T}^{(m)} = \Omega^{-4} T^{(m)}. \qquad (1.160)$$

Now consider the stress-energy tensor of a perfect fluid,

$$T_{ab}^{(m)} = \left(P^{(m)} + \rho^{(m)}\right) u_a u_b + P^{(m)} g_{ab} :$$

the corresponding tensor in the rescaled world is

$$\tilde{T}_{ab}^{(m)} = \left(\tilde{P}^{(m)} + \tilde{\rho}^{(m)}\right) \tilde{u}_a \tilde{u}_b + \tilde{P}^{(m)} \tilde{g}_{ab}, \qquad (1.161)$$

where the four-velocity $\tilde{u}^a$ of the fluid satisfies

$$\tilde{g}_{ab} \tilde{u}^a \tilde{u}^b = -1. \qquad (1.162)$$

Together with the metric rescaling (1.141), this normalization gives

$$\tilde{u}^a = \Omega^{-1} u^a, \qquad \tilde{u}_a = \Omega\, u_a. \qquad (1.163)$$

By comparing eqs. (1.159) and (1.161) and using eq. (1.163), we obtain

$$\left(\tilde{P}^{(m)} + \tilde{\rho}^{(m)}\right) \tilde{u}_a \tilde{u}_b + \tilde{P}^{(m)} \tilde{g}_{ab} = \Omega^{-2} \left[\left(P^{(m)} + \rho^{(m)}\right) u_a u_b + P^{(m)} g_{ab}\right], \qquad (1.164)$$

and we can identify the transformation properties of the energy density and pressure of the fluid under the conformal transformation (1.141),

$$\tilde{\rho}^{(m)} = \Omega^{-4} \rho^{(m)}, \qquad \tilde{P}^{(m)} = \Omega^{-4} P^{(m)}. \qquad (1.165)$$

If, in the Jordan frame, the fluid has a barotropic equation of state of the form

$$P^{(m)} = (\gamma - 1)\rho^{(m)} \qquad (1.166)$$

with γ =constant, then the same equation of state is valid in the Einstein frame, thanks to the relations (1.165) between $\rho^{(m)}, P^{(m)}$ and their conformal cousins $\tilde{\rho}^{(m)}, \tilde{P}^{(m)}$. However, this property does not hold true for a general barotropic equation of state $P = P(\rho)$ which is not of the form (1.166).

In the case of FLRW metrics the usual Jordan frame conservation equation for ordinary matter

$$\frac{d\rho^{(m)}}{dt} + 3H\left(P^{(m)} + \rho^{(m)}\right) = 0 \qquad (1.167)$$

is modified in the Einstein frame to

$$\frac{d\tilde{\rho}^{(m)}}{dt} + 3\tilde{H}\left(\tilde{P}^{(m)} + \tilde{\rho}^{(m)}\right) = \frac{d(\ln\Omega)}{d\phi}\dot{\phi}\left(3\tilde{P}^{(m)} - \tilde{\rho}^{(m)}\right), \qquad (1.168)$$

as follows from eq. (1.157).

11.2 Brans-Dicke theory

The conformal factor of the conformal transformation (1.141) that brings the gravitational sector of the BD action (1.25) into the Einstein form is easy to spot and is given by [290]

$$\Omega = \sqrt{G\phi}. \qquad (1.169)$$

Then the scalar field redefinition

$$\tilde{\phi}(\phi) = \sqrt{\frac{2\omega + 3}{16\pi G}} \ln\left(\frac{\phi}{\phi_0}\right) \qquad (1.170)$$

(where $\phi \neq 0$, $\omega > -3/2$ and $\phi_0^{-1} = G$) brings the scalar field kinetic energy density into canonical form. The resulting *Einstein frame* action is (see Appendix A):

$$S = \int d^4x \left\{ \sqrt{-\tilde{g}} \left[\frac{\tilde{R}}{16\pi G} - \frac{1}{2}\tilde{g}^{ab}\tilde{\nabla}_a\tilde{\phi}\tilde{\nabla}_b\tilde{\phi} - U\left(\tilde{\phi}\right) \right] \right.$$
$$\left. + \exp\left(-8\sqrt{\frac{\pi G}{2\omega + 3}}\,\tilde{\phi}\right) \mathcal{L}^{(m)}[\tilde{g}] \right\}, \qquad (1.171)$$

where $\tilde{\nabla}_a$ is the covariant derivative operator of the rescaled metric $\tilde{g}_{ab}$ and

$$U\left(\tilde{\phi}\right) = V\left[\phi\left(\tilde{\phi}\right)\right] \exp\left(-8\sqrt{\frac{\pi G}{2\omega + 3}}\,\tilde{\phi}\right). \qquad (1.172)$$

The possibility of performing the conformal transformation is sometimes seen as a reason to restrict the range of values of the Jordan frame parameter to $\omega > -3/2$. Note that the Jordan frame scalar has dimensions of G^{-1}, while the Einstein frame scalar $\tilde{\phi}$ has dimensions of $G^{-1/2}$ and can be measured in Planck masses. If ϕ becomes constant (in the limit to general relativity), the Jordan and the Einstein frame coincide.

By looking only at the action (1.171) one could say that in the Einstein frame, gravity is described by general relativity, but there are two important differences. First, the free scalar $\tilde{\phi}$ acting as a source of gravity in the right hand side of the field equations is always present, i.e., one cannot consider solutions of the vacuum field equations $\tilde{R}_{ab} = 0$ in the Einstein frame as one can do in general relativity. The field $\tilde{\phi}$ permeates spacetime and cannot be eliminated, a feature that is reminiscent of the cosmological origin of ϕ (the inverse gravitational coupling determined by distant matter) in the original BD theory. This is in spite of the fact that formally the gravitational field is only described by the metric tensor $\tilde{g}_{ab}$ in the Einstein frame. The conformal transformation moves part of the Jordan frame gravitational variables (ϕ) into Einstein frame matter ($\tilde{\phi}$).

The second difference between Einstein frame BD theory and general relativity is that the matter part of the Lagrangian is now multiplied by the exponential factor in eq. (1.171), which describes an anomalous coupling of this ordinary matter to the scalar $\tilde{\phi}$ — the word "anomalous" is used here because the coupling does not occur in general relativity. This coupling is responsible for the fact that the energy-momentum tensor $\tilde{T}_{ab}^{(m)}$ of ordinary matter in the Einstein frame satisfies eq. (1.157) instead of the usual conservation equation $\nabla^b T_{ab}^{(m)} = 0$. Other consequences are the modification of the geodesic equation and of the equation of geodesic deviation, and the violation of the equivalence principle in the Einstein frame. A similar anomalous coupling occurs in string theories and it is useful to understand the features of this phenomenon, which are discussed below. The corrections to the equation of geodesic deviation are discussed in Chapter 3 in the context of scalar-tensor gravitational waves.

Under the conformal transformation (1.141), the stress-energy tensor $T_{ab}^{(m)}$ scales according to

$$\tilde{T}_{(m)}^{ab} = \Omega^s \, T_{(m)}^{ab} \,, \qquad \tilde{T}_{ab}^{(m)} = \Omega^{s+4} \, T_{ab}^{(m)} \,, \qquad (1.173)$$

where s is an appropriate conformal weight. The conservation equation $\nabla^b T_{ab}^{(m)} = 0$ transforms in four spacetime dimensions as [892]

$$\tilde{\nabla}_a \left(\Omega^s T_{(m)}^{ab} \right) = \Omega^s \, \nabla_a T_{(m)}^{ab} + (s+6) \, \Omega^{s-1} \, T_{(m)}^{ab} \nabla_a \Omega - \Omega^{s-1} g^{ab} \, T^{(m)} \nabla_a \Omega \,. \qquad (1.174)$$

It is convenient to choose the conformal weight $s = -6$ which yields, consistently with eq. (1.160),

$$\tilde{T}^{(m)} \equiv \tilde{g}^{ab} \tilde{T}_{ab}^{(m)} = \Omega^{-4} \, T^{(m)} \,. \qquad (1.175)$$

Hence, $\tilde{T}^{(m)}$ vanishes if and only if $T^{(m)} = 0$. Eq. (1.174) becomes

$$\tilde{\nabla}_a \tilde{T}_{(m)}^{ab} = -\tilde{T}^{(m)} \, \tilde{g}^{ab} \, \tilde{\nabla}_a \left(\ln \Omega \right) \,. \qquad (1.176)$$

Since for BD theory the conformal factor of the transformation (1.141) is $\Omega = \sqrt{G\phi}$,

$$\tilde{\nabla}_a \tilde{T}_{(m)}^{ab} = -\frac{1}{2\phi} \, \tilde{T}^{(m)} \, \tilde{\nabla}^b \phi \qquad (1.177)$$

or, in terms of the Einstein frame scalar [887],

$$\tilde{\nabla}_a \tilde{T}_{(m)}^{ab} = -\sqrt{\frac{4\pi G}{2\omega+3}} \, \tilde{T}^{(m)} \, \tilde{\nabla}^b \tilde{\phi} \,. \qquad (1.178)$$

One can now derive the geodesic equation from eq. (1.178). Consider a dust fluid with zero pressure described by the stress-energy tensor

$$\tilde{T}_{ab}^{(m)} = \tilde{\rho}^{(m)} \, \tilde{u}_a \, \tilde{u}_b \qquad (1.179)$$

and $\tilde{T}^{(m)} = -\tilde{\rho}^{(m)}$. Eq. (1.178) then yields

$$\tilde{u}_a \, \tilde{u}_b \, \tilde{\nabla}^b \tilde{\rho}^{(m)} + \tilde{\rho}^{(m)} \, \tilde{u}_a \, \tilde{\nabla}^b \tilde{u}_b + \tilde{\rho}^{(m)} \, \tilde{u}_c \, \tilde{\nabla}^c \tilde{u}_a - \sqrt{\frac{4\pi G}{2\omega+3}} \, \tilde{\rho}^{(m)} \, \tilde{\nabla}_a \tilde{\phi} = 0 \,. \qquad (1.180)$$

By introducing an affine parameter λ along the fluid worldlines with tangent $\tilde{u}^a$, eq. (1.180) is rewritten as

$$\tilde{u}_a \left(\frac{d\tilde{\rho}^{(m)}}{d\lambda} + \tilde{\rho}^{(m)} \, \tilde{\nabla}^c \tilde{u}_c \right) + \tilde{\rho}^{(m)} \left(\frac{d\tilde{u}_a}{d\lambda} - \sqrt{\frac{4\pi G}{2\omega+3}} \, \tilde{\nabla}_a \tilde{\phi} \right) = 0 \,, \qquad (1.181)$$

equivalent to
$$\frac{d\tilde{\rho}^{(m)}}{d\lambda} + \tilde{\rho}^{(m)}\, \tilde{\nabla}^c \tilde{u}_c = 0 \tag{1.182}$$
and
$$\frac{d\tilde{u}^a}{d\lambda} = \sqrt{\frac{4\pi G}{2\omega+3}}\, \tilde{\nabla}^a \tilde{\phi}\,. \tag{1.183}$$

The geodesic equation is then modified according to [887, 221, 223]

$$\frac{d^2 x^a}{d\lambda^2} + \tilde{\Gamma}^a_{bc}\frac{dx^b}{d\lambda}\frac{dx^c}{d\lambda} = \sqrt{\frac{4\pi G}{2\omega+3}}\, \tilde{\nabla}^a \tilde{\phi} \tag{1.184}$$

in the Einstein frame. There is a fifth force proportional to the gradient of $\tilde{\phi}$ that couples in the same way to any massive test particle. Because of this coupling, scalar-tensor theories in the Einstein frame are non-metric theories. The universality of free fall — all bodies fall with the same acceleration in a gravitational field, independently of their mass and composition — is violated by the fifth force correction to the geodesic equation due to the spacetime dependence of $\tilde{\nabla}^a \tilde{\phi}$. All metric theories of gravity, instead, satisfy the weak equivalence principle [909].

As expected, the equation of null geodesics is not affected by the conformal transformation: no fifth force correction appears in the Einstein frame for massless test particles. In fact, the equation of null geodesics is derived from the Maxwell equations in the high frequency limit of the geometric optics approximation, and Maxwell's equations are conformally invariant in four spacetime dimensions. Alternatively, we note that the electromagnetic field energy-momentum tensor has vanishing trace $T = 0$ and therefore the conservation equation $\nabla^b T_{ab} = 0$ is unaffected by the conformal transformation (1.141), together with the geodesic equation for a null dust described by eq. (1.179) when $\tilde{u}_a \tilde{u}^a = 0$.

A similar correction to the geodesic equation appears in low energy string theory [841, 417], in which the BD field is replaced by the string dilaton. The anomalous coupling between matter and the ϕ-field of BD theory has a counterpart in string theories, in which a similar coupling is responsible for violations of the equivalence principle. The violation must be kept small for the theory to pass the tests imposed by the available experiments in the Solar System [909], and is regarded as a low-energy manifestation of compactified theories [841, 221, 259, 223]. In string theory there is the additional complication that the coupling of the string dilaton is not universal: the latter couples with different strengths to bodies of different internal nuclear structure which carry a dilatonic charge q. The BD field, instead, couples universally to all forms of matter represented by a stress-energy tensor with nonvanishing trace, as described by eq. (1.157). Formally, one could replace the dilatonic charge q with the factor $2\sqrt{\pi G/(2\omega+3)}$ to draw a parallel between the two theories. However, while in string theory it is possible to eliminate the coupling by setting the dilatonic charge q to zero, one cannot get rid of the composition-independent coupling of Einstein frame BD theory.

11.3 Kaluza-Klein cosmology

In Section 1.5 it is shown that Kaluza-Klein theory reduces to a BD cosmology after compactification of the d extra dimensions, under the assumption that the metric on the submanifold M is a FLRW one and that the metric on the submanifold K is Riemannian and diagonal. The BD parameter $\omega = -(d-1)/d$ thus obtained is incompatible with the experimental bounds. Kaluza-Klein cosmology is therefore reformulated in the Einstein frame to make it viable [220, 221, 225, 914, 120, 774, 202, 727, 924, 925]. The conformal transformation (1.169) and the scalar field redefinition are the same as for BD theory, with the minor difference that the Jordan frame scalar of Kaluza-Klein theory is dimensionless — a memory of its geometrical origin. The rescaled metric, called the Einstein or Pauli metric is

$$\tilde{g}_{ab} = \phi\, g_{ab} \tag{1.185}$$

and the new scalar field is

$$\tilde{\phi} = \sqrt{\frac{d+2}{16\pi G\, d}}\, \ln\phi\,. \tag{1.186}$$

Note the dimensional factor $(16\pi G)^{-1/2}$ in eq. (1.186), which is often forgotten in the literature. The transformation (1.185) and (1.186) leads to the Einstein frame action

$$\begin{aligned} S = \int d^4x\, \sqrt{-\tilde{g}} &\left\{ \frac{\tilde{R}}{16\pi G} - \frac{1}{2}\tilde{g}^{ab}\tilde{\nabla}_a\tilde{\phi}\tilde{\nabla}_b\tilde{\phi} - U\left(\tilde{\phi}\right) \right. \\ &\left. + \lambda\left[\det(\rho_{\alpha\beta}) - 1\right] \right\}, \end{aligned} \tag{1.187}$$

where λ is a Lagrange multiplier introduced to enforce the constraint $\det(\rho_{\alpha\beta}) = 1$ on the metric $\rho_{\alpha\beta}$ defined on the submanifold K of the extra dimensions, usually assumed to be an Einstein space. $U(\tilde{\phi})$ is given by

$$U\left(\tilde{\phi}\right) = \frac{R_K}{16\pi G}\exp\left(-\alpha\,\tilde{\phi}\right) + \frac{\hat{\Lambda}}{16\pi G}\exp\left(-\beta\,\tilde{\phi}\right), \tag{1.188}$$

$$\alpha = 4\sqrt{\pi G}\left(\frac{d+2}{d}\right)^{1/2}, \qquad (1.189)$$

$$\beta = 4\sqrt{\pi G}\left(\frac{d}{d+2}\right)^{1/2}. \qquad (1.190)$$

It is often remarked that exponential potentials for scalar fields are the signature of compactified extra dimensions. R_K is the curvature of the metric ρ_{ab} and $\hat{\Lambda}$ is the higher-dimensional cosmological constant — the inclusion of which in Kaluza-Klein theory is debatable, according to the discussion in Section 1.5. The presence of the scalar of geometrical origin in Kaluza-Klein cosmology is interesting in discussions of the early universe, in particular during inflation. If one adopts the common view that inflation is a necessary ingredient for a successful cosmological model, based on the fact that inflation is the only causal mechanism that generates density perturbations with correlations beyond the horizon [495, 565], then one can think of testing Kaluza-Klein cosmology by looking for inflationary solutions to it. The test is feasible since the potential is completely determined by eqs. (1.188)-(1.190). However, the constants in these equations may change in supergravity or string versions of Kaluza-Klein theory due to quantum corrections. It is straightforward to compare the predictions of the simple Kaluza-Klein theory exposed here with the available data on temperature fluctuations in the cosmic microwave background [365, 366].

The Friedmann equation

$$\frac{\ddot{a}}{a} = -\frac{4\pi G}{3}\left(\tilde{\rho} + 3\tilde{P}\right), \qquad (1.191)$$

where the energy density and pressure of the minimally coupled scalar $\tilde{\phi}$ are

$$\tilde{\rho} = \frac{1}{2}\left(\dot{\tilde{\phi}}\right)^2 + U\left(\tilde{\phi}\right), \qquad (1.192)$$

$$\tilde{P} = \frac{1}{2}\left(\dot{\tilde{\phi}}\right)^2 - U\left(\tilde{\phi}\right), \qquad (1.193)$$

yields

$$\frac{\ddot{a}}{a} = -\frac{8\pi G}{3}\left[\left(\dot{\tilde{\phi}}\right)^2 - U(\tilde{\phi})\right]. \qquad (1.194)$$

The minimal condition for an inflationary solution is acceleration of the cosmic expansion, $\ddot{a} > 0$, which requires $U(\tilde{\phi}) > 0$. In practice, there is sufficient inflation if a slow rolling regime is established, in which $U \gg \left(\dot{\tilde{\phi}}\right)^2$. The parameters R_K and $\hat{\Lambda}$ can have either sign and d is also a parameter of the theory. There are several cases to consider.

- **Case 1)** $R_K = 0$, $\hat{\Lambda} \leq 0$: one has $U \leq 0$ and no inflationary solutions.

- **Case 2):** $R_K = 0$; $\hat{\Lambda} > 0$: the potential reduces to the single exponential $U = U_0 \exp\left(-\beta \tilde{\phi}\right)$, well known to be associated with power-law inflation [3, 594]. The potential

$$U\left(\tilde{\phi}\right) = U_0 \exp\left(-\sqrt{\frac{16\pi G}{p}}\,\tilde{\phi}\right) \qquad (1.195)$$

yields the solution

$$a(t) = a_0 t^p, \qquad \phi = \sqrt{\frac{p}{4\pi G}} \ln\left[\sqrt{\frac{8\pi G U_0}{p(3p-1)}}\,t\right], \qquad (1.196)$$

which is inflationary if $p > 1$, and is also an attractor in phase space. The spectral index of density perturbations in power-law inflation can be computed exactly [569] and is $n = 1 - 2/p$. Hence,

$$n = 1 - \frac{2d}{d+2} \leq \frac{1}{3}, \qquad (1.197)$$

a result incompatible with the *COBE* four years data at 1σ $n = 1.2 \pm 0.3$ [114, 440, 479, 54].

- **Case 3)** $R_K < 0$ and $\hat{\Lambda} \leq 0$ yield $U \leq 0$ and inflation does not occur.

- **Case 4)** $R_K < 0$ and $\hat{\Lambda} > 0$: the double exponential potential is studied in Ref. [309] with the conclusion that if $(8\pi G)^{-1/2}\,\beta > 0.5$ the resulting spectrum of density perturbations is ruled out by *COBE*. Since $d \geq 1$ implies $(8\pi G)^{-1/2}\,\beta \geq 0.82$, this Kaluza-Klein scenario is not viable.

- **Case 5)** $R_K > 0$ and $\hat{\Lambda} \neq 0$: this case is ruled out by an analysis similar to the previous one. It is interesting to note that in the limit of a large number of extra dimensions advocated by certain authors [26, 68, 542, 4, 5, 849], the potential is approximated by a single exponential $U = \left(R_K + \hat{\Lambda}\right)\exp\left(-\sqrt{16\pi G}\,\tilde{\phi}\right)$, which has $p = 1$ and describes a coasting universe with scale factor linear in time instead of accelerating.

In conclusion, the simple Kaluza-Klein theory considered here does not provide viable scenarios of inflation, a feature that is regarded as a serious shortcoming of any theory of the early universe. However, the values of the parameters α and β in the potential should not be taken literally, as they may change in string theories leading to a very similar action in the gravitational sector [224]. More sophisticated Kaluza-Klein cosmologies can be obtained by starting in the higher-dimensional world with Einstein-Cartan theory including torsion, instead of simple vacuum general relativity [225]. The effective four-dimensional theory is similar to the one discussed in this section, but it contains two scalar fields instead of a single one. These versions of Kaluza-Klein theory are able to generate viable inflationary scenarios [365]. It is remarkable that data from the cosmic microwave background and large scale structure surveys can then constrain the number of compactified extra dimensions.

11.4 Scalar-tensor theories

Let us consider now a scalar-tensor theory described in the Jordan frame by the action

$$S = \int d^4x \sqrt{-g} \left\{ \frac{1}{16\pi G} \left[\phi R - \frac{\omega(\phi)}{\phi} g^{ab} \nabla_a \phi \nabla_b \phi \right] - V(\phi) \right. $$
$$\left. + \alpha_m \mathcal{L}^{(m)} \right\}, \tag{1.198}$$

where the coupling constant α_m of ordinary matter is explicitly introduced. The conformal factor (1.169) is unchanged while the new scalar field is now defined by the differential relation

$$d\tilde{\phi} = \sqrt{\frac{2\omega(\phi)+3}{16\pi G}} \frac{d\phi}{\phi} \tag{1.199}$$

instead of eq. (1.170) (see Appendix B for the case of D spacetime dimensions). The Einstein frame action is

$$S = \int d^4x \sqrt{-\tilde{g}} \left\{ \frac{\tilde{R}}{16\pi G} - \frac{1}{2} \tilde{g}^{ab} \tilde{\nabla}_a \tilde{\phi} \tilde{\nabla}_b \tilde{\phi} - U\left(\tilde{\phi}\right) + \tilde{\alpha}_m\left(\tilde{\phi}\right) \mathcal{L}^{(m)} \right\}, \tag{1.200}$$

where

$$U\left(\tilde{\phi}\right) = \frac{V\left[\phi(\tilde{\phi})\right]}{(G\phi)^2} \tag{1.201}$$

and

$$\tilde{\alpha}_m\left(\tilde{\phi}\right) = \frac{\alpha_m}{(G\phi)^2}. \tag{1.202}$$

Eq. (1.200) is the action for Einstein gravity with a canonical scalar field with positive definite kinetic energy density, and with the important difference that the factor $\Omega^{-4} = (G\phi)^{-2}$ appears in front of the Lagrangian density of ordinary matter. This prefactor can be interpreted by saying that the fundamental coupling constants (here represented by α_m) vary with time due to their ϕ-dependence. The anomalous coupling described by the prefactor Ω^{-4} is responsible for the non-conservation of $\tilde{T}_{ab}^{(m)}$ (eq. (1.178)) and for the violations of the equivalence principle already discussed.

The conformal transformation technique can be used to generate exact solutions of scalar-tensor theory beginning from known solutions of general relativity [466, 109, 591, 869, 870, 871, 872, 873, 874, 81] or to derive approximate solutions of linearized theory [70]. However, a warning is mandatory about the physical relevance of these solutions in the presence of a potential $V(\phi)$: the solutions obtained with the conformal mapping rarely correspond to a physical potential in the new frame, even starting with a physically well motivated potential in the other frame. For example, consider BD theory in the Jordan frame and allow the BD scalar ϕ to acquire a mass m, as described by the potential $V(\phi) = m^2\phi^2/2$. By using eq. (1.201), the Einstein frame potential becomes

$$U = \frac{1}{2}\left(\frac{m}{G}\right)^2, \tag{1.203}$$

equivalent to a cosmological constant. A given functional form of the self-interaction potential $V(\phi)$ in the Jordan frame corresponds to a very different form of $U\left(\tilde{\phi}\right)$ in the Einstein frame.

Vice-versa, one wants to know what kind of potential $V(\phi)$ must be present in the Jordan frame in order to produce a mass term $U\left(\tilde{\phi}\right) = m^2\tilde{\phi}^2/2$ in the Einstein frame. Eqs. (1.170) and (1.201) yield

$$V(\phi) = U\left(\tilde{\phi}\right)\phi^2 = \frac{m^2 G}{32\pi}(2\omega + 3)\left[\phi\ln\left(\frac{\phi}{\phi_1}\right)\right]^2, \tag{1.204}$$

where ϕ_1 is a constant. This potential would be difficult to justify in the Jordan frame and is not simply related to a mass term for the Jordan frame scalar ϕ. Thus, while one can use exact solutions in the Einstein frame to generate solutions in the Jordan frame, these solutions correspond to unphysical potentials.

11.5 Generalized scalar-tensor theories

The action of generalized scalar-tensor theories

$$S = \frac{1}{16\pi} \int d^4x \sqrt{-g}\left\{ f(\phi, R) - \frac{\eta}{2} g^{ab}\nabla_a\phi\nabla_b\phi \right\}, \quad (1.205)$$

where we omit the matter part of the action for simplicity, can be brought into its Einstein frame form by a conformal transformation which was rediscovered in particular cases by many authors after its original introduction [908, 845, 778, 826, 79, 609, 441, 244, 894]. Quantum corrections originate R^2 terms in the Lagrangian density of Einstein theories: these quadratic Lagrangians are formally reduced to general relativity [476, 845, 908, 372, 122], a result later generalized to supergravity, Lagrangians with terms $\Box^k R$ with $k \geq 1$, Lagrangians polynomial in R [198, 894, 636], and Weyl theory [905, 296].

The conformal rescaling (1.141) with conformal factor

$$\Omega = \left[16\pi G \left| \frac{\partial f}{\partial R} \right| + \text{constant} \right]^{1/2} \quad (1.206)$$

and the scalar field redefinition

$$\tilde{\phi} = \frac{1}{\sqrt{8\pi G}} \sqrt{\frac{3}{2}} \ln\left[\sqrt{32\pi} G \left| \frac{\partial f}{\partial R} \right| \right] \quad (1.207)$$

reduce the action (1.205) to

$$\begin{aligned} S = \alpha \int d^4x \sqrt{-\tilde{g}} &\left\{ \frac{\tilde{R}}{16\pi G} - \frac{1}{2}\tilde{g}^{ab}\tilde{\nabla}_a\tilde{\phi}\tilde{\nabla}_b\tilde{\phi} \right. \\ &\left. - \frac{\eta\alpha}{2} \exp\left[-\sqrt{\frac{16\pi G}{3}}\,\tilde{\phi} \right] - U(\phi, \tilde{\phi}) \right\}, \end{aligned} \quad (1.208)$$

where [609, 498]

$$\alpha = \text{sign}\left(\frac{\partial f}{\partial R} \right), \quad (1.209)$$

$$\begin{aligned} U(\phi, \tilde{\phi}) = \alpha\, e^{-8\sqrt{\frac{\pi G}{3}}\,\tilde{\phi}} &\left[\frac{\alpha R(\phi, \tilde{\phi})}{16\pi G} \exp\left(\sqrt{\frac{16\pi G}{3}}\,\tilde{\phi} \right) \right. \\ &\left. - F(\phi, \tilde{\phi}) \right], \end{aligned} \quad (1.210)$$

$$F(\phi, \tilde{\phi}) = f\left[\phi, R(\phi, \tilde{\phi}) \right]. \quad (1.211)$$

The resulting theory is a non-linear σ-model with canonical gravity and two scalar fields ϕ and $\tilde{\phi}$. These two scalars reduce to a single one if $f\left(\phi, \tilde{\phi}\right)$ is linear in R,

$$f(\phi, R) = f(\phi) R - V(\phi) , \qquad (1.212)$$

in which case the Einstein frame action is

$$S = \frac{|f|}{f} \int d^4x \sqrt{-\tilde{g}} \left[\frac{\tilde{R}}{16\pi G} - \frac{1}{2} \tilde{g}^{ab} \tilde{\nabla}_a \tilde{\phi} \tilde{\nabla}_b \tilde{\phi} - U\left(\tilde{\phi}\right) \right] \qquad (1.213)$$

and

$$\tilde{\phi} = \frac{1}{\sqrt{8\pi G}} \int d\phi \left[\frac{2\eta f(\phi) + 6 \left(df/d\phi\right)^2}{4 f^2(\phi)} \right]^{1/2} , \qquad (1.214)$$

$$U\left(\tilde{\phi}\right) = \frac{\mathrm{sign}(f)\, V(\phi)}{(16\pi G f)^2} , \qquad (1.215)$$

in which it is understood to express ϕ as a function of $\tilde{\phi}$. In the case of non-linear theories of gravity, the field equations are of a higher order than second. Reduction to the simpler second order Einstein equations presents advantages when searching for solutions.

12. Singularities of the gravitational coupling

In the scalar-tensor theory described by the action (1.84), it is customary to interpret

$$G_{eff}(\phi) \equiv \frac{1}{8\pi f(\phi)} \qquad (1.216)$$

as an effective gravitational coupling, at least in the context of cosmology. In the weak field limit, the effective coupling measured in a Cavendish experiment assumes a different form — see Section 2.2.1. The effective coupling diverges if the scalar field values ϕ are such that $f(\phi) = 0$ and G_{eff} changes sign when ϕ crosses this boundary. Moreover, the conformal transformation (1.B.2) to the Einstein frame degenerates when $f(\phi) = 0$. If $f(\phi)$ becomes negative, there is the possibility that also

$$f_1(\phi) \equiv f(\phi) + \frac{3}{2} \left(\frac{df}{d\phi} \right)^2 \qquad (1.217)$$

vanish[6]. The set of points at which $f_1(\phi) = 0$ constitutes a singularity in the scalar field redefinition (1.B.3) that is necessary to define canonical variables for the Einstein frame version of the scalar-tensor theory.

The first kind of singularity is usually avoided in certain scalar-tensor theories by requiring $f(\phi) > 0$ in order to guarantee that gravity is attractive and that the graviton carries positive energy. For example in BD theory, in which $f(\phi) = \phi$, one usually requires $\phi > 0$. However, in other scalar-tensor theories the scalar is allowed to assume values that make $f(\phi)$ negative, which are not *a priori* forbidden by the classical dynamics. For example, this is the case in the nonminimally coupled theory described by $f(\phi) = \kappa^{-1} - \xi\phi^2$, often associated in the literature with a situation where $\kappa_{eff} \equiv \kappa\left(1 - \kappa\xi\phi^2\right)^{-1} < 0$, a regime that is sometimes explicitly called *antigravity* [454, 580, 824, 718, 486]. Other times, this regime is masked by the fact that the field equations are written as

$$G_{ab} = \kappa_{eff} T_{ab}^{(eff)}[\phi] , \qquad (1.219)$$

where $T_{ab}^{(eff)}[\phi]$ is an effective energy-momentum tensor for the scalar ϕ. When the time-time component of this equation is considered, κ_{eff} is allowed to become negative but the Hamiltonian constraint is satisfied because the effective energy density of the scalar $T_{ab}^{(eff)}[\phi] u^a u^b$, where u^a is the four-velocity of FLRW comoving observers, also becomes negative.

The singularity $f(\phi) = 0$ in κ_{eff} corresponds to a Cauchy horizon that is not hidden inside an event horizon and signals the loss of predictability. In fact, the Cauchy problem is well-posed in the Einstein frame, but not in the Jordan frame, and the vanishing of f precludes the possibility of defining the Einstein frame metric. Although not all the solutions of general relativity correspond to globally hyperbolic spacetimes (e.g., plane waves with parallel rays [702]), many authors feel so uncomfortable with the idea of a naked Cauchy horizon that they require $f(\phi) > 0$. An added disadvantage of the $f = 0$ singularity is that the covariant and gauge-invariant analysis of cosmological perturbations (see Chapter 6) becomes invalid and the equations for the perturbations become singular when $f \to 0$. Nevertheless, the dynamics do not disallow values of ϕ such that $f(\phi) = 0$.

---

[6] In $D > 2$ spacetime dimensions one defines f_1 as (cf. Appendix B)

$$f_1(\phi) \equiv f(\phi) + \frac{D-1}{D-2}\left(\frac{df}{d\phi}\right)^2 . \qquad (1.218)$$

It is widely believed that inflation occurred in the early universe leading to a spatially flat FLRW universe and erasing initial anisotropies. Hence it is meaningful to consider an anisotropic universe before inflation begins. This is the best context for understanding the singularities $f(\phi) = 0$ and $f_1(\phi) = 0$. When a simple Bianchi I model is considered in scalar-tensor gravity, the set of points $f(\phi) = 0$ turns out to be a spacetime singularity with divergent Kretschmann scalar $R_{abcd}R^{abcd}$ [824, 8, 7].

Let us turn our attention now to the second kind of singularity $f_1(\phi) = 0$. First, for this singularity to occur $f(\phi)$ must be negative and hence requiring $f(\phi)$ to always be positive eliminates this singularity also. Second, even if negative values of f are permitted, the singularity $f_1(\phi) = 0$ may be forbidden by the classical dynamics (this is the case, e.g., of open or critically open FLRW universes under reasonable assumptions — see below). The $f_1 = 0$ singularity in eq. (1.B.3) precludes the possibility of reformulating the scalar-tensor theory in the Einstein frame. Although this does not *a priori* imply pathologies in the dynamics, the $f_1 = 0$ singularity leaves one uncomfortable for the following reasons:

1. The Cauchy problem is not well-posed in the Jordan frame;

2. When quantizing linearized scalar-tensor gravity, the Einstein frame and not the Jordan frame metric perturbation is identified with the physical graviton and carries positive energy. A problem in defining the Einstein frame then introduces a problem in the quantization.

Again, the singularity $f_1 = 0$ is best understood by studying an anisotropic Bianchi model. It can be shown [8, 7] that a spacetime singularity appears when $f_1(\phi) = 0$ in the scalar-tensor theory (1.84), unless two conditions are met: f and $df/d\phi$ simultaneously vanish and f_1 has as zeros only the zeros of f. The singularities are avoided in induced gravity with a massive scalar, described by $f(\phi) = \epsilon\phi^2$ and $V(\phi) = m^2\phi^2/2$ if $\epsilon > 0$ and $\phi > 0$.

Let us prove now that the $f_1 = 0$ singularity is dynamically avoided in a closed or critically open FLRW universe under reasonable assumptions. In fact, the Hamiltonian constraint in such a universe,

$$3f\left(H^2 + \frac{K}{a^2}\right) = \rho^{(m)} + \frac{\omega}{2}\left(\dot\phi\right)^2 + V - 3H\dot f \qquad (1.220)$$

can be rewritten as

$$\left(H + \frac{\dot f}{2f}\right)^2 = \left(\frac{\dot f}{2f}\right)^2 + \frac{\omega}{6}\frac{\left(\dot\phi\right)^2}{f} + \frac{\rho^{(m)}}{3f} - \frac{K}{a^2} + \frac{V}{3f}. \qquad (1.221)$$

The left hand side of this equation is non-negative. Let us assume that $f + \frac{3}{2}\left(\frac{df}{d\phi}\right)^2 \leq 0$ (which implies that $f < 0$). Then the first two terms on the right hand side add up to

$$\left(\frac{\dot{f}}{2f}\right)^2 + \frac{\omega}{6}\frac{(\dot{\phi})^2}{f} = \frac{1}{6}\left(\frac{\dot{\phi}}{f}\right)^2\left[\omega f + \frac{3}{2}\left(\frac{df}{d\phi}\right)^2\right]$$

$$< \frac{1}{6}\left(\frac{\dot{\phi}}{f}\right)^2 (\omega - 1) f \leq 0 , \qquad (1.222)$$

where the last inequality is satisfied if $\omega \geq 1$, as is usually assumed, and the first two terms on the right hand side of eq. (1.221) give non-positive contributions. Now, by assuming that $V \geq 0$, $K = 0$ or $+1$, and $\rho^{(m)} \geq 0$, one also obtains

$$\left(\rho^{(m)} + V\right)\frac{1}{3f} - \frac{K}{a^2} \leq 0 , \qquad (1.223)$$

with the equality holding only for a vacuum, spatially flat universe with a free BD-like field. Hence the left hand side of eq. (1.221) is non-negative, while the right hand side is negative, an absurdity caused by assuming that $f + \frac{3}{2}\left(\frac{df}{d\phi}\right)^2 \leq 0$. Under the assumptions listed above, the singularity $f_1(\phi) = 0$ is dynamically forbidden.

APPENDIX 1.A: Conformal transformation for BD theory

By using eqs. (1.146) and (1.150), the fact that the conformal factor for BD theory is $\Omega = \sqrt{G\phi}$, and the property

$$\nabla_a \phi = \sqrt{\frac{16\pi G}{2\omega + 3}}\, \phi \nabla_a \tilde{\phi} , \qquad (1.A.1)$$

one obtains

$$\frac{\nabla^c \nabla_c \Omega}{\Omega} = \frac{\nabla^c \nabla_c \phi}{2\phi} - \frac{\nabla^c \phi \nabla_c \phi}{4\phi^2} \qquad (1.A.2)$$

and, using eq. (1.151),

$$R = G\phi \tilde{R} + \frac{3\nabla^c \nabla_c \phi}{\phi} - \frac{3}{2}\frac{\nabla^c \phi \nabla_c \phi}{\phi^2} . \qquad (1.A.3)$$

The integrand in the BD action (1.25) is

$$I \equiv \frac{\sqrt{-g}}{16\pi}\left(\phi R - \frac{\omega}{\phi} g^{ab}\nabla_a \phi \nabla_b \phi - V\right) + \sqrt{-g}\, \mathcal{L}^{(m)} ; \qquad (1.A.4)$$

the use of eqs. (1.146) and (1.A.3) yields

$$I = \frac{\sqrt{-\tilde{g}}}{16\pi}\left(\frac{\tilde{R}}{G} + \frac{3\nabla^c\nabla_c\phi}{G^2\phi^2} - \frac{3}{2G^2\phi^3}\nabla^c\phi\nabla_c\phi - \frac{\omega}{G^2\phi^3}\nabla^c\phi\nabla_c\phi - \frac{V}{G^2\phi^2}\right)$$
$$+ \sqrt{-\tilde{g}}\,(G\phi)^{-2}\,\mathcal{L}^{(m)}\,. \tag{1.A.5}$$

By using eq. (1.A.1), one computes the term in the integrand

$$\frac{\nabla^c\phi\nabla_c\phi}{G^2\phi^3} = \frac{16\pi}{2\omega+3}\,\tilde{\nabla}^c\tilde{\phi}\,\tilde{\nabla}_c\tilde{\phi}\,, \tag{1.A.6}$$

and substituting back in eq. (1.A.4), one has

$$I = \sqrt{-\tilde{g}}\left[\frac{\tilde{R}}{16\pi G} - \frac{1}{2}\tilde{g}^{ab}\tilde{\nabla}_a\tilde{\phi}\tilde{\nabla}_b\tilde{\phi} - U\left(\tilde{\phi}\right)\right] + \frac{3\sqrt{-g}}{16\pi}\Box\phi + \sqrt{-\tilde{g}}\,(G\phi)^{-2}\,\mathcal{L}^{(m)}\,, \tag{1.A.7}$$

where

$$U\left(\tilde{\phi}\right) = \frac{V\left[\phi\left(\tilde{\phi}\right)\right]}{(G\phi)^2} \tag{1.A.8}$$

and the expression $3\sqrt{-g}\,\Box\,\phi/(16\pi)$ contributes a vanishing boundary term. In fact,

$$\int_{V^{(4)}} d^4x\,\sqrt{-g}\,\nabla^c\nabla_c\phi = \int_{V^{(4)}} d^4x\,\partial_c\left(\sqrt{-g}\,\partial^c\phi\right) = \int_{\partial V} d^3x\,n_c\left(\sqrt{-g}\,\partial^c\phi\right)\,, \tag{1.A.9}$$

where n^c is the unit normal to the 3-dimensional boundary ∂V of the 4-dimensional region $V^{(4)}$. The variation of this integral vanishes because the dynamical variables are kept fixed on the boundary ∂V.

The expression (1.170) of the scalar field $\tilde{\phi}$ as a function of ϕ then yields

$$U\left(\tilde{\phi}\right) = V\left[\phi\left(\tilde{\phi}\right)\right]\exp\left(-8\sqrt{\frac{\pi G}{2\omega+3}}\,\tilde{\phi}\right)\,. \tag{1.A.10}$$

The Einstein frame action is then

$$S = \int d^4x\,\sqrt{-\tilde{g}}\left\{\left[\frac{\tilde{R}}{16\pi G} - \frac{1}{2}\tilde{g}^{ab}\tilde{\nabla}_a\tilde{\phi}\tilde{\nabla}_b\tilde{\phi} - U\left(\tilde{\phi}\right)\right] + e^{-8\sqrt{\frac{\pi G}{2\omega+3}}\,\tilde{\phi}}\,\mathcal{L}^{(m)}\right\}\,. \tag{1.A.11}$$

APPENDIX 1.B: Conformal transformation in D dimensions

Consider, in the case of $D > 2$ spacetime dimensions, the scalar-tensor theory described by the action

$$S = \int d^Dx\,\sqrt{-g}\left[f(\phi)R - \omega(\phi)\nabla^c\phi\nabla_c\phi - V + \alpha_m\,\mathcal{L}^{(m)}\right]\,; \tag{1.B.1}$$

the conformal transformation to the Einstein frame is given by the metric rescaling

$$g_{ab} \longrightarrow \tilde{g}_{ab} = f(\phi)^{\frac{2}{D-2}}\,g_{ab}\,, \tag{1.B.2}$$

APPENDIX 1.B: Conformal transformation in D dimensions

and by the scalar field redefinition

$$d\tilde{\phi} = \frac{d\phi}{f(\phi)}\sqrt{f(\phi) + \frac{D-1}{D-2}\left(\frac{df}{d\phi}\right)^2}. \tag{1.B.3}$$

The new scalar field potential in the Einstein frame is given by

$$U\left(\tilde{\phi}\right) = \frac{V\left[\phi\left(\tilde{\phi}\right)\right]}{f^{\frac{D}{D-2}}}. \tag{1.B.4}$$

Chapter 2

EFFECTIVE ENERGY-MOMENTUM TENSORS AND CONFORMAL FRAMES

1. The issue of the conformal frame

The conformal transformations presented in the previous chapter establish a mathematical equivalence between scalar-tensor theories and Einstein's general relativity, with the modifications dictated by the anomalous coupling of the Einstein frame scalar to ordinary matter and its consequences. *A priori*, the mathematical equivalence need not be a physical one. The literature using conformal transformations in cosmology and in alternative theories of gravity is plagued by confusion and ambiguity on the issue of the physical equivalence of the two conformal frames, or which frame should be regarded as the physical one. As a consequence, the conformal transformation technique is often misused and its results are given erroneous interpretations. Incompatible points of view occurring in the literature support the following ideas [613, 363]:

- (*first viewpoint*) that the Jordan and the Einstein frame are physically equivalent,

- (*second viewpoint*) that the two frames are inequivalent — the Jordan frame is the physical one, and the Einstein frame is unphysical,

- (*third viewpoint*) that the two frames are inequivalent — the Jordan frame is unphysical, and the Einstein frame is physical.

"Physical frame" means that the theory formulated in that set of variables (*frame*) is theoretically consistent and able to make predictions that can, in principle, be tested with experiments. Often an author's stand is not sharply stated and the issue is left under a veil of obscurity. Sometimes, the view of an author shifts within a certain paper, or calculations mix the two conformal frames without distinction. An analogous problem is raised in the context of low-energy string theories [287, 183, 194, 27]. For example, the bosonic string theory, described in the *string frame* (analogous to the Jordan frame) by the action [173]

$$S = \int d^n x \, \frac{\sqrt{-g}}{2k_n^2} \, e^{-2\varphi} \left[R - 4 g^{ab} \nabla_a \varphi \nabla_b \varphi - \frac{1}{12} H_{abc} H^{abc} - \frac{n-26}{3 l_S^2} \right] \tag{2.1}$$

can be mapped into its Einstein frame version by a conformal transformation and scalar field redefinition, obtaining

$$S = \int d^n x \, \frac{\sqrt{-\tilde{g}}}{2k_n^2} \left[\tilde{R} - \frac{4}{n-2} \tilde{g}^{ab} \tilde{\nabla}_a \tilde{\varphi} \tilde{\nabla}_b \tilde{\varphi} - \frac{1}{12} e^{\frac{-8\tilde{\varphi}}{n-2}} H_{abc} H^{abc} - \left(\frac{n-26}{3 l_S^2} \right) e^{\frac{4\tilde{\varphi}}{n-2}} \right]. \tag{2.2}$$

While at high energies string theory is conformally invariant, this is not so in the low energy world; like scalar-tensor theories, the issue of the conformal frame deserves attention. The problem of the conformal frame for scalar-tensor theories is a long-standing one, and the concern over it is echoed even among philosophers of science [901]. In the following sections we present separately the three common viewpoints on conformal frames, and the discussion that has arisen due to misinterpretation.

1.1 The first viewpoint

The viewpoint that the Einstein and the Jordan frame are physically equivalent is correct and perfectly consistent, at least at the classical level. This viewpoint can be traced back to Dicke's [290] original paper introducing the conformal transformation technique for BD theory. While technically correct and clear, Dicke's paper is often forgotten or misinterpreted. The two conformal frames are physically equivalent provided that one accepts the idea that, in the Einstein frame, the units of mass, time, and length are varying (they must scale with appropriate powers of the BD scalar ϕ). In other words, physics must be conformally invariant and the symmetry group of gravitational theory should be enlarged to include not only the group of diffeomorphisms, but also conformal transformations. A conformal transformation is seen as a change of scale in the units of mass, time and length, and this change of scale depends on the spacetime point. Since the laws of physics must be invariant under a transformation of units (the choice of units being arbitrary), conformal invariance follows. While this point of view is certainly tenable, the novelty is that the transformation of units depends on the spacetime point: instead of a system of units rigidly attached to spacetime, the Einstein frame exhibits a system of units that change with $\Omega(\phi)$, and hence with the spacetime point. This idea is reminiscent of the basic feature of Weyl theory [906] that the length of a vector changes during parallel transport through the spacetime manifold, together with the vector's direction.

Since the line element

$$ds^2 = g_{ab}\, dx^a dx^b \tag{2.3}$$

must be invariant under a point-dependent rescaling of units, $d\tilde{s}^2 = ds^2$, and g_{ab} transforms according to $g_{ab} \to \tilde{g}_{ab} = \Omega^2 g_{ab}$, then the units of length and time transform according to [290, 644]

$$dt \longrightarrow d\tilde{t} = \Omega\, dt\,, \tag{2.4}$$

$$dx^i \longrightarrow d\tilde{x}^i = \Omega\, dx^i\,, \tag{2.5}$$

while the unit of mass scales in the opposite way,

$$m \longrightarrow \tilde{m} = \Omega^{-1}\, m\,. \tag{2.6}$$

Being the ratio of space and time, the speed of light in vacuo c is invariant, local Lorentz-invariance is preserved, and also the Planck constant with dimensions $[h] = [ML^2T^{-1}]$ is unchanged. Energy, with dimensions $[E] = [Mc^2]$, scales like a mass. The Planck mass, length, and time

$$m_{pl} = \sqrt{\frac{\hbar c}{G}}\,, \qquad l_{pl} = \sqrt{\frac{G\hbar}{c^3}}\,, \qquad t_{pl} = \sqrt{\frac{G\hbar}{c^5}}\,, \tag{2.7}$$

are constant in the Einstein frame and variable in the Jordan frame. If one accepts this point of view, then the Jordan and Einstein frame versions of scalar-tensor theories are merely two physically equivalent representations of the same theory. In the Jordan frame, the effective gravitational coupling $G_{eff} = \phi^{-1}$ varies, while h, c, the masses of elementary particles, and the coupling constants of physics are true constants, and the theory is metric. In the Einstein frame instead, G, h, and c are constants, while the masses of elementary particles and the coupling "constants" vary with time, due to the fact that the units of time, length, and mass themselves vary with ϕ. For example, let m_p be the (constant) proton mass in the Jordan frame; in the Einstein frame, the ϕ-dependent mass of the proton is $\tilde{m}_p = \Omega^{-1} m_p$, and the ratio of $\tilde{m}_p$ to a mass unit $\tilde{m}_u$ arbitrarily chosen and varying with ϕ in the same way as $\tilde{m}_p$ is constant:

$$\frac{\tilde{m}_p}{\tilde{m}_u} = \frac{\Omega^{-1} m_p}{\Omega^{-1} m_u} = \frac{m_p}{m_u}\,, \tag{2.8}$$

the value that this ratio assumes in the Jordan frame. Hence, the measurement of the proton mass does not distinguish between the two frames.

Also, the coupling "constants" vary with ϕ. It is instructive to consider, for example, BD theory with a matter sector consisting of a massive Klein-Gordon field ψ (not to be confused with the gravitational BD field ϕ). The action is

$$S = \int d^4x \sqrt{-g}\, \mathcal{L} = \int d^4x \sqrt{-g} \left(\mathcal{L}^{(g)} + \alpha_{KG}\, \mathcal{L}^{(KG)} \right), \tag{2.9}$$

where

$$\mathcal{L}^{(g)} = R\phi - \frac{\omega}{\phi} g^{cd} \nabla_c \phi \nabla_d \phi, \tag{2.10}$$

$$\mathcal{L}^{(KG)} = -\frac{1}{2} \left(g^{cd} \nabla_c \psi \nabla_d \psi + m^2 \psi^2 \right), \tag{2.11}$$

and where $\alpha_{KG} = 16\pi G$ is the Klein-Gordon coupling constant. By performing the conformal transformation with conformal factor (1.169) and the scalar field redefinition (1.170), one obtains

$$\begin{aligned}\sqrt{-g}\,\mathcal{L} &= \sqrt{-\tilde{g}} \left\{ \tilde{\mathcal{L}}^{(g)} + \tilde{\alpha}_{KG}(\phi)\, \tilde{\mathcal{L}}^{(KG)}[\tilde{g}_{ab}, \psi] \right\} \\ &= \sqrt{-\tilde{g}} \left[\tilde{R} - \frac{1}{2} \tilde{g}^{cd} \tilde{\nabla}_c \tilde{\phi} \tilde{\nabla}_d \tilde{\phi} - \alpha_{KG} \Omega^{-2} \left(\tilde{g}^{cd} \tilde{\nabla}_c \psi \tilde{\nabla}_d \psi \right.\right. \\ &\quad \left.\left. + \tilde{m}^2 \psi^2 \right) \right], \end{aligned} \tag{2.12}$$

where

$$\tilde{\mathcal{L}}^{(g)} = \tilde{R} - \frac{1}{2} \tilde{g}^{cd} \tilde{\nabla}_c \tilde{\phi}\, \tilde{\nabla}_d \tilde{\phi}, \tag{2.13}$$

$$\tilde{\alpha}_{KG}\left(\tilde{\phi}\right) = \Omega^{-2}\, \alpha_{KG} = 16\pi G\, \exp\left[-8 \sqrt{\frac{\pi G}{2\omega + 3}}\, \tilde{\phi} \right], \tag{2.14}$$

$$\tilde{m}\left(\tilde{\phi}\right) = \frac{m}{\Omega} = \frac{m}{\sqrt{G\phi}} = m\, \exp\left[-4 \sqrt{\frac{\pi G}{2\omega + 3}}\, \tilde{\phi} \right]. \tag{2.15}$$

In the Einstein frame, the coupling constant and the mass of the Klein-Gordon field ψ acquire a dependence from the BD scalar. This situation is general, the only exceptions being forms of matter that satisfy conformally invariant evolution equations. The Maxwell field in four spacetime dimensions is such a system: if one adds to the action (2.9) the action for the electromagnetic field

$$S^{(e.m.)} = \int d^4x \sqrt{-g}\, \alpha_{e.m.}\, \mathcal{L}^{(e.m.)}, \tag{2.16}$$

where $\alpha_{e.m.} = 4$ and
$$\mathcal{L}^{(e.m.)} = -\frac{1}{4} F_{ab} F^{ab} \tag{2.17}$$

(F_{ab} is the antisymmetric Maxwell tensor), it is straightforward to verify conformal invariance:

$$\begin{aligned}
\sqrt{-g}\,\mathcal{L}^{(e.m.)} &= -\frac{1}{4}\sqrt{-g}\,g^{ac}g^{bd}F_{ab}F_{cd} \\
&= -\frac{1}{4}\left(\Omega^{-4}\sqrt{-\tilde{g}}\right)\left(\Omega^2\,\tilde{g}^{ac}\right)\left(\Omega^2\,\tilde{g}^{bd}\right)F_{ab}F_{cd} = \\
&= -\frac{1}{4}\sqrt{-\tilde{g}}\,\tilde{g}^{ac}\tilde{g}^{bd}\tilde{F}_{ab}\tilde{F}_{cd}\,,
\end{aligned} \tag{2.18}$$

with $\tilde{F}_{ab} = F_{ab}$. With the exception of conformally invariant matter, the coupling constants of physics (fine structure constant α, Fermi coupling constant G_F, et cetera) vary with Ω and ϕ.

The implementation of the idea that physics should be conformally invariant conflicts with current terminology. For example, consider a massive, conformally coupled Klein-Gordon field, which satisfies the equation

$$\Box \psi - \frac{R}{6}\psi - m^2 \psi = 0 \tag{2.19}$$

in the Jordan frame. The presence of the mass m breaks the conformal invariance that is enjoyed by eq. (2.19) when $m = 0$; and this is standard terminology. However, following Dicke, one can generalize the concept of conformal invariance by allowing the mass to change with the field ϕ. By using the relation

$$g^{ab}\nabla_a\nabla_b\psi - \frac{R}{6}\psi = \Omega^3\left[\tilde{g}^{ab}\tilde{\nabla}_a\tilde{\nabla}_b\tilde{\psi} - \frac{\tilde{R}}{6}\tilde{\psi}\right], \tag{2.20}$$

where $\tilde{\psi} = \Omega^{-1}\psi$, eq. (2.19) yields

$$\widetilde{\Box}\tilde{\psi} - \frac{\tilde{R}}{6}\tilde{\psi} - \tilde{m}^2\tilde{\psi} = 0\,, \tag{2.21}$$

where $\tilde{m} = m/\Omega$, in agreement with eq. (2.6). Thus, if one accepts that masses scale according to this law, eq. (2.19) is conformally invariant in Dicke's sense, although not according to the standard terminology.

Theories have been proposed in which proper lengths, times and masses scale with a fundamental scalar field (e.g., [466, 88, 176, 177, 105, 749, 488, 490, 489, 491, 492, 107, 106, 108]) and in this context, the Jordan and Einstein frame are physically equivalent in the sense described above. Unfortunately, in the literature on scalar-tensor cosmology many authors ignore the features of the Einstein frame and refer measurements in the Einstein frame to a rigid system of units instead of using units that vary with ϕ. This corresponds to adopting the relation $d\tilde{s}^2 = \Omega^2 \, ds^2$ between line elements, instead of $d\tilde{s}^2 = ds^2$. In such cases, the theory used is not physically equivalent to the Jordan frame formulation of scalar-tensor theories, but is a completely new theory related to the original one formulated in the Jordan frame only by a mathematical transformation. It is rather obvious that, if one does not adopt Dicke's [290] point of view, the Jordan and Einstein frame are physically inequivalent. First, in general physics is not conformally invariant with the standard terminology (think of the massive Klein-Gordon field described by eq. (2.19) and used as an example). Second, conformally related metrics are not physically equivalent. For example, consider the FLRW metric with flat spatial sections given by the line element

$$ds^2 = a^2(\eta) \left(-d\eta^2 + dx^2 + dy^2 + dz^2 \right) , \qquad (2.22)$$

or by $g_{ab} = \Omega^2 \, \eta_{ab}$, where η_{ab} is the Minkowski metric, η is the conformal time defined by $dt = a \, d\eta$ in terms of the comoving time t, (x, y, z) are comoving Cartesian coordinates, and $\Omega = a(\eta)$. The metric g_{ab} is physically very different from the flat metric η_{ab}: g_{ab} describes a time-dependent expanding or contracting universe with spacetime curvature, possibly big bang or big crunch singularities, cosmological redshift and matter, while none of these features are present in Minkowski space. It is only if one adopts Dicke's [290] point of view that the two metrics are physically equivalent. Then the expansion of the universe can be described by assuming that the units of length and time shrink, while the line element is the Minkowskian one.

Similarly, when the Jordan and the Einstein frame are used in the cosmological literature, most often a rigid system of units is used in the Einstein frame. While this may be legitimate from the mathematical point of view, it leads to a theory that is not physically equivalent to the original one in the Jordan frame. It may still make sense to consider this new theory in the Einstein frame to model the universe or black holes, but the physical inequivalence of the two theories must be kept in mind at all times. In this book we consider the Einstein frame version of scalar-tensor theories with rigid units, due to its popularity and to the large amount of literature devoted to it. However, it should be noted that this popularity is largely due to the fact that Dicke's original paper and the physical equivalence of the two frames (with running units in the Einstein frame) are forgotten or misinterpreted.

The physical inequivalence of the two frames manifests itself in various phenomena: for example, massive test particles follow geodesic worldlines in the Jordan frame, but deviate from geodesics in the Einstein frame due to a fifth force correction. The latter, in turn, arises from the direct coupling of $\tilde{\phi}$ to ordinary (non-conformally invariant) matter. Ordinary matter is covariantly conserved in the Jordan frame but not in the Einstein frame, except for matter with $T^{(m)} = 0$. The energy conditions (see Section 2.1.6) are satisfied in the Einstein frame and violated in the Jordan frame. As a consequence, singularities present in the Einstein frame can be avoided in the Jordan frame [529, 725]. The Hilbert and Palatini actions for scalar-tensor theories are equivalent in the Einstein frame, but are inequivalent in the Jordan frame [869, 875].

The Arnowitt-Deser-Misner mass of an asymptotically flat spacetime, being a global quantity, does not scale according to the simple law $M_{ADM} \to \tilde{M}_{ADM} = \Omega^{-1} M_{ADM}$ [193, 194]. Generalizations of the Brown-York and of the Hawking-Horowitz quasi-local energy to scalar-tensor theories have been shown to be conformally invariant if the scale factor Ω is a monotonic function of ϕ [145].

While the discussion above points out the differences between the Jordan and the Einstein frame with fixed units of mass, time and length in a purely classical context, these differences are even more significant in the quantum context. In cosmology, a very relevant topic is the quantization of the scalar field fluctuations in the early universe, which generate density perturbations that, in turn, seed the formation of galaxies and clusters at a later epoch. Quantization and the conformal transformation have been found not to commute [402, 668, 333]. Suppose, for example, that an adiabatic vacuum state of the scalar field ϕ is used for quantization in the Jordan frame — the conformally transformed state of the Einstein frame field $\tilde{\phi}$ is not, in general, a vacuum state. The transformation between the fields ϕ and $\tilde{\phi}$ is nonlinear and the two states cannot be related by a Bogoliubov transformation. It is true however that, as long as one does not go beyond first order, scalar and tensor perturbations can be quantized in slow-roll inflation with de Sitter spaces as attractors by using the conformal transformation to the Einstein frame [498, 524, 503, 504].

In the following section we denote by *Einstein frame formulation* of a scalar-tensor theory the version of the theory formulated by assuming a rigid system of units of time, mass and length, against the spirit of Dicke's paper and according to mainstream literature. The issue then becomes: which of the two conformal frames is physical?

1.2 The second viewpoint

The viewpoint 2) that the Jordan frame is the physical one, and that the Einstein frame (with fixed units) is unphysical cannot be supported in general. We provide a counterexample based on linearized BD theory [361, 111]. Consider scalar-tensor gravitational waves described in the Jordan frame of BD theory by

$$g_{ab} = \eta_{ab} + h_{ab}, \qquad (2.23)$$

$$\phi = \phi_0 + \varphi, \qquad (2.24)$$

where the perturbations h_{ab} and φ satisfy

$$O(h_{cd}) = O\left(\frac{\varphi}{\phi_0}\right) = O(\epsilon), \qquad (2.25)$$

with ϵ a smallness parameter and ϕ_0 a constant. The Jordan frame linearized field equations in a region outside sources are

$$R_{ab} = \frac{\partial_a \partial_b \varphi}{\phi_0}, \qquad (2.26)$$

$$\eta^{ab} \partial_a \partial_b \varphi = 0. \qquad (2.27)$$

The scalar field perturbation is expanded as a Fourier integral of monochromatic plane waves

$$\varphi_{\vec{k}} = \varphi_0 \cos(k_a x^a), \qquad (2.28)$$

where the amplitude φ_0 is constant and $\eta_{ab} k^a k^b = 0$. To lowest order in ϵ, the effective energy density of the scalar waves with respect to an observer with (timelike) four-velocity ξ^a is

$$T_{ab} \xi^a \xi^b = -(k_a \xi^a)^2 \frac{\varphi_{\vec{k}}}{\phi_0}, \qquad (2.29)$$

where

$$T_{ab}[\varphi] = \frac{\partial_a \partial_b \varphi}{\phi_0} \qquad (2.30)$$

is identified with the stress-energy tensor of the scalar modes. For a monochromatic wave (2.28), the effective energy density (2.29) does not have definite sign and oscillates with the frequency of $\varphi_{\vec{k}}$. The effective energy density is not quadratic in the first derivatives of the field φ, but depends on its second derivatives. As a consequence, the effective energy density of the scalar modes is of order ϵ, while the contribution of the tensor modes is given by Isaacson's effective stress-energy tensor [511] and is of order ϵ^2.

If one considers the emission of nearly monochromatic gravitational waves by a binary stellar system, one obtains the paradoxical result that the gravitational waves emitted carry away *negative* energy from the binary. In fact, the energy emitted is dominated by the energy of the scalar modes (of order ϵ), while the contribution of the tensor modes (of order ϵ^2) is negligible. While it is true that the sign of the emitted energy is reversed in the next half period, and that the emitted flux averages to zero over many wave periods, it is disturbing that the negative flux persists over macroscopic times which may be of the order of months. For example, this is the case in the system μ-Sco with period $\simeq 3 \cdot 10^5$ s.

On the other hand, linearized BD theory in the Einstein frame does not suffer from this problem and the effective energy density of the scalar modes is positive definite. In the Einstein frame, the metric and scalar field are decomposed according to

$$\tilde{g}_{ab} = \eta_{ab} + \tilde{h}_{ab}, \qquad (2.31)$$

$$\tilde{\phi} = \tilde{\phi}_0 + \tilde{\varphi}, \qquad (2.32)$$

where $\tilde{\phi}_0$ is constant and $\mathrm{O}\left(\tilde{h}_{ab}\right) = \mathrm{O}\left(\tilde{\varphi}/\tilde{\phi}_0\right) = \mathrm{O}\left(\epsilon\right)$. The expansion is consistent with the conformal transformation (1.169) and (1.170) which, in conjunction with eqs. (2.31) and (2.32) yields

$$\tilde{\phi} = \sqrt{\frac{2\omega + 3}{16\pi G}} \frac{\varphi}{\phi_0} + C + \mathrm{O}\left(\epsilon^2\right) \qquad (2.33)$$

where C is an integration constant. Hence, the identification

$$\tilde{\phi}_0 = C, \qquad \tilde{\varphi}_0 = \sqrt{\frac{2\omega + 3}{16\pi G}} \frac{\varphi_0}{\phi_0}, \qquad (2.34)$$

follows. The conformal transformation and the metric decompositions (2.23) and (2.31) also yield

$$\tilde{h}_{ab} = h_{ab} + \frac{\varphi}{\phi_0} \eta_{ab} + \mathrm{O}\left(\epsilon^2\right), \qquad (2.35)$$

which explicitly shows how the Einstein frame gravitational waves recombine the Jordan frame tensor and scalar fields, and also makes it clear that the metric perturbations have the same order of magnitude in the two frames.

In the Einstein frame, the linearized field equations for $\tilde{\varphi}$ and

$$\overline{\tilde{h}}_{ab} \equiv \tilde{h}_{ab} - \frac{1}{2}\eta_{ab}\,\eta^{cd}\,\tilde{h}_{cd} \qquad (2.36)$$

are

$$\tilde{g}^{cd}\tilde{\nabla}_c\tilde{\nabla}_d\overline{\tilde{h}}_{ab} = 0 , \qquad (2.37)$$

$$\eta^{cd}\partial_c\partial_d\,\tilde{\varphi} = 0 . \qquad (2.38)$$

Again, the solutions of eq. (2.38) are expressed as Fourier integrals of monochromatic plane waves

$$\tilde{\varphi}_{\vec{p}} = \tilde{\varphi}_0 \cos\left(p_c\,x^c\right) \qquad (2.39)$$

with $\eta_{ab}\,p^b p^c = 0$. The effective stress-energy tensor associated with the Einstein frame scalar is

$$\tilde{T}_{ab}[\tilde{\varphi}] = \partial_a\tilde{\varphi}\,\partial_b\tilde{\varphi} - \frac{1}{2}\eta_{ab}\,\eta^{cd}\,\partial_c\tilde{\varphi}\,\partial_d\tilde{\varphi} , \qquad (2.40)$$

which is the canonical form of a scalar field energy-momentum tensor [892]. For the monochromatic plane wave (2.39), an observer with (timelike) four-velocity v^c experiences the total gravitational wave energy density

$$\tilde{T}_{ab}\,v^a v^b = \left\{\tilde{T}_{ab}\left[\varphi\right] + \tilde{T}_{ab}^{(eff)}[\tilde{h}_{cd}]\right\} v^a v^b$$

$$= \left[v_c\,p^c\,\tilde{\varphi}_0 \sin\left(p_d\,x^d\right)\right]^2 + \tilde{T}_{ab}^{(eff)}[\tilde{h}_{cd}]\,v^a v^b \geq 0 . \qquad (2.41)$$

The non-canonical form (2.30) of the stress-energy tensor associated with φ_0 in the Jordan frame is the origin of the problem of negative energy fluxes. In the Einstein frame — in which the theory reduces to Einstein gravity with a canonical scalar field — there are no negative energies. Note that in this example the association of an effective energy to linearized gravitational waves around a flat background is unambiguous, while the situation is not so clear in the full theory. In conclusion, there are situations in which the Jordan frame version of scalar-tensor theories is not viable, while its Einstein frame counterpart (with fixed units) is viable and free of ambiguities.

1.3 The third viewpoint

The third viewpoint asserting that the Jordan frame version of scalar-tensor theories is unphysical and that the Einstein frame version is the only physical one is usually based on the following argument. The Jordan frame version of the theory has a kinetic energy density of the scalar field ϕ that may be negative or with an indefinite sign. In the Einstein frame, instead, the kinetic energy density of $\tilde{\phi}$ is positive definite. A negative kinetic energy is usually taken as a sign that the theory does not have a stable ground state and that the system decays toward a lower and lower energy state *ad infinitum*. The existence of a ground state that is stable against small fluctuations is a necessary requirement for a viable classical theory of gravity [833, 319, 2]. A negative energy is often associated with the formulation of the theory in terms of unphysical variables. The example in the previous subsection using linearized gravitational waves in BD theory supports this argument: there, the energies of scalar and tensor waves are identifiable without ambiguities. Any energy term describing the interaction between scalar and tensor modes is absent to the first order in ϵ in the Jordan frame (the equations (2.26) and (2.27) for the two kinds of perturbations are decoupled to this order). Hence, the problem of the negative energy radiated by a binary stellar system originates from the indefiniteness of the sign of the effective kinetic energy of ϕ, which is the only term of order ϵ.

However, in the full theory it is not trivial to define the energy of the Jordan frame scalar ϕ, because ϕ and g_{ab} are directly coupled. Any identification of energies would presumably include an energy density for ϕ, an energy density for g_{ab}, and interaction terms describing the energy exchange between the two fields. Despite much effort, nothing is yet known about the energy density of g_{ab}, and the problem of energy localization for the gravitational field is not solved even in general relativity. The argument of negative kinetic energy density of ϕ leading to runaway solutions is therefore not compelling, since ϕ is only a subsystem of a larger physical system comprising g_{ab}, and one should consider the *total* energy density of the system instead of the energy of ϕ.

Inspection of the BD field equation (1.29) naturally leads to the (somewhat naive) identification of

$$T_{ab}^{(I)}[\phi] = \frac{\omega}{\phi^2}\left(\nabla_a\phi\nabla_b\phi - \frac{1}{2}g_{ab}\nabla^c\phi\nabla_c\phi\right)$$
$$+ \frac{1}{\phi}\left(\nabla_a\nabla_b\phi - g_{ab}\Box\phi\right) - \frac{V}{2\phi}g_{ab} \qquad (2.42)$$

with the effective stress-energy tensor of the BD field (e.g., [711, 775, 762, 639, 895]). This identification reflects the prejudice that the BD field equations should be read as effective Einstein equations. While this approach is useful in certain applications, it is ultimately questionable because the BD field equations are definitely different from those of general relativity.

In the case of a FLRW metric

$$ds^2 = -dt^2 + a^2(t)\left[\frac{dr^2}{1-Kr^2} + r^2\left(d\theta^2 + \sin^2\theta\, d\varphi^2\right)\right], \quad (2.43)$$

$T_{ab}^{(I)}[\phi]$ assumes the form of a perfect fluid with energy density and pressure given, in comoving coordinates, by

$$\rho_\phi^{(I)} = T_{00}^{(I)}[\phi] = \frac{\omega}{2}\left(\frac{\dot\phi}{\phi}\right)^2 - 3H\frac{\dot\phi}{\phi} + \frac{V}{2\phi}, \quad (2.44)$$

$$P_\phi^{(I)} = \frac{T_{ii}^{(I)}[\phi]}{g_{ii}} = \frac{\omega}{2}\left(\frac{\dot\phi}{\phi}\right)^2 + \frac{\ddot\phi}{\phi} + 2H\frac{\dot\phi}{\phi} - \frac{V}{2\phi}, \quad (2.45)$$

and the time-time component of the field equation (1.29) reduces to

$$H^2 = \frac{8\pi}{3}\frac{\rho^{(m)}}{\phi} + \frac{\rho_\phi^{(I)}}{3} - \frac{K}{a^2}, \quad (2.46)$$

where $\rho^{(m)}$ is the energy density of ordinary matter and K the curvature index of the metric. Consider pure BD gravity in vacuum ($T_{ab}^{(m)} = 0$): then $\rho_\phi^{(I)} = 3\left(H^2 + K/a^2\right) \geq 0$ for spatially flat ($K = 0$) and closed ($K = +1$) FLRW universes and, with the definition (2.42) of effective energy-momentum tensor for ϕ, the energy density of ϕ is non-negative. There are no runaway solutions and the Jordan frame version of BD theory appears to be perfectly legitimate in this case.

There are two points to learn from this example:
a) the discussion of the sign of the energy requires prior identification of an energy-momentum complex for the BD-like scalar, which is not unique or free of ambiguities, and
b) once a reasonable definition of $T_{ab}[\phi]$ is found, the argument of negative kinetic energy density for ϕ is incomplete (kinetic energy is not the only form of energy associated with ϕ) and cannot be used to rule out the Jordan frame version of the theory in general.

Other arguments are raised in the literature to support the use of the Einstein frame versus its Jordan counterpart. These include the argument that in the Jordan frame scalar and tensor modes are mixed, a feature that makes the Jordan frame inconvenient for the formulation of the Cauchy problem, which is only well-posed in the Einstein frame [845, 254, 256, 257]. It is difficult to quantize the scalar field fluctuations in the Jordan frame, and rather easy to do it in the Einstein frame [767, 543, 402, 668, 333, 322]. Indeed, it is the Einstein frame, not the Jordan frame, metric perturbation that must be identified with the physical graviton when one quantizes the linearized version of scalar-

Effective energy-momentum tensors and conformal frames 67

tensor gravity, as is done for general relativity in the covariant perturbation approach [816, 223, 254]. Let us consider explicitly the case of BD theory: the Jordan frame metric and BD field are decomposed according to

$$g_{ab} = \eta_{ab} + h_{ab}, \qquad \phi = \phi_0 + \varphi. \tag{2.47}$$

It is not h_{ab} that can be quantized and leads to the correct propagator, but rather the tensor field

$$\alpha_{ab} \equiv h_{ab} + \eta_{ab}\frac{\varphi}{\phi_0}. \tag{2.48}$$

By performing the usual conformal transformation to the Einstein frame and the scalar field redefinition

$$g_{ab} \longrightarrow \tilde{g}_{ab} = \sqrt{G\phi}\, g_{ab}, \tag{2.49}$$

$$\phi \longrightarrow \tilde{\phi} = \sqrt{\frac{2\omega+3}{16\pi G}} \ln\left(\frac{\phi}{C}\right) \equiv \tilde{\phi}_0 + \tilde{\varphi}, \tag{2.50}$$

one obtains, to first order,

$$\tilde{g}_{ab} = \eta_{ab} + \tilde{h}_{ab} = \sqrt{G\phi_0}\left(\eta_{ab} + h_{ab} + \eta_{ab}\frac{\varphi}{\phi_0}\right) + O(2), \tag{2.51}$$

$$\tilde{\phi} = \tilde{\phi}_0 + \tilde{\varphi} = \sqrt{\frac{2\omega+3}{16\pi G}}\,(\varphi + \text{const.}) + O(2), \tag{2.52}$$

where

$$\tilde{\varphi} = \sqrt{\frac{2\omega+3}{16\pi G}}\,\varphi, \tag{2.53}$$

and

$$\tilde{h}_{ab} = h_{ab} + \frac{\varphi}{\phi_0}\eta_{ab} = \alpha_{ab}, \tag{2.54}$$

where the last equation is obtained by dropping the irrelevant constant $\sqrt{G\phi_0}$ in front of the metric. Hence α_{ab} coincides to first order with the conformally transformed metric perturbation in the Einstein frame. One introduces

$$\tilde{\Theta}_{ab} \equiv \tilde{h}_{ab} - \frac{1}{2}\eta_{ab}\tilde{h}, \tag{2.55}$$

where $\tilde{h} \equiv \eta^{ab}\tilde{h}_{ab}$ and $\tilde{\Theta} = -\tilde{h}$ and the linearized field equations in the transverse gauge $\partial^b\tilde{\Theta}_{ab} = 0$ are

$$\eta^{cd}\partial_c\partial_d\tilde{\Theta}_{ab} = -\frac{16\pi}{\phi_0}T_{ab}, \tag{2.56}$$

$$\Box \varphi = \frac{8\pi T}{(2\omega + 3)\phi_0} \,. \tag{2.57}$$

As already noted, the kinetic terms of $\tilde{\phi}$ and α_{ab} have standard forms in the Einstein frame but not in the Jordan frame. The propagators of the graviton and of the scalar field are found by first looking for the Green function $G(x', x)$ which solves the field equation with a four-dimensional delta as source,

$$\eta^{c'd'}(x')\, \partial_{c'} \partial_{d'} G(x', x) = -\delta^{(4)}(x', x) \,. \tag{2.58}$$

The Fourier expansions of $G(x', x)$ and $\delta^{(4)}(x', x)$

$$G(x', x) = \frac{1}{(2\pi)^2} \int d^4k \, \hat{G}(k)\, e^{-ik_\alpha(x'^\alpha - x^\alpha)} \,, \tag{2.59}$$

$$\delta^{(4)}(x', x) = \frac{1}{(2\pi)^2} \int d^4k \, e^{-ik_\alpha(x'^\alpha - x^\alpha)} \,, \tag{2.60}$$

when substituted into eq. (2.58), yield

$$\hat{G}(k) = -\frac{1}{k^2} \,. \tag{2.61}$$

Hence,

$$\tilde{\Theta}_{ab} = \frac{16\pi}{\phi_0 k^2} T_{ab} \,, \tag{2.62}$$

$$\tilde{\varphi} = \frac{8\pi T}{(2\omega + 3)\phi_0 k^2} \,, \tag{2.63}$$

and

$$\tilde{h}_{ab} = \frac{16\pi}{\phi_0^2 k^2} P_{abcd} T^{cd} \,, \tag{2.64}$$

where

$$P_{abcd} = \frac{1}{2}\left(\eta_{ac}\eta_{bd} + \eta_{bc}\eta_{ad} - \eta_{ab}\eta_{cd}\right) \,; \tag{2.65}$$

and, in analogy to the photon propagator of quantum electrodynamics, the propagators of the graviton and of the scalar are $\frac{i}{k^2} P_{abcd}$ and $\frac{i}{k^2}$ (e.g., [816]). The propagators of the scalar and tensor gravitational degrees of freedom are well defined in the Einstein frame, but a similar derivation cannot be given in the Jordan frame, in which the eigenstates of the elicity — and hence the concepts of field and particle — are not correctly defined.

Other arguments proposed to support the use of the Einstein versus the Jordan frame include the following: in supergravity with ten or eleven dimensions, a conformal transformation is needed to identify the physical fields [776, 200, 292], and the conformal factor presented here

Effective energy-momentum tensors and conformal frames 69

from the classical point of view is recovered by imposing that the supersymmetry transformation of eleven-dimensional supergravity takes an $SU(8)$ covariant form [285]. In the study of boson stars in scalar-tensor theory, several definitions of the star's mass are possible, leading to different results in the Jordan frame, but to the same mass in the Einstein frame [857].

1.4 Other viewpoints

Many authors have tried to reach a conclusion about the issue of the conformal frame by using the equivalence principle as a guide [150, 242, 243, 844, 780, 613]. This approach can be misleading, since the equivalence principle could well be violated in nature, provided that the violations are sufficiently small to escape detection by the current experiments and thus satisfy the available constraints. As already discussed, in the Jordan frame the stress-energy tensor of ordinary matter is covariantly conserved, with massive test particles following geodesics of the Jordan metric. Thus the equivalence principle is satisfied (however, quantum corrections spoil the equivalence principle even in the Jordan frame [223]). On the contrary, in the Einstein frame, the stress-energy tensor $T_{ab}^{(m)}$ is not conserved unless matter is conformally invariant and eq. (1.178) holds. Here massive test particles deviate from geodesic worldlines and the equivalence principle is violated. Some authors find this violation of the equivalence principle unacceptable *a priori*. The anomalous coupling of the scalar ϕ and the fifth force arising in the Einstein frame is better seen as a manifestation of compactified theories or of new gravitational physics and not rejected *a priori*; the equivalence principle could be a prejudice.

In order to save both the equivalence principle and the Einstein frame formulation of scalar-tensor theories, the introduction of matter that is nonminimally coupled in the Jordan frame has been proposed [613], described by the action term

$$S^{(m)} = \int d^4x \sqrt{-g}\, (G\phi)^2 \mathcal{L}^{(m)}, \qquad (2.66)$$

so that the conformal transformation produces a minimally coupled $\tilde{T}_{ab}$ in the Einstein frame, according to

$$\sqrt{-g}\, (G\phi)^2 \mathcal{L}^{(m)} = \sqrt{-\tilde{g}}\, \mathcal{L}^{(m)}. \qquad (2.67)$$

This mathematical trick is very artificial and unjustified from the physical point of view. If the Einstein frame is the physical one (a point derived from a misinterpretation of Dicke's [290] conformal transformation), there is *a priori* no problem in accepting the anomalous coupling of ϕ and its consequences. If instead, one accepts the idea of running units of mass, time and length in the Einstein frame, then the two frames are physically equivalent and the fifth force proportional to the gradient $\tilde{\nabla}^c \phi$ in the Einstein frame can be reduced (up to constants) to the gradient of the varying mass $\tilde{\nabla}^c \tilde{m}$ of a massive particle. The fact that the mass is varying is responsible for the deviation of the particle from a geodesic worldline: massless particles always have zero gradient $\tilde{\nabla}^c \tilde{m}$ and follow geodesics.

In the literature one also finds the idea that a form of matter couples to the Jordan frame metric, while the dark energy responsible for the present acceleration of the universe (see Chapter 8) couples to the Einstein frame metric [483]. In the light of these considerations, this *ad hoc* procedure cannot be regarded as an acceptable one.

1.5 Einstein frame or Jordan frame?

If one disregards the physical equivalence of the two conformal frames, as is done in most of the cosmological literature, and chooses fixed units in the Einstein frame, then the issue of which conformal frame is physical becomes contentious. As demonstrated in the previous sections, there are situations in which the Jordan frame version of the theory is unviable and the Einstein frame version is acceptable; and there are also situations in which the Jordan frame is physical. When the Einstein frame version of the theory (with a rigid system of units) is accepted, one cannot avoid wondering where the original physical motivations of the Jordan frame version of the theory went. The theory is radically changed by the conformal transformation if running units are not adopted in the Einstein frame. Indeed, several authors see the necessity of a conformal transformation as the statement that general relativity is the only relativistic theory of gravity [127, 613, 612, 812]. As seen above, the argument of negative kinetic energy densities cannot be applied in general to decide in favour of the Einstein frame.

So, what is the verdict? Einstein frame with fixed units or Jordan frame? At present, a satisfactory answer doesn't exist: the Einstein frame version with fixed units is always well-behaved in terms of energies, while the Jordan frame version is not always viable. However, the former may completely lack the physical motivation of the latter. Thus, we see that there is no valid reason to discard the Jordan frame version of a theory when it is physically acceptable.

1.6 Energy conditions in relativistic theories

In this section we summarize the point-wise energy conditions of relativistic gravity referred to in the following chapters. There are slight differences in the definitions of energy conditions adopted in the literature [468, 892]. The energy conditions satisfied by a specified form of matter are formulated in terms of the stress-energy tensor T_{ab} describing it. In FLRW cosmology it is meaningful to refer to an homogeneous and isotropic fluid characterized by the stress-energy tensor

$$T_{ab} = (P+\rho)\, u_a\, u_b + P g_{ab}\,, \qquad (2.68)$$

and we summarize the energy conditions for such a fluid as follows.

- The **weak energy condition (WEC)** is satisfied if

$$T_{ab}\, t^a\, t^b \geq 0 \quad \text{for all timelike vectors } t^a \qquad (2.69)$$

or, for the fluid (2.68),

$$\rho \geq 0 \quad \text{and} \quad \rho + P \geq 0\,. \qquad (2.70)$$

- The **dominant energy condition (DEC)** is satisfied if WEC is satisfied and, in addition, $T^{ab} t_a$ is a null or timelike vector (i.e., $T_{ab} T^b{}_c t^a t^c \leq 0$) for any timelike vector t^a. For the fluid (2.68) this means that

$$\rho \geq |P|\,, \qquad (2.71)$$

or that the speed at which energy flows in this form of matter does not exceed c.

- The **null energy condition (NEC)** consists of

$$T_{ab}\, l^a\, l^b \geq 0 \quad \text{for all null vectors } l^a\,; \qquad (2.72)$$

for the fluid (2.68), this means

$$\rho + P \geq 0\,. \qquad (2.73)$$

Violations of the null energy condition are studied in the context of macroscopic traversable wormholes [381, 61, 64] and occur in cosmology if the expansion of the universe is superaccelerated ($\dot{H} > 0$); in this case energy is called *superquintessence* or *phantom energy*.

- The **null dominant energy condition (NDEC)** consists of

$$T_{ab}\, l^a\, l^b \geq 0 \quad \text{and} \quad T^{ab} l_b \text{ is null or timelike for any null vector } l^a\,. \qquad (2.74)$$

This condition is the same as the DEC but here l^a is a null instead of timelike vector; for the fluid (2.68), this means

$$\rho \geq |P| \quad \text{or} \quad \rho = -P. \tag{2.75}$$

- The **strong energy condition (SEC)** consists of

$$T_{ab} t^a t^b \geq -\frac{1}{2} T \quad \text{for any timelike vector} \quad t^a \tag{2.76}$$

or, for the fluid (2.68),

$$\rho + P \geq 0 \quad \text{and} \quad \rho + 3P \geq 0. \tag{2.77}$$

This condition ensures that gravity is always attractive, and is violated by a positive cosmological constant with $P^{(\Lambda)} = -\rho^{(\Lambda)}$ and during inflation or a quintessence-dominated era with $P < -\rho/3$.

Quantum systems are expected to violate all of the energy conditions, including positivity of the energy density on short timescales (e.g., [384]). However, even at the classical level scalar-tensor theories and the theory of a scalar field nonminimally coupled to the Ricci curvature of spacetime can violate all of the energy conditions (see Refs. [61, 384, 63, 884, 110]).

1.7 Singularity theorems and energy conditions

Next we consider the scalar-tensor theory without self-interaction described by the Jordan frame action

$$S = \int d^4x \, \frac{\sqrt{-g}}{16\pi} \left[R\phi - \frac{\omega(\phi)}{\phi} g^{cd} \nabla_c \phi \nabla_d \phi \right] + \sqrt{-g}\, \mathcal{L}^{(m)}. \tag{2.78}$$

As previously discussed, the conformal transformation $g_{ab} \to \tilde{g}_{ab} = G\phi\, g_{ab}$ and the scalar field redefinition

$$\tilde{\phi} = \int \frac{d\phi}{\phi} \sqrt{\frac{2\omega(\phi) + 3}{16\pi G}} \tag{2.79}$$

recast the action into the Einstein-Hilbert form, but with nonminimal coupling between $\tilde{\phi}$ and ordinary matter,

$$S = \int d^4x \, \sqrt{-\tilde{g}} \left[\frac{\tilde{R}}{16\pi} - \frac{1}{2} \tilde{g}^{cd} \tilde{\nabla}_c \tilde{\phi} \tilde{\nabla}_d \tilde{\phi} + \frac{\mathcal{L}^{(m)}}{(G\phi)^2} \right]. \tag{2.80}$$

If one chooses $\phi < 0$ initially in the Jordan frame, one has $G_{eff} < 0$ in the Jordan frame; hence, the restriction $\phi > 0$ is usually imposed.

Similarly, $\omega > -3/2$ is chosen in order to avoid a negative kinetic energy for the Einstein frame scalar $\tilde{\phi}$. If $\phi < 0$ and $\omega > -3/2$ are chosen, the weak energy condition is violated and the Hawking-Penrose singularity theorems [469, 468, 892] do not apply in the original Jordan frame. A similar situation occurs if $\phi > 0$ and $\omega < -3/2$. However, if $\phi > 0$ and $\omega > -3/2$, the weak energy condition holds for the field $\tilde{\phi}$ in the Einstein frame (with fixed units), and the singularity theorems are valid in this frame, if ordinary matter also satisfies the energy conditions [467]. In the Jordan frame, the strong and dominant energy conditions [892] may be violated even if the weak energy condition is not. This situation opens the door to the possibility of circumventing the Hawking-Penrose singularity theorems. Hence, it is possible to find solutions without a big bang initial singularity.

The violation of the weak energy condition in the Jordan frame is also responsible for the violation of the second law of black hole thermodynamics [892] — the surface area of a black hole never decreases — in Jordan frame BD theory. In addition, the violation of the weak energy condition is responsible for the breakdown of the apparent horizon theorem, that is, an apparent horizon cannot lie outside an event horizon [775].

The absence of singularities in one frame and their occurrence in the conformally rescaled world often is a source of marvel to authors who maintain Dicke's [290] statement that the two conformal frames are equivalent, but incorrectly use fixed units in the Einstein frame (e.g., [529, 725]). If the two frames are equivalent, how can the solution of the field equations exhibit a big bang singularity in one frame, while the conformally transformed counterpart is non-singular? The contradiction vanishes when variable units of mass, time and length are adopted in the Einstein frame. Consider a FLRW metric in the Jordan frame: in the Einstein frame, one has the same line element

$$ds^2 = d\tilde{s}^2 = -d\tilde{t}^2 + \tilde{a}^2(\tilde{t}) \left[\frac{dr^2}{1 - Kr^2} + r^2 \left(d\theta^2 + \sin^2\theta \, d\varphi^2 \right) \right], \quad (2.81)$$

where $d\tilde{t} = \Omega \, dt$, $\tilde{a} = \Omega \, a$ and $\Omega = \sqrt{G\phi}$ as in any scalar-tensor theory. To decide whether a solution is singular in the Einstein frame, it is not sufficient to consider the behaviour of $\tilde{a}(\tilde{t})$ as $\tilde{t} \to 0$. Rather, one has to study the ratio of a typical physical length $\tilde{a}(\tilde{t}) \left| d\vec{x} \right|$ to the unit of length $\tilde{l}_u(\phi) = \Omega \, l_0$, where l_0 is a fixed unit in the Jordan frame and $|d\vec{x}|$ is a coordinate distance:

$$\frac{\tilde{a}(\tilde{t})\,|d\vec{\tilde{x}}|}{\tilde{l}_u(\phi)} = \frac{\Omega\, a(t)\,|d\vec{x}|}{\Omega\, l_0} = \frac{a(t)\,|d\vec{x}|}{l_0}. \qquad (2.82)$$

Hence, $\tilde{a}\,|d\vec{\tilde{x}}|/\tilde{l}_u \to 0$ if and only if $a\,|d\vec{x}|/l_0 \to 0$: i.e., a singularity occurs in the Einstein frame if and only if it is present in the Jordan frame. To complete the argument, one has to show that $\tilde{t} \to 0$ is equivalent to $t \to 0$, i.e., that the singularity occurs at a finite time in both frames. This is easily accomplished by considering the ratio of $\tilde{t}$ to the varying unit of time $\tilde{t}_u(\phi) = \Omega\, t_0$, where t_0 is the fixed unit in the Jordan frame. One has

$$\frac{\tilde{t}}{\tilde{t}_u} = \frac{\int_0^t \Omega(\phi)\, dt}{\Omega(\phi)\, t_0} \simeq \frac{t}{t_0} \qquad \text{as } t \to 0\,; \qquad (2.83)$$

hence $\tilde{t} \to 0 \Leftrightarrow t \to 0$. In terms of energy density, eq. (1.165) yields $\tilde{\rho} = \Omega^{-4}\rho$, while the Einstein frame unit of energy density is

$$\tilde{\rho}_u(\phi) \simeq \frac{\tilde{m}}{\tilde{l}^3} = \frac{\Omega^{-1} m_0}{\Omega^3 l_0^3} = \Omega^{-4} \rho_0\,, \qquad (2.84)$$

where $\rho_0 \simeq m/l_0^3$ is the unit of energy density in the Jordan frame. Again, in the Einstein frame, one must not consider $\tilde{\rho}$, but the ratio

$$\frac{\tilde{\rho}}{\tilde{\rho}_u} = \frac{\Omega^{-4}\rho}{\Omega^{-4}\rho_0} = \frac{\rho}{\rho_0}. \qquad (2.85)$$

A measurement of the cosmological density yields the same result in the two frames and $\tilde{\rho}/\tilde{\rho}_u \to \infty \Leftrightarrow \rho/\rho_0 \to \infty$. When constant units are adopted in the Einstein frame, it is not surprising that singularities occur in one frame but not in the other; then the two frames are not equivalent.

2. Effective energy-momentum tensors

In Section 1.2, the Jordan frame BD field equations (1.29) are written as

$$G_{ab} = \frac{8\pi}{\phi} T_{ab}^{(m)} + T_{ab}^{(I)}[\phi]\,, \qquad (2.86)$$

where

$$\begin{aligned} T_{ab}^{(I)}[\phi] &\equiv \frac{\omega}{\phi^2}\left(\nabla_a\phi\nabla_b\phi - \frac{1}{2}g_{ab}\nabla^c\phi\nabla_c\phi\right) \\ &+ \frac{1}{\phi}(\nabla_a\nabla_b\phi - g_{ab}\,\Box\,\phi) - \frac{V}{2\phi}g_{ab} \end{aligned} \qquad (2.87)$$

is interpreted as an effective stress-energy tensor for the field ϕ. The second bracket on the right hand side of eq. (2.87) is rather unconventional in scalar field theory: instead of being quadratic in the first derivatives of ϕ, this term is linear in the second derivatives. This feature causes the trouble with linearized gravitational waves discussed in Section 2.1.2 and tempts some authors to question the role of $T_{ab}^{(I)}$ as an energy-momentum complex for ϕ [768]. The stress-energy tensor (2.87) has, in general, the structure of an imperfect fluid

$$T_{ab}^{(I)}[\phi] = \left(P_\phi^{(I)} + \rho_\phi^{(I)}\right) u_a u_b + 2q_{(a}^{(I)} u_{b)} + P_\phi^{(I)} g_{ab} + \Pi_{ab}^{(I)}, \quad (2.88)$$

where

$$u^a = \frac{\nabla^a \phi}{\sqrt{-\nabla_c \phi \nabla^c \phi}} \quad (2.89)$$

is the four-velocity field of the cosmological fluid associated with ϕ, $\rho_\phi^{(I)}$ and $P_\phi^{(I)}$ are its energy density and pressure, $q_{(I)}^a$ is the heat flux vector, and $\Pi_{ab}^{(I)}$ is the anisotropic stress tensor [711]. It is assumed that the gradient $\nabla^c \phi$ does not become a null vector.

There is the possibility of introducing a different energy-momentum tensor for ϕ by rewriting the field equations (2.86) as

$$\phi G_{ab} = 8\pi T_{ab}^{(m)} + \phi T_{ab}^{(I)}[\phi], \quad (2.90)$$

then adding and subtracting on the right hand side the term G_{ab}/G_0 (where G_0 is the present day value of the gravitational coupling measured in the Solar System). One obtains (e.g., [855])

$$G_{ab} = 8\pi G_0 \left[T_{ab}^{(m)} + T_{ab}^{(II)}[\phi]\right], \quad (2.91)$$

where

$$T_{ab}^{(II)}[\phi] = \frac{\phi}{8\pi} T_{ab}^{(I)}[\phi] + \left(\frac{1 - G_0 \phi}{8\pi G_0}\right) G_{ab}. \quad (2.92)$$

The alternative stress-energy tensor $T_{ab}^{(II)}[\phi]$ contains a "geometrical" contribution proportional to the Einstein tensor. The form (2.92) of $T_{ab}^{(II)}[\phi]$ is analogous to that of the energy-momentum tensor of a scalar field conformally coupled to the Ricci curvature introduced by Callan, Coleman and Jackiw [172] to renormalize the $\lambda \varphi^4$ theory in a curved space. This theory (see Chapter 7) is described by the action [210, 172]

$$S = \int d^4 x \sqrt{-g} \left[\frac{1}{2}\left(\frac{1}{\kappa} - \frac{\phi^2}{6}\right) R - \frac{1}{2} \nabla^c \varphi \nabla_c \varphi - \lambda \varphi^4\right]. \quad (2.93)$$

The theory is generalized to arbitrary scalar field potentials and generic values of the coupling constant ξ between ϕ and R [136]. The field equations for this scalar field theory can be written in a form that exhibits a ϕ-dependent gravitational coupling

$$G_{eff}[\varphi] = \frac{G}{1 - 8\pi G \xi \varphi^2} , \qquad (2.94)$$

or in the Callan-Coleman-Jackiw approach [172] in which the gravitational coupling is a true constant and with a different T_{ab} (called *improved energy momentum tensor* of φ). The formulation *à la* Callan-Coleman-Jackiw is familiar to physicists working in quantum field theory in curved spaces, but its analogue in BD and scalar-tensor theories apparently went unnoticed for a long time.

The BD field equations in the form (2.91) exhibit a gravitational coupling that is strictly constant, achieved by using (2.92) instead of (2.87) as the effective energy-momentum tensor for ϕ. Since the nonminimally coupled scalar field theory (2.93) can also be regarded as a scalar-tensor theory (see eqs. (7.6)-(7.9)), the introduction of the improved energy momentum tensor suggests regarding $T_{ab}^{(II)}[\phi]$ as the effective energy-momentum tensor of scalar-tensor gravity. This point of view, used in studies of quintessence in scalar-tensor gravity [855], is reinforced by the fact that $T_{ab}^{(II)}(\phi)$ obeys a conservation equation while $T_{ab}^{(I)}(\phi)$ does not, as

$$\nabla^b T_{ab}^{(I)}(\phi) = G_{ab} \nabla^b \phi . \qquad (2.95)$$

A third effective energy-momentum tensor for the scalar ϕ is obtained by writing the field equations as

$$G_{ab} = \frac{8\pi}{\phi} \left[T_{ab}^{(m)} + T_{ab}^{(III)}[\phi] \right] , \qquad (2.96)$$

where

$$T_{ab}^{(III)}[\phi] = \frac{\phi}{8\pi} T_{ab}^{(I)}[\phi] . \qquad (2.97)$$

2.1 Time-dependence of the gravitational coupling

It is often said that the Jordan frame version of scalar-tensor theories is constrained by the experimental limits on the time variation of the gravitational coupling (e.g., [909, 910, 83]). The time variation of the gravitational coupling is also used to constrain models of the universe dominated in the present era by a BD-like scalar (*quintessence* models — see Chapter 8) [211, 643, 855, 709, 50]. The experimental limits are usually quoted as [909]

$$\left|\frac{\dot{G}}{G}\right| < 10^{-10} \, \text{yr}^{-1} \,, \tag{2.98}$$

with recent improvements pushing the upper bound down to 10^{-11} yr^{-1} [431] or even $1.6 \cdot 10^{-12}$ yr^{-1} [450]. The constraint depends on the specific solution of the scalar-tensor theory. For example, eq. (2.104) below does not contain an explicit dependence on ω but the ω-dependence is contained in any specific solution $\phi(t)$, possibly with additional dependence on other parameters such as integration constants.

From the action (1.25) or from the field equations (1.29), one reads the gravitational coupling of BD theory as

$$G_{eff}(\phi) = \frac{1}{\phi}. \tag{2.99}$$

In practice, the gravitational coupling effective when two mass distributions attract each other — as determined in a Cavendish-like experiment — is slightly different. The analysis of the BD spherically symmetric solution in the context of the Parametrized Post-Newtonian (PPN) formalism [909] yields the effective coupling for Solar System experiments[1] [677]

$$G_{eff} = \frac{2(\omega + 2)}{2\omega + 3} \frac{1}{\phi}. \tag{2.102}$$

This effective coupling is constant is Barker's theory [67] in which the coupling function is

$$\omega(\phi) = \frac{4 - 3G_0\phi}{2(G_0\phi - 1)}. \tag{2.103}$$

However, the constant effective gravitational coupling (2.102) is only defined for spherically symmetric solutions, and not for cosmological

---

[1] For the generalized scalar-tensor theory described by the action

$$S = \int d^4x \sqrt{-g} \left[\frac{F(\phi)}{2} R - \frac{\omega(\phi)}{2} \nabla^c \phi \nabla_c \phi - V(\phi) + \mathcal{L}^{(m)} \right] \tag{2.100}$$

the effective gravitational coupling in a Cavendish experiment is given by [152, 254, 909]

$$G_{eff}(\phi) = \frac{2\omega F + (2dF/d\phi)^2}{8\pi F \left[2\omega F + 3(dF/d\phi)^2\right]}. \tag{2.101}$$

This expression can also be derived from cosmological perturbation theory [138].

ones. Note that while all that is required for the effective coupling (2.99) to be positive is that $\phi > 0$, positivity of the coupling (2.102) requires either $\phi > 0$ and ω in the range $\omega < -2$, $\omega > -3/2$, or $\phi < 0$ and $-2 < \omega < -3/2$ which corresponds to negative coupling from the point of view of cosmology.

The use of either eq. (2.99) or eq. (2.102) yields the percent variation of the gravitational coupling in BD theory as

$$\frac{\dot{G}_{eff}}{G_{eff}} = -\frac{\dot{\phi}}{\phi}, \qquad (2.104)$$

where the time variation is due to the cosmological evolution of ϕ. It is customary to express $\dot{G}_{eff}/G_{eff}$ as

$$\frac{\dot{G}_{eff}}{G_{eff}} = \sigma H_0, \qquad (2.105)$$

where H_0 is the present value of the Hubble parameter. This relation expresses the expectation that G_{eff} changes on a cosmological time scale. It is the quantity $\dot{G}_{eff}/G_{eff}$ that is constrained by Solar System experiments.

The requirement that scalar-tensor theories in the Jordan frame satisfy the experimental constraints on $\dot{G}_{eff}/G_{eff}$ is based on the interpretation of ϕ as the inverse effective coupling (eq. (2.99) or eq. (2.102)). This interpretation, in turn, is based on writing the field equations in the traditional form (1.29). However, the latter is only one possibility: when one writes the field equations in the form (2.91) with the effective energy-momentum tensor of ϕ à la Callan-Coleman-Jackiw [172], the gravitational coupling is strictly constant. This procedure casts doubts on the validity of the process of constraining scalar-tensor theories using the time variation of G_{eff}. Fortunately, the Solar System constraints on the coupling function ω are usually more stringent than the limits based on the time variation of G_{eff}. The Solar System bounds cannot be eliminated by rewriting the field equations in a different form because physically, the massless BD scalar mediates an infinite range force that affects local experiments and cannot be removed.

In the Einstein frame version of scalar-tensor theories, instead, the gravitational coupling is strictly constant.

2.2 Conservation equations for the various $T_{ab}^{(J)}[\phi]$

In this subsection we discuss the conservation properties of the Jordan frame effective stress-energy tensors $T_{ab}^{(J)}[\phi]$ ($J = I, II$, or III) for the

Jordan frame

Consider the energy-momentum tensor $T_{ab}^{(I)}[\phi]$ defined by eq. (2.87). By using the field equations (2.86) and the contracted Bianchi identities $\nabla^b G_{ab} = 0$, one immediately concludes that

$$\nabla^b \left(\frac{T_{ab}^{(m)}}{\phi} \right) + \frac{1}{8\pi} \nabla^b T_{ab}^{(I)}[\phi] = 0 \,. \tag{2.106}$$

Since ϕ does not couple directly to ordinary matter in the Jordan frame action (1.25) and the Lagrangian density $\mathcal{L}^{(m)}$ does not depend on ϕ, $T_{ab}^{(m)}$ is covariantly conserved[2] and therefore, in the presence of matter, $T_{ab}^{(I)}[\phi]$ satisfies

$$\nabla^b T_{ab}^{(I)}[\phi] = \frac{8\pi}{\phi^2} T_{ab}^{(m)} \nabla^b \phi \,. \tag{2.107}$$

Thus $T_{ab}^{(I)}[\phi]$ is covariantly conserved in vacuum ($T_{ab}^{(m)} = 0$), but not in the presence of ordinary matter.

Now let us consider the alternative stress-energy tensor $T_{ab}^{(II)}[\phi]$ in the Callan-Coleman-Jackiw form (2.92); from the field equations (2.91) and the contracted Bianchi identities, one concludes that

$$\nabla^b T_{ab}^{(II)}[\phi] + \nabla^b T_{ab}^{(m)} = 0 \,. \tag{2.108}$$

Since ordinary matter and ϕ are not directly coupled in the action (1.25), it follows that $\nabla^b T_{ab}^{(m)} = 0$ and $T_{ab}^{(II)}[\phi]$ is covariantly conserved also in the presence of matter,

$$\nabla^b T_{ab}^{(II)}[\phi] = 0 \,. \tag{2.109}$$

This feature constitutes a further motivation for using $T_{ab}^{(II)}[\phi]$ instead of $T_{ab}^{(I)}[\phi]$ [855].

---

[2]The conservation equation $\nabla^b T_{ab}^{(m)} = 0$ follows from the invariance of the matter part of the action under diffeomorphisms.

Table 2.1. A comparison of the three possible approaches to Jordan frame Brans-Dicke theory. The field equations are listed in the first line, and the stress-energy tensor and the conservation law it obeys appear in the second and third line, respectively.

(I) Standard approach	(II) Approach á la CCJ	(III) Mixed Approach
$G_{ab} = \frac{8\pi}{\phi} T_{ab}^{(m)} + T_{ab}^{(I)}$	$G_{ab} = 8\pi G_0 \left(T_{ab}^{(m)} + T_{ab}^{(II)} \right)$	$G_{ab} = \frac{8\pi}{\phi} \left(T_{ab}^{(m)} + T_{ab}^{(III)} \right)$
$T_{ab}^{(I)} = \frac{\omega}{\phi^2} \left(\nabla_a \phi \nabla_b \phi - \frac{1}{2} g_{ab} \nabla^c \phi \nabla_c \phi \right)$ $+ \frac{1}{\phi} \left(\nabla_a \nabla_b \phi - g_{ab} \Box \phi \right)$ $- \frac{V}{2\phi} g_{ab}$	$T_{ab}^{(II)} = \frac{\phi}{8\pi} T_{ab}^{(I)} + \frac{1 - G_0 \phi}{8\pi G_0} G_{ab}$	$T_{ab}^{(III)} = \frac{\phi}{8\pi} T_{ab}^{(I)}$
$\nabla^b T_{ab}^{(I)} = \frac{8\pi}{\phi^2} T_{ab}^{(m)} \nabla^b \phi$	$\nabla^b T_{ab}^{(II)} = 0$	$\nabla^b T_{ab}^{(III)} = \left(T_{ab}^{(m)} + T_{ab}^{(III)} \right) \nabla^b \left(\ln \frac{\phi}{\phi_0} \right)$

Finally, the alternative tensor $T_{ab}^{(III)}[\phi]$ introduced by eqs. (2.96) and (2.97) satisfies

$$\nabla^b T_{ab}^{(III)} = \left(T_{ab}^{(m)} + T_{ab}^{(III)}[\phi]\right)\nabla^b\left(\ln\frac{\phi}{\phi_0}\right). \tag{2.110}$$

• **Einstein frame**

In the Einstein frame, the energy-momentum tensor of the scalar $\tilde{\phi}$ assumes the canonical form for a scalar field [892]

$$\tilde{T}_{ab}\left[\tilde{\phi}\right] = \tilde{\nabla}_a\tilde{\phi}\,\tilde{\nabla}_b\tilde{\phi} - \frac{1}{2}\tilde{g}_{ab}\,\tilde{g}^{cd}\,\tilde{\nabla}_c\tilde{\phi}\,\tilde{\nabla}_d\tilde{\phi}\,; \tag{2.111}$$

the conservation equation in the Einstein frame is

$$\tilde{\nabla}^b \tilde{T}_{ab}[\tilde{\phi}] = 0\,. \tag{2.112}$$

Due to the anomalous coupling of $\tilde{\phi}$ to ordinary matter, $\tilde{T}_{ab}^{(m)}$ is not conserved in general, but it obeys eq. (1.178).

Having clarified the different possibilities for the energy-momentum tensor of the scalar and the issue of the conformal frame, we proceed to discuss gravitational waves in scalar-tensor cosmology.

Chapter 3

GRAVITATIONAL WAVES IN SCALAR-TENSOR COSMOLOGY

1. Introduction

The detection of gravitational waves is anticipated to provide a powerful tool to discriminate between general relativity and alternative theories of gravity. An array of detectors could distinguish between massless tensor modes with spin two — the only ones present in general relativity — and modes of different spin, such as the spin zero modes of the scalar gravitational field in scalar-tensor gravity [887, 307, 308, 909]. Regarding the astrophysics related to the generation of gravitational waves, the most obvious feature is that scalar gravitational radiation can be monopole radiation to lowest order, in contrast to the quadrupole radiation predicted as the lowest order in Einstein's theory. However, there are factors that may decrease the chance of detection of monopole radiation [254]. Here our main interest is in cosmological gravitational waves and the discussion is limited to a brief introduction to gravitational waves in scalar-tensor gravity. Since gravitational waves in the Jordan frame version of scalar-tensor theories are widely discussed in the literature (e.g., [909]), but their Einstein frame counterpart with fixed units of mass, time and length (which is the popular version in the literature) is not, we focus on the latter. After general consideration, we turn our attention to gravitational waves of cosmological origin and to effects that could potentially lead to the discovery of scalar gravitational waves. The production of gravitational waves during inflation in the context of scalar-tensor gravity is discussed in Chapter 6 together with the analogous generation of density perturbations.

2. Einstein frame scalar-tensor waves

Scalar-tensor gravitational waves in the Einstein frame are seldom discussed in the literature, which focuses on the Jordan frame version of scalar-tensor gravity [909]. On the other hand, scalar-tensor gravity formulated in the Einstein frame with fixed units of mass, time and length is a recurrent subject in modern literature and should be augmented with the appropriate treatment of gravitational waves. Another motivation for study is the similarity between scalar-tensor theories and low-energy string theories formulated in the Einstein frame, which contain dilatonic waves as well as tensor modes, and which mimic (with certain differences) Einstein frame scalar-tensor gravity.

2.1 Gravitational waves in the Einstein frame

In the Einstein frame, linearized gravitational waves outside sources are described by

$$\tilde{g}_{ab} = \eta_{ab} + \tilde{h}_{ab} , \qquad (3.1)$$

$$\tilde{\phi} = \tilde{\phi}_0 + \tilde{\varphi} , \qquad (3.2)$$

where $\mathrm{O}\left(\tilde{h}_{ab}\right) = \mathrm{O}\left(\tilde{\varphi}/\tilde{\phi}_0\right) = \mathrm{O}\left(\epsilon\right)$, with ϵ a smallness parameter. Note that by performing the conformal transformation (1.169) and (1.170) from the Jordan frame one obtains

$$\tilde{g}_{ab} = G\left(\phi_0 + \varphi\right)\left(\eta_{ab} + h_{ab}\right) = G\phi_0 \left(1 + \frac{\varphi}{\phi_0}\right)\left(\eta_{ab} + h_{ab}\right) , \qquad (3.3)$$

where h_{ab} and φ are the Jordan frame perturbations. The constant factor $G\phi_0$ in eq. (3.3) can be absorbed in a redefinition of the coordinates. Hence, in terms of Jordan frame quantities, the Einstein frame metric perturbations are

$$\tilde{h}_{ab} = h_{ab} + \frac{\varphi}{\phi_0}\,\eta_{ab} ; \qquad (3.4)$$

the last equation shows how spin two and spin zero modes are rearranged by the conformal transformation. The linearized field equations outside sources in the Einstein frame are

$$\tilde{R}_{ab} - \frac{1}{2}\tilde{g}_{ab}\tilde{R} = 8\pi G \left[\tilde{T}_{ab}\left[\tilde{\varphi}\right] + T_{ab}^{(eff)}\left[\tilde{h}_{ab}\right] \right] , \qquad (3.5)$$

$$\eta^{cd}\,\partial_c\partial_d\,\tilde{\varphi} = 0 , \qquad (3.6)$$

where

$$\tilde{T}_{ab}\left[\tilde{\varphi}\right] = \partial_a\tilde{\varphi}\,\partial_b\tilde{\varphi} - \frac{1}{2}\,\eta_{ab}\,\eta^{cd}\,\partial_c\tilde{\varphi}\,\partial_d\tilde{\varphi} , \qquad (3.7)$$

and $T_{ab}^{(eff)}\left[\tilde{h}_{ab}\right]$ is the Isaacson [511] effective energy-momentum tensor of spin two modes. One uses the auxiliary tensor

$$\tilde{\Theta}_{ab} \equiv \tilde{h}_{ab} - \frac{1}{2}\eta_{ab}\tilde{h}, \qquad (3.8)$$

where $\tilde{h} \equiv \eta^{ab}\tilde{h}_{ab}$ and $\tilde{\theta} \equiv \eta^{ab}\tilde{\Theta}_{ab} = -\tilde{h}$. It is straightforward to verify that, in terms of Jordan frame quantities, one has

$$\tilde{\Theta}_{ab} = h_{ab} - \frac{1}{2}\eta_{ab} h - \frac{\varphi}{\phi_0}\eta_{ab}. \qquad (3.9)$$

It is possible to find a gauge in which, to first order,

$$\partial^b \tilde{\Theta}_{ab} = 0, \qquad \eta^{cd}\partial_c \partial_d \tilde{\Theta}_{ab} = 0, \qquad (3.10)$$

and use the remaining gauge freedom to set equal to zero the time-time and time-space components of $\tilde{\Theta}_{ab}$ to first order,

$$\tilde{\Theta}_{0a} = 0. \qquad (3.11)$$

2.2 Corrections to the geodesic deviation equation

Section 1.11.2 shows how the equation of timelike geodesics in the Einstein frame receives a correction, interpreted as the effect of a fifth force and violating the weak equivalence principle. A necessary tool in the analysis of resonant gravitational wave detectors is the equation of geodesic deviation [839] which receives a correction as well in the Einstein frame [887, 417, 111] (see also [801, 611]). The derivation of the geodesic deviation equation parallels that of general relativity [839, 892].

Consider a smooth one-parameter family of worldlines $\{\gamma_s(\lambda)\}$ of massive test particles; λ is an affine parameter along these worldlines, and s is a parameter that labels the curves in the family. Let $t^a = (\partial/\partial\lambda)^a$ and $s^a = (\partial/\partial s)^a$ be the (dimensionless) tangent four-vector to the curve γ_s and the deviation vector, respectively. t^a and s^a are independent coordinate vector fields on the two-dimensional manifold spanned by the curves $\{\gamma_s(\lambda)\}$ and therefore they commute:

$$\frac{\partial t^a}{\partial s} = \frac{\partial s^a}{\partial \lambda}, \qquad \frac{Dt^a}{Ds} = \frac{Ds^a}{D\lambda}, \qquad (3.12)$$

where

$$\frac{DA^a}{D\lambda} \equiv t^b \nabla_b A^a, \qquad \frac{DA^a}{Ds} \equiv s^b \nabla_b A^a \qquad (3.13)$$

for any vector field A^a. The relative four-acceleration of two neighbouring curves is

$$a^a = \frac{D^2 s^a}{D\lambda^2} = \frac{D}{D\lambda}\left(\frac{Ds^a}{D\lambda}\right) = \frac{\partial}{\partial\lambda}\left(\frac{Ds^a}{D\lambda}\right) + \Gamma^a_{bc}\left(\frac{Ds^b}{D\lambda}\right)t^c$$

$$= \frac{\partial}{\partial\lambda}\left(\frac{\partial s^a}{\partial\lambda} + \Gamma^a_{bc} s^b t^c\right) + \Gamma^a_{bc} t^c \left(\frac{\partial s^b}{\partial\lambda} + \Gamma^b_{de} s^d t^e\right). \quad (3.14)$$

By using eq. (3.12), one obtains

$$a^a = \frac{\partial^2 t^a}{\partial s \partial \lambda} + \Gamma^a_{bc,d} t^d s^b t^c + \Gamma^a_{bc} \frac{\partial s^b}{\partial \lambda} t^c + \Gamma^a_{bc} s^b \frac{\partial t^c}{\partial \lambda} + \Gamma^a_{bc} t^c \frac{\partial t^b}{\partial s} + \Gamma^a_{bc}\Gamma^b_{de} s^d t^c t^e. \quad (3.15)$$

Now, one uses the corrected geodesic equation (1.184) to derive

$$\frac{\partial^2 t^a}{\partial s \partial \lambda} = -\Gamma^a_{bc,d} s^d t^b t^c - \Gamma^a_{bc} \frac{\partial t^b}{\partial s} t^c - \Gamma^a_{bc} t^b \frac{\partial t^c}{\partial s} - \Gamma^a_{bc} s^c \frac{\partial t^b}{\partial \lambda}$$

$$- \Gamma^a_{bc}\Gamma^b_{de} t^d t^e s^c + \sqrt{\frac{4\pi G}{2\omega+3}} \frac{D}{Ds}\left(\partial^a \widetilde{\phi}\right). \quad (3.16)$$

When eq. (3.16) is substituted into eq. (3.15) one obtains

$$a^a = -\Gamma^a_{bc,d} s^d t^b t^c - \Gamma^a_{bc} \frac{\partial t^b}{\partial s} t^c - \Gamma^a_{bc} t^b \frac{\partial t^c}{\partial s} - \Gamma^a_{bc} s^c \frac{\partial t^b}{\partial \lambda} - \Gamma^a_{bc}\Gamma^b_{de} t^d s^c t^e$$

$$+ \sqrt{\frac{4\pi G}{2\omega+3}} \frac{D}{Ds}\left(\partial^a \widetilde{\phi}\right) + \Gamma^a_{bc,d} t^d s^b t^c + \Gamma^a_{bc} \frac{\partial s^b}{\partial \lambda} t^c + \Gamma^a_{bc} s^b \frac{\partial t^c}{\partial \lambda}$$

$$+ \Gamma^a_{bc} t^c \frac{\partial t^b}{\partial s} + \Gamma^a_{bc}\Gamma^b_{de} s^d t^c t^e. \quad (3.17)$$

Using again eq. (3.12) and the symmetry of Γ^a_{bc}, and renaming some of the indices yields[1] [887, 111]

$$a^a = \widetilde{R}_{bcd}{}^a s^b t^c t^d + \sqrt{\frac{4\pi G}{2\omega+3}} \frac{D}{Ds}\left(\partial^a \widetilde{\phi}\right). \quad (3.18)$$

A similar correction to the geodesic deviation equation is found in string theory [417] but it depends on the dilatonic charge. This feature of string theories results in the fact that two gravitational wave detectors

---

[1] The analysis of Ref. [887] includes the possibility that the field $\widetilde{\phi}$ has a mass and a finite range.

built with materials of different compositions would respond differently to dilatonic waves. Eq. (3.18) can be used to analyze the relative motion of two nearby bodies in the presence of scalar-tensor gravitational waves, or the response of a laser interferometer [801, 417, 111].

3. Gravitational lensing by scalar-tensor gravitational waves

It is well known that in general relativity, gravitational waves deflect light rays propagating through them and cause a frequency shift. The effects of scintillation of a light source and lensing by gravitational waves has received a considerable amount of attention in the literature [907, 921, 911, 922, 124, 125, 148, 339, 338, 554, 328, 330, 329, 331, 332, 340, 621, 458, 526, 255, 548, 726]. Gravitational waves as perturbations of the caustics of ordinary gravitational lenses (i.e., mass concentrations) have also been studied [627, 22, 23, 550, 395, 66], and the observational search for scintillation induced by gravitational waves has also begun [715, 714, 716, 146]. It appears that lensing by gravitational waves from astrophysical sources is either so infrequent that it has only a very small probability of detection, or its magnitude is so small as to defy observation [255]. Therefore, it is interesting to ask whether the lensing effect has the same, or a larger, order of magnitude in scalar-tensor theories in which scalar modes go hand in hand with tensor modes. We present the analysis in both the Jordan and the Einstein frame (with fixed units of length, mass and time), in order to contrast the different physics in the two frames and to make explicit once again the effect of negative energy density in the unphysical Jordan frame.

3.1 Jordan frame analysis

We begin the analysis in the Jordan frame version of BD gravity [340, 147] by considering the amplification of a light beam crossing a region of space filled with scalar-tensor gravitational waves. The amplification effect is due to the focusing of null rays by the energy present in the gravitational waves. The latter are assumed to vanish at infinity, where the light source and the observer are located.

This phenomenon was discussed in the context of general relativity and soon discarded as unimportant because the magnitude of the amplification is only of second order in the wave amplitude, and hence too small to be observable [921, 922, 124, 125]. In the Jordan frame version of scalar-tensor theories, instead, the effect is of first order in the wave amplitudes and has the potential to be physically interesting. The difference with respect to general relativity is due to the spin zero modes, which are absent in general relativity and behave as matter in the beam of light.

The amplification of a beam crossing gravitational waves is best studied [754, 921, 124] by considering a congruence of null rays that travel from a distant light source to an observer, in the geometric optics approximation. If k^a is the field of four-vectors tangent to these rays, the optical scalars and the (complex) shear of the congruence are

$$\theta_l = \frac{1}{2} \nabla_a k^a , \tag{3.19}$$

$$|\sigma|^2 = \frac{1}{2} \left(\nabla_{(b} k_{a)} \right) \left(\nabla^{(b} k^{a)} \right) - \theta_l^2 , \tag{3.20}$$

$$\omega_l^2 = \frac{1}{2} \left(\nabla_{[b} k_{a]} \right) \left(\nabla^{[b} k^{a]} \right) , \tag{3.21}$$

$$\sigma = (c^a)^* c^b \nabla_b k_a = \frac{1}{\sqrt{2}} \left[\left(\nabla_{(b} k_{a)} \right) \left(\nabla^{(b} k^{a)} \right) - \frac{1}{2} (\nabla_c k^c)^2 \right]^{1/2} , \tag{3.22}$$

where c^a is a null complex vector that satisfies the relations

$$c^a c_a = 0 , \quad c^a k_a = 0 , \quad c^a (c_a)^* = -1 . \tag{3.23}$$

The evolution of the optical scalars as one moves along the curves is given by the propagation equations

$$\frac{d\theta_l}{d\lambda} = -\theta_l^2 - |\sigma|^2 + \omega_l^2 - \frac{1}{2} R_{ab} k^a k^b , \tag{3.24}$$

$$\frac{d\sigma}{d\lambda} = -2\theta_l \sigma - C_{abcd}(c^a)^* k^b (c^c)^* k^d \equiv -2\theta_l \sigma - C(\lambda) , \tag{3.25}$$

$$\frac{d\omega_l}{d\lambda} = -2\omega_l \theta_l , \tag{3.26}$$

where $C^a{}_{bcd}$ is the Weyl tensor and $R_{ab} = \partial_a \partial_b \varphi / \phi_0 + O(\epsilon^2)$. Let us consider the situation in which the metric and scalar field are given by

$$g_{ab} = \eta_{ab} + h_{ab} , \quad \phi = \phi_0 + \varphi , \tag{3.27}$$

with $O(h_{ab}) = O(\varphi/\phi_0) = O(\epsilon)$, and tensor and scalar modes are Fourier integrals of monochromatic components. The elementary spin zero waves are of the form

$$\varphi_{\vec{l}} = \varphi_0 \exp \left(\pm i l^a x_a \right) , \tag{3.28}$$

where the amplitude φ_0 is constant and l^a is a constant null vector of the background, $\eta_{ab} l^a l^b = 0$. A more realistic scenario would involve the

Gravitational waves

use of a cosmological metric instead of η_{ab}, but the main features of the scintillation induced in the light beam can be grasped using eq. (3.27), in analogy to the approach used in the theory of standard gravitational lenses [781]. The situation in general relativity corresponds to no scalar field and to the vanishing of the Ricci tensor in eq. (3.24) outside sources of gravitational waves. For a point-like source of light, and to lowest order in ϵ, the solution of the propagation equations is then of the form

$$\theta_l = \frac{1}{\lambda} + \delta\theta , \qquad (3.29)$$

$$\sigma = \delta\sigma , \qquad (3.30)$$

$$\omega_l = 0 , \qquad (3.31)$$

with $\delta\theta$ and $\delta\sigma$ small [754]. We use the convention that $\lambda = 0$ at the light source and assume that the deflection of light rays is small (paraxial approximation). For a simple monochromatic scalar wave (3.28), the propagation equations reduce to the equations for the perturbations

$$\frac{1}{\lambda^2}\frac{d}{d\lambda}\left(\lambda^2 \delta\theta_l\right) = -\delta\theta^2 - |\delta\sigma|^2 + \frac{1}{2}(l_a k^a)^2 \frac{\varphi}{\phi_0} , \qquad (3.32)$$

$$\frac{1}{\lambda^2}\frac{d}{d\lambda}\left(\lambda^2 \delta\sigma\right) = -2\,\delta\theta\,\delta\sigma - C(\lambda) . \qquad (3.33)$$

In general relativity, the only difference is that the last term on the right hand side of eq. (3.32) is absent and therefore $\delta\theta = \mathrm{O}\left(\epsilon^2\right)$ [921, 124]. In scalar-tensor gravity, this term dominates the right hand side and hence the expansion θ_l of the congruence of null geodesics is

$$\delta\theta = \mathrm{O}\left(\frac{\varphi}{\phi_0}\right) = \mathrm{O}\left(\epsilon\right) \qquad (3.34)$$

when $l_a k^a \neq 0$. The case $l_a k^a = 0$ corresponds to gravitational waves propagating in the same direction as the null rays, a situation that gives zero deflection in scalar-tensor gravity as well as in general relativity. The amplification of a bundle of null rays, described by the expansion scalar θ_l, is a first order effect in the amplitude of gravitational waves in Jordan frame scalar-tensor theories, contrary to the case of Einstein's theory. One would then expect the amplification effect to be many orders of magnitude larger than in general relativity, and to be potentially observable. While in general relativity lensing is only due to the shear of the tensor modes, the Jordan frame scalar of scalar-tensor gravity behaves as matter in the beam.

Note that in general relativity, if there is matter in the beam, the sign of the Ricci term $-R_{ab}k^a k^b/2$ in eq. (3.24) is always negative as a consequence of the fact that matter satisfies the weak energy condition and $R_{ab}k^a k^b \geq 0$. Hence the old *adagio* "matter in the beam always focuses"[2]. Instead, in eq. (3.32) of scalar-tensor gravity, the sign of the Ricci term $(l_a k^a)^2 \varphi/(2\phi_0)$ oscillates in time with the period of the scalar wave (3.28). This is an aspect of the fact that the effective energy density of the scalar field has oscillating sign in the Jordan frame.

3.2 Einstein frame analysis

In the Einstein frame, the trajectories of photons are null geodesics — as in the Jordan frame. One decomposes the metric and scalar field according to eqs. (3.1) and (3.2) and expresses the solutions of the linearized field equations (3.5) and (3.6) as Fourier integrals of monochromatic plane waves of the form (2.39). The energy-momentum tensor of the scalar field in the Einstein frame assumes the canonical form

$$\tilde{T}_{ab}[\tilde{\varphi}] = \partial_a \tilde{\varphi} \, \partial_b \tilde{\varphi} - \frac{1}{2} \eta_{ab} \, \eta^{cd} \, \partial_c \tilde{\varphi} \, \partial_d \tilde{\varphi} \, . \qquad (3.35)$$

Now, the term $-\tilde{R}_{ab} k^a k^b$ in the right hand side of the Raychaudhuri equation (3.24) — which is responsible for the relatively large (first order in ϵ) amplification effect in the Jordan frame — turns out to be of second order, and comparable to the contribution from tensor modes [358],

$$\tilde{R}_{ab} k^a k^b = 4\pi \left[p_a k^a \tilde{\varphi}_0 \sin(p_c x^c) \right]^2 + \tilde{T}_{ab}^{(eff)}[\tilde{h}_{cd}] k^a k^b \, . \qquad (3.36)$$

This expression of order ϵ^2 is quadratic in the first derivatives of both $\tilde{\varphi}$ and $\tilde{h}_{ab}$ and is positive-definite.

In the Einstein frame, the scalar waves always satisfy the weak energy condition and focus the null rays, but the effect is of the same order of magnitude as in general relativity and is exceedingly small from the observational point of view. The difference between the magnitude of the amplification effect in the Jordan and the Einstein frame is due solely to the difference between the respective effective energy-momentum tensors for φ and $\tilde{\varphi}$: linear in the second derivatives of φ in the Jordan frame, quadratic in the second derivatives of $\tilde{\varphi}$ in the Einstein frame. The analysis of the focusing effect shows once again the physical inequivalence of the two conformal frames when fixed units of mass, time, and length are used in the Einstein frame.

---

[2]The fact that, for matter in general relativity, $-R_{ab}k^a k^b \leq 0$ is a key ingredient in the proof of the singularity theorems of general relativity [469, 468, 892].

3.3 Propagation of light through a gravitational wave background

In any theory of gravity, long wavelength gravitational waves generated at an early time can overlap to produce a cosmological background with random phases, polarizations, and directions of propagation. In scalar-tensor theories, this background is composed of both spin two and spin zero modes.

The propagation of light from distant sources (e.g., quasars at cosmological distances) through a cosmological background of gravitational waves has been studied in Einstein gravity. Since each gravitational wave produces a deflection of null rays of order ϵ (where ϵ is the characteristic amplitude of the wave), the problem arises as to whether or not the numerous random deflections due to a stochastic background of waves cumulate with the distance D traveled by the photons. Were such a secular or "D-effect" present, it could lead to observationally significant deflection angles and frequency shifts since D can be a cosmological distance. Hence, the effect was studied in conjunction with redshift anomalies and periodicities reported in galaxy groups and clusters [732, 264] and in relation to proper motions of quasars [458].

A priori, one would expect that a photon traveling through a random background of gravitational waves experiences N scatterings and that the deflections due to each wave add stochastically, leading to a D-effect proportional to $\sqrt{N}$ [911, 621]. However, this simplistic picture is incorrect because it neglects the fact that the speed of propagation of the signals is equal to the speed of the random inhomogeneities of the medium, which are highly non-stationary [921, 922, 264, 125, 584, 585, 148, 526]. The root mean square deflection of a light ray moving in a medium with inhomogeneities due to a certain field depends on the spin of the field, as well as the velocity of these inhomogeneities. Thus, one expects that scalar-tensor theories — which contain spin two and spin zero modes — would yield a result different from the one found in Einstein's theory.

The analysis of general inhomogeneities of spin zero, one, and two in a medium is well known [584] and the results can be immediately adapted to the case of scalar modes propagating at the speed of light, and corresponding to zero mass and infinite range for the BD-like scalar $\tilde{\phi}$. For a photon that has an unperturbed path parallel to the z-axis one obtains the result [526, 358] that the root mean square deflections are

$$(\delta k^x)_{rms} = (\delta k^y)_{rms} = \left(\frac{\tilde{\varphi}}{\tilde{\phi}_0}\right)_{rms} \left[\ln\left(\frac{2\pi D}{\lambda_{gw}}\right)\right]^{1/2}, \quad (3.37)$$

$$(\delta k^z)_{rms} = 0, \tag{3.38}$$

where λ_{gw} is the typical gravitational wave wavelength. The dependence of $(\delta k^i)_{rms}$ ($i = x, y$) on D is qualitatively different from the effect found in general relativity but the difference is not numerically significant. For example, for $\lambda_{gw} = 5$ cm (corresponding to a 1 K cosmic gravitational wave background) and $D = 500$ Mpc, one has

$$(\delta k^i)_{rms} \approx 8 \left(\frac{\tilde{\varphi}}{\tilde{\phi}_0} \right)_{rms}. \tag{3.39}$$

This is an improvement of less than one order of magnitude over the result found in general relativity. It seems that every aspect of lensing by gravitational waves has approximately the same order of magnitude in general relativity and in Einstein frame scalar-tensor theories when fixed units are adopted in this frame.

Chapter 4

EXACT SOLUTIONS OF SCALAR-TENSOR COSMOLOGY

1. Introduction

Exact solutions of a physical theory are required in order to gain insight into the physical content and predictions of the theory and, in the case of cosmology, to provide workable scenarios of the unperturbed universe. In this chapter, a selection of exact solutions of scalar-tensor cosmology is presented, ranging from classical big bang solutions to more recent ones. No attempt is made to compile a comprehensive catalogue of exact solutions (see Ref. [591] for an early review); instead, example solutions are presented which display features of particular interest in modern cosmology, for particular choices of ordinary matter. The core of this chapter consists of FLRW solutions of BD theory in the Jordan frame. Once these are known, one can use eqs. (1.141) and (1.169) and integrate eq. (1.199) to obtain the corresponding Einstein frame solutions. In practice the reverse path is sometimes followed, i.e., exact solutions of general relativity are mapped back into the Jordan frame to obtain exact solutions in this representation.

The equations of FLRW cosmology in scalar-tensor gravity are introduced in Section 1.7. The metric is given by

$$ds^2 = -dt^2 + a^2(t)\left[\frac{dr^2}{1-Kr^2} + r^2\left(d\theta^2 + \sin^2\theta\, d\varphi^2\right)\right] \quad (4.1)$$

in comoving polar coordinates. Vacuum solutions $(a(t), \phi(t))$ correspond to $T_{ab}^{(m)} = 0$ and have the gravitational scalar field as the only formal "source" of gravity, while non-vacuum solutions are usually derived by assuming that ordinary matter is described by a perfect fluid with stress-energy tensor of the form

$$T_{ab}^{(m)} = \left(P^{(m)} + \rho^{(m)}\right) u_a u_b + P^{(m)} g_{ab} \tag{4.2}$$

and with equation of state

$$P^{(m)} = (\gamma - 1)\rho^{(m)}, \quad \gamma = \text{const.} \tag{4.3}$$

We consider the range of values $0 \leq \gamma \leq 2$, corresponding to

$$-1 \leq \frac{P^{(m)}}{\rho^{(m)}} \leq 1 \tag{4.4}$$

and we treat a cosmological constant Λ as a form of ordinary matter fluid with $\gamma = 0$.

As seen in Section 2.2, in the Jordan frame ordinary matter is minimally coupled to the BD field ϕ and the stress-energy tensor $T_{ab}^{(m)}$ is therefore covariantly conserved. For a perfect fluid of the form (4.2) one then has

$$\frac{d\rho^{(m)}}{dt} + 3H\left(\rho^{(m)} + P^{(m)}\right) = 0. \tag{4.5}$$

If, in addition, the equation of state (4.3) is postulated, eq. (4.5) is immediately integrated to yield

$$\rho^{(m)} = \frac{C}{a^{3\gamma}}, \tag{4.6}$$

where C is an integration constant related to initial conditions. Consider a big bang solution with scale factor such that $a(t) \to 0$ as $t \to 0^+$: clearly, if two barotropic fluids with indices γ_1 and γ_2 (with $0 \leq \gamma_1, \gamma_2 \leq 2$) are present, the fluid with the highest γ will dominate the dynamics near the initial singularity. In particular, a stiff fluid defined by $\gamma = 2$ and $P^{(m)} = \rho^{(m)}$ will dominate any other barotropic fluid in such conditions. Long waves of a scalar field ψ which is massless, free of self-interaction, and minimally coupled to the Ricci curvature, mimic a stiff fluid [73, 895] (see Appendix A).

A caveat about vacuum solutions of scalar-tensor theories is necessary: it is tempting to interpret the field equation (1.29) as an effective Einstein equation for general relativity, with the BD-like scalar as a form of alternative matter and as a source of gravity, and to look for the corresponding solutions. In principle, vacuum solutions of scalar-tensor gravity (i.e., with ϕ as the only form of "matter") do not seem to be physically very significant. In fact, Mach's principle, which constitutes the primary motivation for the introduction of BD theory, states that local gravitational properties (i.e., the local value of the gravitational coupling ϕ^{-1}) are determined by distant matter in the universe. In this sense, the Minkowski solution corresponding to ϕ =constant is not very meaningful from the physical point of view [642]. However, if one sets $V(\phi) = 0$, after using eq. (1.42) to eliminate $\ddot{\phi}$, eqs. (2.44) and (2.45) yield $P_\phi^{(I)} = \rho_\phi^{(I)}$; the BD field behaves as a stiff fluid and dominates the matter fluids present [646].

Since the trace $T^{(m)}$ of the stress-energy tensor of ordinary matter acts as the only scalar field source, one expects to find similarities between vacuum solutions ($T_{ab}^{(m)} = 0$) and solutions corresponding to radiative matter with $T^{(m)} = 0$. Indeed these similarities are found [895, 771].

2. Exact solutions of Brans-Dicke cosmology

One begins with the study of solutions corresponding to a spatially flat FLRW universe. The relevant equations (1.38), (1.41), and (1.42) are presented in Section 1.3. Big bang solutions resembling those known in general relativity have been known for a long time. Many of the works on exact solutions of BD cosmology assume the boundary condition

$$\lim_{t \to 0} a^3(t)\, \dot{\phi}(t) = 0 \qquad (4.7)$$

as the big bang singularity is approached[1].

When the boundary condition (4.7) is imposed, one of the four integration constants $\left(a_0, \dot{a}_0, \phi_0, \dot{\phi}_0\right)$ is eliminated and the dimension of the phase space is reduced by one, with loss of generality of the solutions [895, 639]. In general, or even with the restriction (4.7), the phase space of FLRW cosmology in scalar-tensor theories has a larger dimensionality than the corresponding phase space of general relativity (which requires initial conditions $(a_0, \dot{a}_0)$) and therefore scalar-tensor gravity exhibits a richer variety of solutions than general relativity.

Most of the solutions presented in the following are of the power-law type

$$a(t) \propto t^q, \qquad \phi(t) \propto t^s, \qquad (4.8)$$

with $3q + s \geq 1$. In BD theory, a power-law solution plays a role analogous to that of the inflationary de Sitter attractor in general relativity.

When quintessence models (a non-vacuum situation, see Chapter 8) are studied in scalar-tensor gravity, a power-law solution is very often assumed and the exponents q and s are determined *a posteriori* together with other parameters.

---

[1]The quantity $a^3 \dot{\phi}$ is related to the d'Alembertian of ϕ by eq. (1.36).

When ordinary matter is described by a barotropic fluid as in eqs. (4.2) and (4.3), and in the absence of a scalar field potential, the equation satisfied by the BD scalar reduces to

$$\ddot{\phi} + 3H\dot{\phi} = \frac{8\pi}{2\omega + 3}(4 - 3\gamma)\rho^{(m)}, \quad (4.9)$$

while the cosmic fluid obeys eq. (4.6). If the scale factor is described by a power-law $a = a_0 t^q$, the equation for the BD scalar becomes

$$\ddot{\phi} + \frac{3q}{t}\dot{\phi} = \frac{8\pi(4 - 3\gamma)C}{(2\omega + 3)a_0{}^{3\gamma}} t^{-3\gamma q}. \quad (4.10)$$

A particular integral of this equation is

$$\phi = \phi_1 t^{2-3\gamma q}, \quad (4.11)$$

where

$$\phi_1 = \frac{8\pi C(4 - 3\gamma)}{(2\omega + 3)a_0{}^{3\gamma}(2 - 3\gamma q)[1 + 3q(1 - \gamma)]}. \quad (4.12)$$

The general integral of eq. (4.10) is then

$$\phi = \phi_0 t^s + \phi_1 t^{2-3\gamma q}, \quad (4.13)$$

with $s = 0$ or $3q + s = 1$.

2.1 $K = 0$ FLRW solutions

In the literature more exact solutions are found corresponding to zero curvature index than to positive or negative values of K, reflecting the simpler form of the equations when the spatial curvature vanishes. We begin with the most well known solutions.

2.1.1 The O'Hanlon and Tupper solution

The O'Hanlon and Tupper [681] solution corresponds to vacuum, $V(\phi) = 0$, and to the range of the BD parameter $\omega > -3/2$, $\omega \neq 0, -4/3$:

$$a(t) = a_0 \left(\frac{t}{t_0}\right)^{q_\pm}, \quad (4.14)$$

$$\phi(t) = \phi_0 \left(\frac{t}{t_0}\right)^{s_\pm}, \quad (4.15)$$

with

$$q_\pm = \frac{\omega}{3(\omega + 1) \mp \sqrt{3(2\omega + 3)}} = \frac{1}{3\omega + 4}\left[\omega + 1 \pm \sqrt{\frac{2\omega + 3}{3}}\right], \quad (4.16)$$

$$s_\pm = \frac{1 \mp \sqrt{3(2\omega+3)}}{3\omega+4}, \qquad (4.17)$$

and satisfies $3q + s = 1$. This solution can be derived with different methods (e.g., [681, 639, 639, 545]), has a big bang singularity as $t \to 0$, and does not correspond to the initial condition (4.7). The limit of the solution as $\omega \to +\infty$ is

$$a(t) \propto t^{1/3}, \qquad \phi = \text{const}. \qquad (4.18)$$

This limit does not reproduce the corresponding general relativistic solution which is Minkowski space.

The solution corresponding to the exponents (q_-, s_-) is called the *slow* solution, while the one described by (q_+, s_+) is the *fast* solution. The terminology corresponds to the behaviour of the BD scalar ϕ at early times $t \to 0$. The slow solution describes increasing ϕ and decreasing coupling $G_{eff} \simeq 1/\phi$ as time progresses (for $\omega > -4/3$), while the fast solution is associated with decreasing ϕ and increasing $G_{eff} \simeq 1/\phi$ as functions of t at early times (for $\omega > -4/3$). The fast and slow solutions are interchanged under the duality transformation (1.49) and (1.50) which yields

$$(q_\pm, s_\pm) \longrightarrow (q_\mp, s_\mp). \qquad (4.19)$$

It is often remarked that, in the limit $\omega = -4/3$, the O'Hanlon-Tupper solution approaches the de Sitter space

$$a(t) = a_0 \exp(H t), \qquad (4.20)$$

$$\phi(t) = \phi_0 \exp(-3H t), \qquad (4.21)$$

with H =constant. Strictly speaking, this is not true as one obtains the described behaviour only by choosing simultaneously the values q_+ and s_- of the exponents. However, it is true that eqs. (4.20) and (4.21) provide a solution of the BD equations (1.38), (1.41), and (1.42) for $K = 0$ [681]; this is the only de Sitter solution for $K = 0$ and vacuum. This solution differs from the usual de Sitter space which is obtained in the scalar field cosmology of Einstein theory by keeping the scalar field constant. Examples of de Sitter solutions with a non-constant scalar field are rather common in scalar-tensor theories [451].

2.1.2 The Brans-Dicke dust solution

The Brans-Dicke dust solution [152] corresponds to a pressureless dust fluid ($\gamma = 1$) and to a matter-dominated universe with $V(\phi) = 0$ and

$\omega \neq -4/3$,
$$a(t) = a_0 \, t^q \,, \tag{4.22}$$

$$\phi(t) = \phi_0 \, t^s \,, \tag{4.23}$$

with
$$q = \frac{2(\omega+1)}{3\omega+4} \,, \tag{4.24}$$

$$s = \frac{2}{3\omega+4} \,, \tag{4.25}$$

and satisfies $3q + s = 2$; while
$$\rho^{(m)} = \frac{C}{a^{3\gamma}} = \rho_0 \, t^r \,, \tag{4.26}$$

with
$$r = -3q = \frac{-6(\omega+1)}{3\omega+4} \,, \quad \rho_0 = \frac{C}{a_0^3} \,. \tag{4.27}$$

2.1.3 The Nariai solution

The Nariai solution [662, 454] is a particular power-law solution for a $K = 0$ FLRW universe with $V(\phi) = 0$, $\omega \neq -4\left[3\gamma(2-\gamma)\right]^{-1} < 0$, and the perfect fluid equation of state (4.3). It can be written as

$$a(t) = a_0 \left(1 + \delta \, t\right)^q \,, \tag{4.28}$$

$$\phi(t) = \phi_0 \left(1 + \delta \, t\right)^s \,, \tag{4.29}$$

where
$$q = \frac{2\left[\omega(2-\gamma) + 1\right]}{3\omega\gamma(2-\gamma) + 4} \,, \tag{4.30}$$

$$s = \frac{2(4 - 3\gamma)}{3\omega\gamma(2-\gamma) + 4} \,. \tag{4.31}$$

The exponents q and s satisfy the relation $s + 3\gamma q = 2$. The energy density of ordinary matter scales as

$$\rho^{(m)}(t) = \frac{C}{a^{3\gamma}} = \rho_0 \left(1 + \delta \, t\right)^r \,, \quad r = -3\gamma q \,. \tag{4.32}$$

Sometimes it is convenient to rewrite the exponents q, s, and r as functions of the parameters [545]

$$\alpha \equiv \frac{2(4 - 3\gamma)}{(2\omega+3)(2-\gamma) + 3\gamma - 4} \,, \quad A \equiv \frac{2\omega+3}{12} \,, \tag{4.33}$$

as
$$q = \frac{2}{\alpha + 3\gamma}, \quad s = \frac{2\alpha}{\alpha + 3q}, \quad r = \frac{-6\gamma}{\alpha + 3\gamma}, \quad (4.34)$$

and
$$\delta = \left(\frac{\alpha + 3\gamma}{2}\right) \left\{ \frac{8\pi\rho_0}{3\phi_0 \left[(1 + \alpha/2)^2 - A\alpha^2\right]} \right\}. \quad (4.35)$$

The BD field equations for $K = 0$, $V = 0$, and the equation of state (4.3) are

$$H^2 = \frac{8\pi\rho^{(m)}}{3\phi} + \frac{\omega}{6}\left(\frac{\dot\phi}{\phi}\right)^2 - H\frac{\dot\phi}{\phi}, \quad (4.36)$$

$$\dot H = \frac{-8\pi\rho^{(m)}}{(2\omega + 3)\phi}(\omega\gamma + 2) - \frac{\omega}{2}\left(\frac{\dot\phi}{\phi}\right)^2 + 2H\frac{\dot\phi}{\phi}, \quad (4.37)$$

$$\ddot\phi + 3H\dot\phi = \frac{8\pi\rho^{(m)}}{2\omega + 3}(4 - 3\gamma). \quad (4.38)$$

Particular cases of the Nariai solution include a dust fluid ($\gamma = 1$, $P^{(m)} = 0$, $\alpha = (\omega + 1)^{-1}$, $\omega \neq -4/3$), for which

$$a(t) = a_0 (1 + \delta t)^{\frac{2(\omega+1)}{3\omega+4}}, \quad (4.39)$$

$$\phi(t) = \phi_0 (1 + \delta t)^{\frac{2}{3\omega+4}}, \quad (4.40)$$

$$\rho^{(m)}(t) = \rho_0 (1 + \delta t)^{\frac{-6(\omega+1)}{3\omega+4}}, \quad (4.41)$$

$$\delta = \left(\frac{4\pi\rho_0}{\phi_0} \frac{3\omega + 4}{2\omega + 3}\right)^{1/2}, \quad (4.42)$$

with $3q + s = 2$, which reproduces the Brans-Dicke [152] dust solution. If $\omega = -1$, one has a Minkowski space resulting from the balance between the radiative fluid and the scalar field that grows quadratically with time. The solution is instead expanding if $\omega < -4/3$ or $\omega > -1$, and pole-like if $-4/3 < \omega < -1$. In this case, there are two disconnected branches[2], one expanding for $t < -t_0$ and the other contracting for $t > -t_0$.

---

[2]Pre-big bang cosmology [579] is built upon a class of solutions of low-energy string theory similar to the one presented here.

Another particular case of the Nariai solution is that of a radiative fluid ($\gamma = 4/3$, $P^{(m)} = \rho^{(m)}/3$):

$$a(t) = a_0 \sqrt{1 + \delta t}, \qquad (4.43)$$

$$\phi(t) = \phi_0 = \text{const.}, \qquad (4.44)$$

$$\rho^{(m)}(t) = 3P^{(m)}(t) = \frac{C}{a^4} = \frac{\rho_0}{(1 + \delta t)^2}, \qquad (4.45)$$

$$\delta = \left(\frac{32\pi \rho_0}{3\phi_0}\right)^{1/2}. \qquad (4.46)$$

This solution is independent of ω and corresponds to a vanishing Ricci curvature $R = 6\left(\dot{H} + 2H^2\right)$.

Another case of special interest contained in Nariai's general solution is that of a universe with cosmological constant [625, 552] corresponding to $\gamma = 0$, $P^{(m)} = -\rho^{(m)}$, $\alpha = 4(2\omega + 1)^{-1}$,

$$a(t) = a_0 (1 + \delta t)^{\omega + \frac{1}{2}}, \qquad (4.47)$$

$$\phi(t) = \phi_0 (1 + \delta t)^2, \qquad (4.48)$$

$$\delta = \left[\frac{32\pi \rho_0}{\phi_0} \frac{1}{(6\omega + 5)(2\omega + 3)}\right]^{1/2}. \qquad (4.49)$$

This solution is important for the extended inflationary scenario discussed in the next chapter. It is not the only solution describing a universe dominated by a cosmological constant, but it is an attractor in phase space.

2.1.4 Other solutions with cosmological constant

These solutions [743] describe a spatially flat FLRW solution for $V = 0$, $-3/2 < \omega < -4/3$, with a cosmological constant Λ as the only form of matter. The first one is given by

$$a(t) = a_0 \left\{ \left|(C + 1)\exp\left(2\sqrt{A} t\right) + C - 1\right|^{\omega + 1} \right.$$

$$\left. \left[\frac{\left|\exp\left(\sqrt{A} t\right) - \sqrt{\frac{1-C}{1+C}}\right|}{\exp\left(\sqrt{A} t\right) + \sqrt{\frac{1-C}{1+C}}}\right]^{\pm \sqrt{\frac{2\omega + 3}{3}}} \right\}^{\frac{1}{3\omega + 4}}$$

$$\cdot \exp\left(-\frac{\omega+1}{3\omega+4}\sqrt{A}\,t\right), \qquad (4.50)$$

$$\phi(t) = \phi_0 \left\{ \left|(C+1)\exp\left(2\sqrt{A}\,t\right) + C - 1\right| \right.$$

$$\cdot \left[\frac{\left|\exp\left(\sqrt{A}\,t\right) - \sqrt{\frac{1-C}{1+C}}\right|}{\exp\left(\sqrt{A}\,t\right) + \sqrt{\frac{1-C}{1+C}}}\right]^{\mp\sqrt{3(2\omega+3)}} \right\}^{\frac{1}{3\omega+4}}$$

$$\cdot \exp\left(-\frac{\sqrt{A}\,t}{3\omega+4}\right), \qquad (4.51)$$

where A and C are constants with $|C| < 1$ and

$$A = \frac{2\Lambda(3\omega+4)}{2\omega+3}. \qquad (4.52)$$

Another solution corresponding to the same vacuum energy and to $|C| > 1$ is [743]

$$a(t) = a_0 \left|(C+1)\exp\left(2\sqrt{A}\,t\right) + C - 1\right|^{\frac{\omega+1}{3\omega+4}}$$

$$\cdot \exp\left[\left(-\frac{\omega+1}{3\omega+4}\right)\sqrt{A}\,t \pm \frac{2}{3}\frac{\sqrt{3\,|3\omega+3|}}{3\omega+4}\right.$$

$$\left. \cdot \arctan\left(\exp\left(\sqrt{A}\,t\right)\sqrt{\frac{C+1}{C-1}}\right)\right], \qquad (4.53)$$

$$\phi(t) = \phi_0 \left|(C+1)\exp\left(2\sqrt{A}\,t\right) + C - 1\right|^{\frac{1}{3\omega+4}}$$

$$\cdot \exp\left(-\frac{\sqrt{A}\,t}{3\omega+4} \mp \frac{2\sqrt{3\,|2\omega+3|}}{3\omega+4}\arctan\left(e^{\sqrt{A}\,t}\sqrt{\frac{C+1}{C-1}}\right)\right). \qquad (4.54)$$

When $|\omega| \to +\infty$, these solutions do not go over to the corresponding solutions of general relativity [746].

2.1.5 Generalizing Nariai's solution

Nariai's solution is not the most general power-law solution for a spatially flat FLRW model and it can be generalized [454, 646] by employing the new timelike coordinate τ defined by

$$d\tau = \frac{dt}{a^{3(\gamma-1)}}, \qquad (4.55)$$

which only coincides with t for a dust fluid with $\gamma = 1$. The generalized Nariai solution is given by

$$a(\tau) = a_0 (\tau - \tau_-)^{q_\mp} (\tau - \tau_+)^{q_\pm}, \qquad (4.56)$$

$$\phi(\tau) = \phi_0 (\tau - \tau_-)^{s_\mp} (\tau - \tau_+)^{s_\pm}, \qquad (4.57)$$

where

$$q_\pm = \frac{\omega}{3\left[1 + \omega(2-\gamma) \mp \sqrt{\frac{2\omega+3}{3}}\right]}, \qquad (4.58)$$

$$s_\pm = \frac{1 \pm \sqrt{\frac{2\omega+3}{3}}}{1 + \omega(2-\gamma) \mp \sqrt{\frac{2\omega+3}{3}}}, \qquad (4.59)$$

with a_0, ϕ_0 and $\tau_\pm$ constants and $\omega > -3/2$. Nariai's solution is recovered for $\tau_+ = \tau_-$. If $\tau_+ \neq \tau_-$, the solution (4.56)-(4.59) approaches the O'Hanlon-Tupper vacuum solution (4.14)-(4.17) as $\tau \to \tau_\pm$.

2.1.6 Phase space analysis for $K = 0$ and $V(\phi) = 0$

To obtain a qualitative overall view of the dynamics of the universe it is convenient to approach the problem with dynamical system methods [821, 890]. In this subsection the possibility of a cosmological constant is allowed. The latter can either be described by a constant scalar field potential $V = $ constant $\neq 0$ or by a barotropic fluid with $\gamma = 0$ while $V = 0$. It is this second possibility that is contemplated here.

The general dynamical behaviour of $K = 0$ FLRW solutions without self-interaction and with a barotropic fluid described by eqs. (4.2) and (4.3) is analyzed in Refs. [743, 545, 544, 771, 482]. The BD field equations for a FLRW universe reduce to two coupled autonomous first order equations for the (rescaled) scale factor and BD scalar [544, 771, 482]. The phase space is two-dimensional and this feature is likely to eliminate the possibility of chaotic dynamics [480, 449]. The two parameters ω and γ vary in the parameter space $(\omega, \gamma) \in (-3/2, +\infty) \times [0, 2]$. Many kinds of behaviour of the scale factor $a(t)$ are possible, including *bouncing* models. These are solutions that contract to a minimum size and then expand. Using the terminology of Chapter 8, these solutions exhibit a regime of *superacceleration*, i.e., $\dot{H} > 0$. For example, the expanding branch of a pole-like solution $a(t) = a_0 \, t^q$ with $q < 0$ is superaccelerating ($\dot{H} = -q/t^2 > 0$). Bounce models go from $H < 0$ during contraction to $H > 0$ during expansion, with a corresponding period in which $\dot{H} > 0$. These bounce models are present for $\omega < 0$.

2.1.7 Phase plane analysis for $K = 0$ and $V(\phi) = \Lambda \phi$

A phase space analysis for spatially flat FLRW models with a barotropic fluid and a linear potential is given in Ref. [544] (see Ref. [744] for a study of the dynamics with pure Λ and no fluid). A linear potential is obtained by adding a cosmological constant Λ to the Ricci curvature multiplying the BD field ϕ in the action (1.25). This procedure generates a different result than the usual addition of a constant Λ to the overall Lagrangian density of BD theory.

For values of the BD parameter $\omega > 0$ all initially expanding universes approach de Sitter-like exponential expansion at late times, irrespective of the value of the fluid parameter γ. If $\omega < 0$ this is no longer true. Instead, there are bouncing universes and *vacillating universes* — solutions that expand, slow down, contract for a short period of time and then re-expand again.

Two de Sitter fixed points are always present irrespective of the value of γ. They are given by [81, 744, 544]

$$a^{(\pm)}(t) = a_0 \exp\left\{\pm (\omega + 1)\left[\frac{2\Lambda}{(2\omega + 3)(3\omega + 4)}\right]^{1/2} t\right\}, \quad (4.60)$$

$$\phi^{(\pm)}(t) = \phi_0 \exp\left\{\pm \left[\frac{2\Lambda}{(2\omega + 3)(3\omega + 4)}\right]^{1/2} t\right\}. \quad (4.61)$$

Note that in general (for $\omega \neq -1$) the scalar field is not constant for these de Sitter spaces, unlike the case of general relativity; however, the general relativistic solution is recovered in the $\omega \to \infty$ limit.

The two fixed points (4.60) and (4.61) corresponding to the upper or lower sign respectively, are attractors: their attraction basin is limited to most, but not all, of the solutions that initially expand. They reduce to Minkowski spaces with exponentially expanding or contracting BD field if $\omega = -1$. Although ruled out by Solar System experiments, this case reproduces string theory. Such Minkowski spaces are stable (analogues are known in string cosmology [660, 41]) and correspond to a balance between the BD field and ordinary matter.

There are additional fixed points of the dynamical system for special values of the parameter γ of the barotropic fluid ($\gamma = 0, 1, 4/3$, and 2), of various natures.

If $\omega > 0$, all solutions that start out contracting end up in a big crunch singularity. If instead $-1 < \omega < 0$, there are nonsingular bouncing universes, some of which are attracted by the de Sitter fixed points, depending on the initial conditions [454, 544, 235, 324].

A phase plane analysis of the dynamics with self-interaction potential $V(\phi) = V_0 \phi^{2n}$ for the BD field is contained in Ref. [483].

2.2 $K = \pm 1$ solutions and phase space for $V = 0$

A few exact solutions of BD theory representing spatially closed ($K = +1$) or open ($K = -1$) FLRW universes are known [645, 589, 76, 562, 639, 895]. A phase space analysis of these BD cosmologies is presented in Refs. [545, 771, 482]. By introducing the new timelike coordinate

$$d\tau = \sqrt{\frac{8\pi \rho^{(m)}}{(2\omega + 3)\phi}}\, dt \qquad (4.62)$$

and the dynamical variables

$$x \equiv \frac{1}{2\phi}\frac{d\phi}{d\tau}, \qquad (4.63)$$

$$y \equiv \frac{1}{a}\frac{da}{d\tau}, \qquad (4.64)$$

the BD field equations for a FLRW model of any curvature index reduce to [482]

$$x' = -x^2 - \frac{3(2-\gamma)}{2}xy + \left(2 - \frac{3\gamma}{2}\right), \qquad (4.65)$$

$$y' = \frac{-2(1-3\alpha^2)}{3\alpha^2}x^2 + 3xy + \frac{3\gamma - 2}{2}y^2 - \frac{3(4-3\gamma)\alpha^2 + 3\gamma - 2}{6\alpha^2}. \qquad (4.66)$$

Here

$$\alpha \equiv \frac{1}{\sqrt{2\omega + 3}} \qquad (4.67)$$

and only values of the BD parameter such that $2\omega + 3 > 0$ are considered. The field equations are reduced to an autonomous system of two coupled first order equations, and the phase space is two-dimensional.

By contrast, in general relativity with a single scalar field, the phase space can only be reduced to a two-dimensional manifold when $K = 0$ [386, 38, 451]: chaos is then possible and is indeed reported for non-spatially flat models [174, 137, 175, 140, 241, 851]. Moreover, the phase space of BD theory may be three-dimensional when the self-interaction potential $V(\phi)$ of the BD scalar does not vanish [771]. The $K = 0$ FLRW universes lie on the separatrix described by the equation

$$x + y = \pm \sqrt{\frac{1+x^2}{3\alpha^2}} \,. \tag{4.68}$$

The phase plane is symmetrical under time reversal $\tau \to -\tau$ together with the reflection about the origin $(x, y) \to (-x, -y)$. The fixed points $(x', y') = (0, 0)$ of the dynamical system (4.65) and (4.66) are

$$\left(x_\pm^{(1)}, y_\pm^{(1)} \right) = \frac{\pm 1}{\sqrt{3(2-\gamma)^2 - \alpha^2(4-3\gamma)^2}}$$
$$\cdot \left(\alpha(4-3\gamma) \,,\, 2 - \gamma - \alpha^2(4-3\gamma) \right) , \tag{4.69}$$

and

$$\left(x_\pm^{(2)}, y_\pm^{(2)} \right) = \pm \left(\frac{\sqrt{2-3\gamma}}{2} \,,\, \frac{1}{\sqrt{2-3\gamma}} \right) , \tag{4.70}$$

where the solutions denoted by a '+' are expanding, while those corresponding to a '−' are contracting according to the symmetry mentioned above. It is sufficient to consider the '+' solutions. The critical points $\left(x_\pm^{(1)}, y_\pm^{(1)} \right)$ lie on the $K = 0$ separatrix. It can be shown that, in terms of the original variables, the fixed points correspond to the power-law solutions

$$a(t) = a_0 \, t^q \,, \tag{4.71}$$

$$\phi(t) = \phi_0 \, t^s \,, \tag{4.72}$$

where

$$q^{(i)} = \frac{2y^{(i)}}{2x^{(i)} + 3\gamma \, y^{(i)}} \,, \tag{4.73}$$

$$s = \frac{4x^{(i)}}{2x^{(i)} + 3\gamma \, y^{(i)}} \,. \tag{4.74}$$

The critical point $\left(x_+^{(1)}, y_+^{(1)} \right)$ is the Nariai solution and exists for

$$\alpha < \frac{\sqrt{3}(2-\gamma)}{|4-3\gamma|} \,, \tag{4.75}$$

which translates into constraints on the value of ω. In the limiting case $\omega = -4/\left[3\gamma\left(2-\gamma\right)\right]$, the fixed point degenerates into the de Sitter space $a(t) = a_0\, e^{Ht}$ with $H =$const. In another limiting case with $\gamma = 0$ corresponding to a pure cosmological constant, the fixed point reduces to the solution (4.47)-(4.49) of extended inflation (see Chapter 5).

In terms of the original variables, the fixed point $\left(x_+^{(2)}, y_+^{(2)}\right)$ is the coasting universe

$$a(t) = a_0\, t , \tag{4.76}$$

$$\phi(t) = \phi_0\, t^{2-3/\gamma} , \tag{4.77}$$

and it exists for all values of $\gamma < 2/3$, irrespective of the value of the BD parameter ω. In general relativity this solution only appears for $\gamma = 2/3$. If $\omega > \omega_* \equiv \frac{2}{(2-\gamma)(2-3\gamma)}$ this fixed point is located in the $K > 0$ region, while if $\omega < \omega_*$ it appears in the $K < 0$ region. If $\omega = \omega_*$, the fixed point lies on the $K = 0$ separatrix.

A stability analysis [545, 771, 482] establishes that

- the fixed point $\left(x_+^{(1)}, y_+^{(1)}\right)$ (the expanding Nariai solution) is a late time attractor if $\omega > \omega_*$ and a saddle point if $\omega < \omega_*$. This fixed point corresponds to power-law inflation and is an attractor (for $\omega > \omega_*$) for models with curvature index $K \neq 0$

- the fixed point $\left(x_+^{(2)}, y_+^{(2)}\right)$ is an attractor for $\omega < \omega_*$ and a saddle point for $\omega > \omega_*$

- if $\omega = \omega_*$ the two fixed points coincide and lie on the $K = 0$ separatrix: this point is an attractor point for $K \leq 0$ and a saddle point for $K > 0$.

Compactification of the phase plane using the Poincaré projection shows the fixed points at infinity given by the polar angles

$$\theta^{(3)} = \tan^{-1}\left(-1 + \frac{1}{\sqrt{3\alpha}}\right) , \tag{4.78}$$

$$\theta^{(4)} = \tan^{-1}\left(-1 - \frac{1}{\sqrt{3\alpha}}\right) , \tag{4.79}$$

$$\theta^{(5)} = \frac{\pi}{2} , \tag{4.80}$$

plus the symmetric ones corresponding to time reversal. $\theta^{(3)}$ and $\theta^{(4)}$ lie on the $K = 0$ separatrix.

In terms of the original variables these fixed points correspond to the spatially flat O'Hanlon-Tupper solutions [681] $a \propto t^q$, $\phi \propto \phi^s$ with

$$q = \frac{\sin \theta^{(i)}}{F\left(\theta^{(i)}\right)}, \qquad (4.81)$$

$$s = \frac{2 \cos \theta^{(i)}}{F\left(\theta^{(i)}\right)}, \qquad (4.82)$$

where

$$F(\theta) = 2\cos\theta + \left[3 - 2(\sin\theta + \cos\theta)^2 + \frac{2}{3\alpha^2}\cos^2\theta\right]\sin\theta, \qquad (4.83)$$

with $\theta^{(4)}$ being the *fast* solution (expanding) and $\theta^{(3)}$ the *slow* solution (expanding if $\omega > 0$ and contracting if $\omega < 0$). $\theta^{(5)}$ corresponds to the Milne universe of general relativity $a = a_0 t$, $\phi =$const.

A stability analysis of the fixed points at infinity [545, 771, 482] establishes that:

- the fixed point $\theta^{(5)}$ (expanding Milne universe) is a late time attractor if $\gamma > 2/3$

- whenever the critical point $(x^{(1)}, y^{(1)})$ lies at a finite distance from the origin, the fixed points at infinity $\theta^{(3)}$ and $\theta^{(4)}$ are nodes

- when the critical point $(x^{(1)}, y^{(1)})$ disappears, the nature of the critical points at infinity depends on the value of the parameter γ. If $\gamma < 4/3$ the fixed point $\theta^{(3)}$ is a saddle point, and it is also an attractor for $K = 0$ solutions lying on the separatrix. $\theta^{(4)}$ is a node. This behaviour is reversed when $\gamma > 4/3$.

In the case $K = +1$ certain solutions end in a big crunch singularity and there are also *hesitating universes* that emerge from a big bang singularity, slow down their expansion, and finally expand again; as well as coasting universes with $a(t) \propto t$. There are also vacillating universe solutions emerging from a big bang, contracting and then re-expanding; and also solutions that begin contracting, and only later stop, expand, and reach an inflationary regime with $\ddot{a} > 0$.

For $K = -1$ there are bouncing universes and solutions starting from a big bang singularity and approaching a coasting universe (the $\theta^{(5)}$ attractor) at late times. An important feature is that, for both $K = \pm 1$, there are solutions that are dominated by a cosmological constant ($\gamma = 0$) but do not approach a $K = 0$ solution at late times (in particular, they do not approach de Sitter space) [771]. Hence, the cosmic no-hair theorems of general relativity stating that the inflationary de Sitter space is a late time attractor in phase space for the cosmic dynamics, and that inflation is a generic phenomenon in cosmology leading to a flat universe, do not hold in general in BD theory. This is not surprising in view of the fact that the spatially flat solution of BD cosmology corresponding to a cosmological constant is not de Sitter space, but a power-law solution.

For a fixed value of the parameter ω, the necessary condition for inflation is

$$\gamma < \gamma_* \equiv \frac{2}{3}\left(2 - \sqrt{\frac{2\omega + 3}{2\omega}}\right). \tag{4.84}$$

2.3 Phase space for any K and $V = m^2\phi^2/2$

The case in which the scalar field potential is $V(\phi) = m^2\phi^2/2$, the curvature index is arbitrary, and a barotropic fluid is present, is studied in Ref. [771]. By using the conformal time η and the new variables

$$X \equiv \sqrt{\frac{2\omega + 3}{12}}\,\frac{\phi'}{\phi}, \tag{4.85}$$

$$Y \equiv \frac{a'}{a} + \frac{\phi'}{2\phi}, \tag{4.86}$$

$$Z \equiv \frac{m^2}{2}a^2\phi, \tag{4.87}$$

where a prime denotes differentiation with respect to η, the dynamical system is reduced to the three first order coupled equations

$$X' = -2XY + \frac{4 - 3\gamma}{4A}\left(Y^2 - X^2 - \frac{Z}{6} + K\right), \tag{4.88}$$

$$Y' = \frac{2 - 3\gamma}{2}\left(Y^2 - X^2 + K\right) - 2X^2 + \frac{\gamma}{4}Z, \tag{4.89}$$

$$Z' = 2ZY, \tag{4.90}$$

where

$$A \equiv \frac{1}{2}\sqrt{\frac{2\omega + 3}{3}} \tag{4.91}$$

and where it is assumed that $\omega > -3/2$.

In general, the phase space is three-dimensional, contrary to the case in which the potential for the BD field vanishes [545] or is linear [744, 544] (the former case is recovered by setting $m = 0$, which eliminates the Z-variable). However, the dimensionality of the phase space reduces to two in vacuum ($T_{ab}^{(m)} = 0$), in which case the variable Z can be eliminated. In the presence of a radiative fluid ($\gamma = 4/3$), the variable X can be eliminated and the phase space again reduces to a plane [771].

The existence of critical points depends on the value of the curvature index K and of the parameter γ. Moreover, for vacuum and radiation fluids, there are power-law contracting and expanding attractors, and expanding de Sitter attractors [771].

2.3.1 The Dehnen-Obregon solution

This is a solution of BD cosmology [268] representing a $K = +1$ coasting universe filled with dust fluid ($\gamma = 1$, $P^{(m)} = 0$) and valid for $\omega < -2$ and $V(\phi) = 0$:

$$a(t) = \sqrt{\frac{-2}{2+\omega}}\, t\,, \tag{4.92}$$

$$\phi(t) = \frac{-8\pi}{2\omega+3}\, \rho(t)\, t^2 = \frac{\phi_0}{t} \tag{4.93}$$

with

$$2\pi^2 a^3(t)\, \rho(t) = M\,, \tag{4.94}$$

$$\phi_0 = \frac{-\sqrt{2}\, M\, |\omega+2|^{3/2}}{\pi\, (2\omega+3)}\,, \tag{4.95}$$

where M is a constant representing the total mass of the closed universe. The energy density is

$$\rho(t) = \frac{M\, |\omega+2|^{3/2}}{4\sqrt{2}\, \pi^2}\, \frac{1}{t^3}\,. \tag{4.96}$$

Although at a first sight it seems that this solution has no analogue in general relativity as $\omega \to -\infty$ [268], careful analysis shows that an analogue indeed exists [641, 642].

Other solutions are usually expressed in terms of conformal time and of variables different from the scale factor $a(t)$ and the BD scalar ϕ. A variety of these solutions is known [591, 76, 895, 639, 426, 856, 685].

2.4 Bianchi models

Bianchi model solutions representing homogeneous and anisotropic universes are also known, and the dynamics near the initial singularity have been investigated in detail [663, 664]. Usually these solutions are presented in terms of variables different from the scale factor and the scalar field, which makes their interpretation non-trivial.

The isotropization of Bianchi models of type I, V and IX in BD theory with $V(\phi) = 0$ and the barotropic equation of state (4.3) are studied in Refs. [663, 664, 203].

2.4.1 Bianchi V universes

A Bianchi V model is represented by the metric

$$ds^2 = -dt^2 + a_1^2\, dx^2 + a_2^2\, e^{-2x}\, dy^2 + a_3^2\, e^{-2x}\, dz^2 \qquad (4.97)$$

in synchronous coordinates. In the case in which ordinary matter is described by a perfect fluid with equation of state (4.3) and $V(\phi) = 0$, solutions are found [199] by using a rescaled time coordinate τ and a new scalar field defined by

$$d\tau \equiv a^{-3(\gamma-1)}\, dt\,, \qquad (4.98)$$

$$\psi(\tau) \equiv a^{-3\gamma}\, \phi\,, \qquad (4.99)$$

where $a \equiv (a_1\, a_2\, a_3)^{1/3}$. For a non-radiative fluid with $\gamma \neq 4/3$ a solution is [199]

$$a_1(\tau) = \left\{ \frac{2(2-3\gamma)}{[2+\omega\gamma(2-3\gamma)]\, m_\gamma} \right\}^{\frac{1}{2(4-3\gamma)}} \left(A\tau^2 + B\tau + C \right)^{\frac{1}{2(4-3\gamma)}}, \qquad (4.100)$$

$$a_2(\tau) = a_1 \exp\left[\frac{-2h_2}{\sqrt{\Delta}} \operatorname{arctanh}\left(\frac{-2(4-3\gamma)\,[(4-3\gamma)\, m_\gamma \tau + \tau_0]}{(3\gamma-2)\sqrt{\Delta}} \right) \right]$$
if $\Delta > 0$, $\qquad (4.101)$

$$a_2(\tau) = a_1 \exp\left[\frac{2h_2}{\sqrt{|\Delta|}} \arctan\left(\frac{-2(4-3\gamma)\,[(4-3\gamma)\, m_\gamma \tau + \tau_0]}{(3\gamma-2)\sqrt{|\Delta|}} \right) \right]$$
if $\Delta < 0$, $\qquad (4.102)$

$$a_3(\tau) = \frac{a_1^2}{a_2}, \qquad (4.103)$$

where

$$A = \frac{(4-3\gamma)^2}{2-3\gamma}\, m_\gamma\,, \qquad (4.104)$$

$$B = \frac{2(4-3\gamma)}{2-3\gamma}\tau_0, \tag{4.105}$$

$$C = \left[\frac{-(2-3\gamma)^2\left(h_1^2+h_2^2+h_3^2\right)}{18(\gamma-1)+\omega(2-3\gamma)^2}-\tau_0^2\right]\frac{1}{m_\gamma(3\gamma-2)}, \tag{4.106}$$

where h_i ($i = 1, 2, 3$) and τ_0 are constants satisfying the relation

$$h_1 + h_2 + h_3 = 0 \tag{4.107}$$

and

$$\Delta = B^2 - 4AC = -\frac{8(4-3\gamma)^2}{18(\gamma-1)+(2-3\gamma)^2\omega}. \tag{4.108}$$

The BD field is

$$\phi(\tau) = \left\{\frac{[2+(2-\gamma)(3\gamma-2)\omega]m_\gamma}{2(2-3\gamma)}\right\}^{\frac{-3\gamma}{2(4-3\gamma)}}\left[A\tau^2 + B\tau + C\right]^{\frac{(2-3\gamma)}{2(4-3\gamma)}}. \tag{4.109}$$

If $\Delta = 0$ one has an isotropic solution $a_1 = a_2 = a_3$ as a special case. For $K = -1$, the latter reproduces a solution found by Dehnen and Obregon [268, 269, 561, 204]. The density of ordinary matter obeys the equation

$$\rho = \frac{M_\gamma}{a^{3\gamma}}, \tag{4.110}$$

where M_γ is constant.

For a radiative fluid with $\gamma = 4/3$ one has the solution [199]

$$a_i(\tau) = \alpha_i^{(0)}\sqrt{c_1 e^{-2\tau} + c_2 e^{2\tau} - \frac{4\pi M_r}{3}}$$
$$\cdot \exp\left\{\frac{c_3 - h_i}{2\sqrt{\Delta}}\operatorname{arctanh}\left[\frac{1}{\sqrt{\Delta}}\left(c_2 e^{2\tau} - \frac{2\pi M_r}{3}\right)\right]\right\} \quad \text{if } \Delta > 0,$$

$$a_i(\tau) = \alpha_i^{(0)}\sqrt{c_1 e^{-2\tau} + c_2 e^{2\tau} - \frac{4\pi M_r}{3}}$$
$$\cdot \exp\left\{\frac{h_i - c_3}{2\sqrt{|\Delta|}}\arctan\left[\frac{1}{\sqrt{|\Delta|}}\left(c_2 e^{2\tau} - \frac{2\pi M_r}{3}\right)\right]\right\} \quad \text{if } \Delta < 0,$$

$$a_i(\tau) = \alpha_i^{(0)}\exp\left(\frac{c_3 - h_i}{2c_2 e^{2\tau} - \frac{4\pi M_r}{3}}\right) \quad \text{if } \Delta = 0,$$

where $i = 1, 2$, or 3, α_i, c_1 and c_2 are integration constants, and M_r is a constant given by $\rho^{(m)} a^4 = M_r$. The BD scalar field is

$$\phi(\tau) = \frac{\exp\left\{\frac{-c_3}{\sqrt{\Delta}} \text{arctanh}\left[\frac{1}{\sqrt{\Delta}}\left(c_2 e^{2\tau} - \frac{2\pi M_r}{3}\right)\right]\right\}}{\left(\alpha_1^{(0)} \alpha_2^{(0)} \alpha_3^{(0)}\right)^{2/3}}$$

if $\Delta > 0$,

$$\phi(\tau) = \frac{\exp\left\{\frac{c_3}{\sqrt{|\Delta|}} \arctan\left[\frac{1}{\sqrt{|\Delta|}}\left(c_2 e^{2\tau} - \frac{2\pi M_r}{3}\right)\right]\right\}}{\left(\alpha_1^{(0)} \alpha_2^{(0)} \alpha_3^{(0)}\right)^{2/3}}$$

if $\Delta < 0$, $\omega < -3/2$,

$$\phi(\tau) = \frac{\exp\left(\frac{-c_3}{c_2 e^{2\tau} - \frac{2\pi M_r}{3}}\right)}{\left(\alpha_1^{(0)} \alpha_2^{(0)} \alpha_3^{(0)}\right)^{2/3}}$$

if $\Delta = 0$.

Particular choices of the integration constants c_i reproduce the Bianchi models found in Refs. [592, 589, 752, 645, 76].

3. Exact solutions of scalar-tensor theories

Numerous cosmological solutions of general scalar-tensor theories with ω varying as a function of ϕ are known [103, 128, 398, 14, 13, 712, 81, 270, 76, 82, 179, 9, 639, 895, 426, 856, 853, 183, 184, 83, 532, 531, 685, 367, 368]. Usually, specialized techniques are needed in order to find such solutions. These techniques have been initially studied by Barrow and Maeda [81] and Barrow [76] in the case of isotropic cosmologies that are vacuum- or radiation-dominated, with a vanishing trace of the energy-momentum tensor. The case of isotropic and spatially flat universes dominated by a perfect fluid with equation of state (4.3) and $0 \leq \gamma \leq 4/3$ in general scalar-tensor gravity is studied by Barrow and Mimoso [82]. For arbitrary choices of $\omega(\phi)$, a solution-generating technique for isotropic models of any spatial curvature restricted to vacuum or a radiation fluid, is presented in Ref. [83], together with a method valid for any value of γ but restricted to spatially flat isotropic universes.

APPENDIX 4.A: The equation of state of a scalar field fluid

Anisotropic cosmological models in general scalar-tensor gravity are studied in Refs. [639, 895]. The isotropization of Bianchi models in scalar-tensor theories with $V(\phi) = 0$ and ordinary matter in the form of a barotropic fluid with the equation of state (4.3) is discussed in Ref. [639]. The case of a vacuum with a scalar field potential $V(\phi)$ satisfying

$$\frac{\phi}{V}\frac{dV}{d\phi} = 2 \pm k\sqrt{\frac{3}{2} + \omega(\phi)} \qquad (4.111)$$

(with k =constant) is studied in Refs. [132, 231]. This class of potentials is such that the conformally transformed potential in the Einstein frame is exponential, $U\left(\tilde{\phi}\right) = \exp\left(k\tilde{\phi}\right)$. It has been found that, with the exception of certain Bianchi type IX metrics, all Bianchi models isotropize if $0 < k^2 \leq 2$ and in addition, they inflate if $k^2 < 2$. Bianchi VIIb models with $k > 0$ have also been found to isotropize, but they do not inflate [231].

The existence of point (Noether) symmetries which lead to first integrals of motion in anisotropic scalar-tensor cosmologies is studied in Ref. [577] with the result that, if the cosmological constant is nonzero, BD theory is the only scalar-tensor theory containing such a symmetry. The symmetry survives in the $\Lambda \to 0$ limit, but in this case other scalar-tensor theories may exhibit point symmetries as well [577].

APPENDIX 4.A: The equation of state of a scalar field fluid

A free scalar field ψ that is minimally coupled to the Ricci curvature in a FLRW universe mimics a stiff fluid. In fact, the expression of the canonical energy-momentum tensor for a self-interacting field

$$T_{ab} = \nabla_a \psi \nabla_b \psi - \frac{1}{2} g_{ab} \nabla^c \psi \nabla_c \psi - g_{ab} V(\psi) \qquad (4.A.1)$$

yields the energy density and pressure

$$\rho = \frac{\left(\dot{\psi}\right)^2}{2} + V, \qquad (4.A.2)$$

$$P = \frac{\left(\dot{\psi}\right)^2}{2} - V. \qquad (4.A.3)$$

In the absence of self-interaction, this form of matter has the effective equation of state $P = \rho$ of a stiff fluid.

If ψ depends on the spatial position as well, then

$$\rho = \frac{\left(\dot{\psi}\right)^2}{2} + V + \frac{\left(\vec{\nabla}\psi\right)^2}{2a^2}, \qquad (4.A.4)$$

$$P = \frac{\left(\dot{\psi}\right)^2}{2} - V - \frac{\left(\vec{\nabla}\psi\right)^2}{6a^2}. \qquad (4.\text{A}.5)$$

If ψ is a free field, consider waves of ψ with physical wave number k/a such that $(k/a)^2 \ll H^2, \dot{H}$ (long waves). For these waves, the gradient terms are negligible and the equation of state is again $P \simeq \rho$. Near the initial singularity, instead, the terms containing the gradients dominate and the effective equation of state is $P \simeq -\rho/3$.

Chapter 5

THE EARLY UNIVERSE

1. Introduction

The standard big bang model of the universe is a very successful one supported by three major pieces of evidence:

- the expansion of the universe, which constitutes the most natural interpretation of the redshift of galaxies,

- the existence of a cosmic background of electromagnetic radiation, a relic of the primordial state of high density, pressure and temperature. The predicted spectrum of the cosmic microwave background is that of a blackbody with temperature of 2.7 K, in remarkable agreement with the observations, and

- primordial nucleosynthesis: the relative abundances of H, He, and a few other light elements are predicted and are in agreement with observations.

If one replaces general relativity with BD theory or scalar-tensor gravity in the standard big bang model, one still finds many spatially homogeneous and isotropic solutions that are in qualitative and quantitative agreement with the observed recession of galaxies, spectrum and temperature of the cosmic microwave background, and primordial nucleosynthesis [444, 445, 291, 898]. The only uncertainty is the boundary condition on the BD-like scalar ϕ at the big bang singularity. Primordial nucleosynthesis does impose constraints on scalar-tensor gravity, but these can be satisfied by a number of solutions. By contrast, in other alternative theories of gravity, cosmological observations are sufficient to rule out many solutions [909].

Where the predictions of scalar-tensor theories differ from those of general relativity is during inflation, a very short period of accelerated, superluminal expansion of the universe that precedes the radiation- and matter-dominated eras in the standard big bang model. The correct description of the universe before inflation is unknown: it is possible that inflation interrupts a previous radiation-dominated era à la big bang model, or that the inflationary universe directly emerges from a quantum gravity regime [581, 541, 569]. It is quite possible that general relativity is not the correct description of gravity in the early universe, and indeed quantum gravity must introduce corrections to Einstein's theory when one goes back to a sufficiently early time. Since string theory — the only theory of quantum gravity presently available, and about which very little is known indeed — exhibits many analogies with scalar-tensor theories, it is meaningful to explore the dynamics of the early universe in the context of scalar-tensor gravity. A direct approach to extended inflation using compactified string theories is subject to large uncertainties [412].

2. Extended inflation

The extended inflationary scenario [552] brings a new lease on life to one of the peculiar features of Guth's [455] old inflationary scenario, namely the idea that the universe undergoes a spontaneous first order phase transition from a metastable vacuum. In old inflation, de Sitter-like expansion proceeds due to the fact that the inflaton field ψ is trapped in a false vacuum state: a situation described as *supercooling* of the universe. The field ψ is anchored at the value $\psi = 0$, its constant potential energy density $V(0) > 0$ is equivalent to a cosmological constant $\Lambda = 8\pi G V(0)$, and it induces an exponential expansion of the universe $a(t) = a_0 \, e^{Ht}$. Inflation is supposed to end due to tunneling of ψ from the false to the true vacuum, with spontaneous nucleation of bubbles of true vacuum. Let η_V be the nucleation rate of true vacuum bubbles per Hubble time and per Hubble volume; one has

$$\eta_V = \frac{\Gamma}{H^4}, \qquad (5.1)$$

where Γ is the nucleation rate per unit time and H^{-1} is the Hubble radius. In most field theories, Γ is a constant determined by the shape of the potential barrier separating the true and false vacua. In old inflation based on general relativity H is constant, corresponding to exponential expansion of the universe, and the nucleation rate η_V is also constant. In order to complete the phase transition from false to true vacuum ending old inflation, there must be enough bubbles nucleating per Hubble time

The early universe 117

and volume, i.e., it must be $\eta_V \sim 1$. On the other hand, the requirement that there be a sufficient amount of inflation to solve the problems of the standard big bang model (approximately 60 e-folds) imposes the constraint $\eta_V \ll 1$, which can be understood by noting that if η_V is too large, nucleation of true vacuum bubbles stops inflation too early.

Due to the incompatibility of these constraints, the old inflationary scenario was abandoned and replaced by other scenarios based on general relativity in which the expansion of the universe is only approximately exponential:

$$a(t) = a_0 \exp\left(H(t)\right) \qquad (5.2)$$

where

$$H(t) = H_0 + H_1 t + ..., \qquad (5.3)$$

and where the constant term H_0 dominates the expansion in the slow-roll approximation. In these scenarios, slow-roll inflation occurs due to the fact that the potential $V(\psi)$ has a shallow section mimicking a cosmological constant, and inflation is terminated by a second order phase transition. It is assumed that the potential becomes steep and the inflaton starts moving fast on it, thus breaking the slow-roll approximation. The inflaton rolls down the steep potential towards its minimum $V = 0$, overshoots it, and oscillates around it. The potential energy of ψ is then dissipated in oscillations around its minimum, which generate particles in the *reheating* regime, in which the temperature of the universe is raised again after the cooling due to the inflationary expansion.

2.1   The original extended inflationary scenario

The first order phase transition that ends old inflation has a chance to work in the context of BD theory, contrary to what happens in general relativity. As noted in the previous chapter, in BD gravity the solution of the field equations corresponding to a cosmological constant as the only (or dominant) form of matter is not de Sitter space, but the power-law solution (4.47) and (4.48) [625]

$$a(t) \propto t^{\omega+1/2}, \qquad (5.4)$$

$$\phi(t) \propto t^2, \qquad (5.5)$$

which is a special case of the Nariai solution (4.28)-(4.31) and is inflationary if $\omega > 1/2$. This form of inflation is called *power-law inflation* and was originally introduced in the context of general relativity with an exponential scalar field potential [3, 594]. In this case, the Hubble parameter $H = (\omega + 1/2)\, t^{-1}$ is not constant and the nucleation rate of true vacuum bubbles is not constant either,

$$\eta_V = \frac{\Gamma}{H^4} \propto t^4 \, . \tag{5.6}$$

This is exactly what is needed to cure the problem of the old inflationary scenario: η_V is small at early times during which only a few bubbles of true vacuum nucleate and inflation proceeds (albeit at a rate smaller than exponential) for a sufficient number of e-folds. Later on, nucleation of bubbles of true vacuum becomes more efficient, the cosmic expansion slows down, the phase transition is completed, and the false vacuum energy $V(0)$ disappears as tunneling proceeds. The latent heat of the transition is dissipated through collisions of the rapidly moving bubbles, leading to the production of a thermal bath of particles at high temperature and of gravitational waves. This spontaneous *graceful exit* is more appealing than a second order phase transition that occurs because of the shape of the potential $V(\psi)$, that has to be assumed *ad hoc*.

However, extended inflation has a serious flaw called the *big bubble problem* [897, 553]. The nucleation rate η_V, although small, is not exactly zero at early times and a few true vacuum bubbles do nucleate in this early epoch. While they are still small in comparison to the Hubble radius, they do not feel the spacetime curvature and their dynamics are essentially the same as if they were in flat spacetime. In this regime, these early bubbles expand at the speed of light; then their size becomes comparable to the Hubble radius and they begin expanding at the inflationary cosmic rate. The crucial point is that bubbles nucleated during early times do reach a cosmological size: because there are only a few of them and they end up being so large, they don't thermalize like the many small bubbles nucleated later during inflation do. They leave a significant imprint in the cosmic microwave background, introducing inhomogeneities that add to, and even dominate, those generated by quantum fluctuations of the inflaton and by gravitational waves. The bubble-induced fluctuations should be detectable today [897, 553, 572, 540]; since they are not observed, a constraint is placed on extended inflation. A calculation of the bubble spectrum [572] leads to the limit on the BD parameter $\omega \leq 20$. This constraint obviously conflicts with the bounds coming from present-day experiments in the Solar System requiring that ω be larger than 500 [731] or even 3300 [910].

2.2 Alternatives

Attempts to rescue the original extended inflationary scenario from the big bubble problem generated a few alternative scenarios [553, 10, 484, 485, 628, 557]. The simplest way to circumvent the present day constraint $\omega > 500$ is by introducing a potential $U(\phi)$ for the BD field (not to be confused with the potential $V(\psi)$ for the inflaton) [553]. This potential begins to dominate the dynamics of ϕ after the end of inflation; it has a minimum at a certain value ϕ_0, and keeps ϕ anchored there, with $\phi_0 = G^{-1}$ being the inverse of the present day value of the gravitational coupling. The effect of keeping ϕ constant is that general relativity is recovered, provided that $\phi_0 \neq 0$ and that $V(\phi_0) \simeq 0$, otherwise a cosmological constant is introduced in the field equations even if ω stays small. This way, bubble nucleation proceeds at a low rate with a low ω, but the Solar System limits on ω are circumvented and the big bubble problem is cured. However, at the level of perturbations, this remedy for extended inflation destroys the scale-invariance of the spectrum of density perturbations [567] — a significant tilt from a Harrison-Zeldovich spectrum violates the observational constraints on the spectral index. The calculated function $n(\omega)$ is monotonically increasing and the constraint $\omega \leq 20$ implies $n \leq 0.8$. This range of values of n violates the observational limit $n = 1.2 \pm 0.3$ coming from the *COBE* four year data at 1σ [114, 440, 479, 54]. Only a higher value of ω could yield a spectral index in agreement with the observations.

A different cure for the big bubble problem of extended inflation consists of allowing extended inflation to terminate via a first order phase transition, followed by a second phase of inflation in the slow-rolling regime. The latter is terminated by a second order phase transition and is advocated in order to erase the bubble perturbations in the cosmic microwave background left behind by the first order phase transition ending the previous stage of extended inflation. This scenario is called *plausible double inflation*. The main motivation of the original extended inflation model, i.e., ending inflation by a spontaneous first order phase transition, is lost in double inflationary models.

In general, the alternative models trying to resurrect extended inflation advocate rather artificial situations. It seems that, in order to make it work, extended inflation requires a significant amount of "conceptual fine-tuning" which is untidy. In this case, it is better to abandon BD theory altogether in favour of scalar-tensor theories.

Extended inflation is also considered in the semiclassical context in a regime in which short wavelength quantum fluctuations of both the inflaton and the BD field act as noise for long wavelength (classical) modes, inducing Brownian motion described by a Fokker-Planck diffusion equation [408, 410, 413, 409, 836, 838, 837]. In this kind of inflation, named *stochastic inflation*, the trajectory of the scalar field peaks around classical values and the problems associated with classical scenarios persist.

3. Hyperextended inflation

The big bubble problem of extended inflation can be circumvented by setting the inflationary scenario into the context of a scalar-tensor theory instead of BD gravity. In this way, the coupling function $\omega(\phi)$ can vary with time, going from relatively small values at the beginning of inflation to large values after the end of inflation, without the need to introduce a mass term or other potential for the BD-like scalar (we remind the reader that a scalar-tensor theory converges to general relativity when $\omega \to \infty$ and $\omega^{-3} \, d\omega/d\phi \to 0$). An inflationary scenario based on scalar-tensor gravity is called *hyperextended inflation*.

The way the big bubble problem is evaded in hyperextended inflation is acceptable in principle, since the function $\omega(\phi)$ could span a large range of values between two epochs as different as the early inflationary universe and the present day. However, while possible, the change in the value of ω is wild and occurs in a short time when compared with the age of the universe. Providing a plausible way of achieving such a change is the basis for criticism [573] of early versions of hyperextended inflation (e.g., that of Ref. [81]).

In the version of hyperextended inflation proposed by Steinhardt and Accetta ([830], see also Ref. [573]), the value of ω is initially large, producing an acceptable spectrum of density perturbations. Later on, ω assumes low values and only a few true vacuum bubbles are nucleated at this time, leaving no significant imprint in the cosmic microwave background. Inflation stops due to the dynamics of the scalar, not due to a first order phase transition and nucleation of bubbles. Significant nucleation of true vacuum bubbles only occurs *after* the end of inflation thanks to tunneling, thus removing the effective vacuum energy due to the potential being anchored to the value $V(\psi = 0)$ corresponding to the supercooled metastable state. This theory is described by an action of the type

$$S = \int d^4x \sqrt{-g} \left[\frac{1}{16\pi} \left(\phi R - \frac{\omega(\phi)}{\phi} \nabla^c \phi \nabla_c \phi - U(\phi) \right) - \frac{1}{2} \nabla^c \psi \nabla_c \psi - V(\psi) \right], \tag{5.7}$$

where the inflaton potential $V(\psi)$ must have a nonzero minimum $V(0)$ corresponding to the metastable state. The dynamics of the inflaton field ψ can again be described in the slow-roll approximation (the same as usual inflationary scenarios ending with a second order phase transition). The slow-roll approximation under these circumstances is studied in detail in Refs. [410, 77, 854]. One of the possibilities appearing in hyperextended gravity is intermediate inflation[1]. There are essentially no large bubbles created and the big bubble problem is thus solved. Working models can be found [442, 557], although this task is not trivial [573, 442]. As is the case for the alternatives to extended inflation discussed in the previous section, the original motivation for extended inflation — ending inflation via a first order phase transition — is lost in hyperextended scenarios.

Finally, one should note that inflation can be implemented not only in the arenas of general relativity and scalar-tensor gravity, but also in the context of higher derivative theories which contain in the action a term proportional to R^2. Indeed, one of the first models of inflation, proposed by Starobinsky [822] is of this kind: a BD version is studied in Refs. [475, 236].

4. Real inflation?

When deciding whether or not inflation occurs in the Einstein frame, care must be exerted in considering not just the behaviour of the scale factor $a(t)$, but rather the ratio of any physical length $a(t)\, l_0$ (where l_0 is a constant comoving length) to the Planck length

$$l_{pl}(t) \equiv \sqrt{\frac{G_{eff}\hbar}{c^3}} = \sqrt{\frac{\hbar}{c^3\,\phi}}\,, \tag{5.8}$$

which also depends on time. There are situations in which $a(t)$ expands with $\ddot{a} > 0$ but $l_{pl}(t)$ grows faster than a and the ratio $a(t)l_0/l_{pl}(t)$ actually decreases. This situation occurs in string-inspired pre-big bang cosmology [579] — which actually corresponds to a BD theory with $\omega = -1$ — and the absence of actual inflation (defined as $(al_0/l_{pl})\ddot{}\, > 0$) constitutes the basis for serious criticism of this theory [246].

While pole-like inflation $a(t) = a_0\,(-t)^p$ with $t < 0$ and $p < 0$ is considered both in pre-big bang cosmology and in scalar-tensor theories, we focus on the Jordan frame power-law solution (4.47) and (4.48) corresponding to a cosmological constant and rewritten as $a = a_0\, t^{\omega+1/2}$, $\phi = \phi_0 t^2$. The ratio $a\, l_0/l_{pl}$ is, in this case,

$$\frac{a(t)\, l_0}{l_{pl}(t)} = a_0\, l_0 \sqrt{\frac{c^3\phi_0}{\hbar}}\, |t|^{\omega+3/2} \tag{5.9}$$

---

[1] Intermediate inflation [75, 658, 84] corresponds to an expansion of the universe which interpolates between exponential and power-law and yields a Harrison-Zeldovich spectrum ($n = 1$), or a *blue spectrum* ($n > 1$) of density perturbations.

and is inflationary if $t > 0$ and $\omega > -1/2$ (power-law inflation) or if $t < 0$ and $\omega < -3/2$ (pole-like inflation). Were one to simply take into account only the behaviour of the scale factor, the conclusion would be reached that there is power-law inflation only when $\omega > 1/2$. In the particular case $\omega = -1$ corresponding to pre-big bang cosmology [579] there is no real inflation [246].

5. Constraints from primordial nucleosynthesis

One of the three major successes of the standard big bang model of the universe is the prediction of primordial nucleosynthesis ([25, 24, 487, 696, 889], see also [893]) and the matching of the predicted and observed abundances of light elements. During the radiation era, when the temperature drops to about 10^9 K due to the cosmic expansion — which happens when the universe is about three minutes old — nucleosynthesis begins with the production of approximately 25% by mass of ^{4}He and small amounts of ^{2}H, ^{3}He and ^{7}Li. There is almost no production of heavier elements (see Refs. [690] and [782] for reviews).

The abundance of ^{4}He and of the other light elements is sensitive to the expansion rate of the universe. Hence one can constrain scalar-tensor gravity by using observational limits on the relative abundance of light elements [444, 445, 291, 631, 300, 11, 799, 196, 852, 797, 769]. The idea is as follows. At nucleosynthesis practically all the neutrons present are used to manufacture helium nuclei, hence the mass fraction $X\left(^4\text{He}\right)$ of helium depends on the ratio of the neutron and proton number densities n_n and n_p at the time of nucleosynthesis,

$$X\left(^4\text{He}\right) = 2 \left. \frac{\frac{n_n}{n_p}}{1 + \frac{n_n}{n_p}} \right|_{nucleosynthesis} . \tag{5.10}$$

Before a certain time t_F (*freeze-out*) the weak interaction maintains neutrons and protons in chemical equilibrium; after freeze-out, this no longer happens and the ratio of neutron to proton number densities is frozen to the value that it had at the time t_F,

$$\left. \frac{n_n}{n_p} \right|_{t_F} = \exp\left(\frac{m_n - m_p}{kT_F}\right) , \tag{5.11}$$

where m_n and m_p are the neutron and proton masses, respectively, k is the Boltzmann constant, and T_F is the temperature at freeze-out. The neutrons decay freely during the time between freeze-out and nucleosynthesis and the duration of this period affects the final abundance of neutrons at nucleosynthesis and the mass fraction of ^{4}He produced. If the cosmic expansion proceeds faster, freeze-out occurs earlier at a higher temperature T_F and the ratio n_n/n_p is closer to unity. In addition, the time between freeze-out and nucleosynthesis is shorter, fewer neutrons decay and more neutrons are available overall, resulting in a higher mass fraction $X\left({}^4\text{He}\right)$. Note that in the right hand side of eq. (5.10), the expression $2x\left(1+x\right)^{-1}$ is a monotonically increasing function of its argument $x \equiv n_n/n_p$. If the expansion of the universe is slower, underproduction of ${}^4\text{He}$ occurs instead.

The deviation from general relativity is quantified by the expansion rate (Hubble parameter) in the scalar-tensor theory H_{ST}, measured in units of the expansion rate H that one would have in general relativity with the same form and amount of matter:

$$\xi_n \equiv \frac{H_{ST}}{H} = H_{ST}\sqrt{\frac{8\pi G}{3}\rho} \qquad (5.12)$$

in a $K=0$ FLRW universe. This ratio is called *speedup factor* and assumes the value unity in general relativity. A calculation [256, 257] yields

$$\xi_n = \frac{1}{\Omega(\phi_n)\sqrt{1+\left[\frac{1}{G\Omega}\frac{d\Omega}{d\phi}|_{\phi_0}\right]^2}}, \qquad (5.13)$$

where $\Omega(\phi) = \sqrt{G\phi}$ is the scale factor of the conformal transformation mapping the scalar-tensor theory into the Einstein frame, and ϕ_n and ϕ_0 are the values of the scalar field at the time of nucleosynthesis and now, respectively. If ξ_n deviates from unity there is underproduction or overproduction of ${}^4\text{He}$ during nucleosynthesis [888, 73, 258]. Most authors agree on the estimate of ξ_n within the range $0.8 \leq \xi_n \leq 1.2$. A more conservative interpretation of the observed abundances leads to the more stringent limits $0.95 \leq \xi_n \leq 1.03$ [196, 797]. The mass fraction of ${}^4\text{He}$ produced during nucleosynthesis is [783, 683, 196]

$$\begin{aligned}X\left({}^4\text{He}\right) &= 0.228 + 0.010\ln\left(10^{10}\frac{n_b}{n_\gamma}\right) + 0.012\left(N_\nu - 3\right) \\ &+ 0.185\left(\frac{\tau_\nu - 889.8\,\text{s}}{889.8\,\text{s}}\right) + 0.327\ln\xi_n,\end{aligned} \qquad (5.14)$$

where n_b and n_γ are the number densities of baryons and photons, respectively, and N_ν is the number of light (mass less than 1 MeV) neutrino species. τ_ν is the neutrino lifetime.

Based on the available experimental limits, a reasonable choice ($N_\nu = 3$, $\tau_\nu < 885$ s) yields [769] $X\left({}^4\text{He}\right) < 0.250$, and this limit in turn yields $\ln\xi_n \leq 0.0797$.

The limits set on ξ_n translate into constraints on the coupling function $\omega(\phi)$. These constraints are not universal and depend on the particular scalar-tensor theory under study. For example, the limit on the present value of ω is $\omega(\phi_0) \geq 10^7$ in a certain class of scalar-tensor theories [769, 258], while other authors report a much more stringent limit $\omega(\phi_0) \geq 10^{20}$ for a different class of theories [300, 797]. The discrepancy is due to the fact that, in the absence of self-interaction of the BD-like scalar field ϕ, two opposite mechanisms regulate the convergence of a scalar-tensor theory to general relativity and the smallness of ξ_n. One mechanism attracts the theory toward general relativity and dominates in certain scalar-tensor theories satisfying particular boundary conditions, leading to a monotonic behaviour of the speedup factor. This rapid convergence yields nucleosynthesis bounds that are much more stringent than in scalar-tensor theories in which the attractor mechanism is not so efficient [798]. A repulsive mechanism counteracts the attractor one and shows its effects in theories with non-monotonic or oscillating speedup factor ξ_n, in which the nucleosynthesis bounds are much less stringent.

When a self-interaction potential $V(\phi)$ is included in the picture, the situation changes because the scalar ϕ can be attracted toward a fixed value corresponding to a minimum of $V(\phi)$, thus effectively returning general relativity. The situation is actually more complicated because the dynamics of the scalar ϕ are regulated not only by $V(\phi)$ but also by the source terms

$$\frac{-8\pi T^{(m)} - \dot{\omega}\dot{\phi}}{2\omega + 3} \tag{5.15}$$

in eq. (1.103). In this case the theory may converge to general relativity but, in certain situations, only asymptotically. For example, this behaviour is reported for a massive BD-like scalar in a $K = 0$ FLRW universe, or for a massless scalar in a $K = -1$ universe [770].

Although the nucleosynthesis bounds depend on the scalar-tensor theory under study, instead of the full theory it makes sense to consider its expansion about the present-day value ϕ_0 of the scalar. Then, a scalar-tensor theory is described, back to the epoch of primordial nucleosynthesis, by only the first two coefficients in the expansion of the conformal factor Ω. One has

$$\ln \Omega = a_1 (\phi - \phi_0) + \frac{a_2}{2} (\phi - \phi_0)^2 + \cdots, \tag{5.16}$$

$$\frac{d(\ln \Omega)}{d\phi} = a_1 + a_2 (\phi - \phi_0) + \cdots, \tag{5.17}$$

and
$$\xi_n = \left[G\phi_n\left(1 + \frac{a_1^2}{G^2}\right)\right]^{-1/2}. \tag{5.18}$$

For comparison, the post-Newtonian parameters of the scalar-tensor theory used in the analysis of Solar System experiments [909] are given by [678, 254]

$$\gamma - 1 \simeq \frac{2\,a_1^2}{1 + a_1^2}, \quad \beta - 1 \simeq -\frac{a_1^2\,a_2}{2\left(1 + a_1^2\right)^2}. \tag{5.19}$$

Chapter 6

PERTURBATIONS

1. Introduction

The perturbations of a FLRW universe are extremely important because they leave an imprint as temperature fluctuations in the cosmic microwave background, which have been detected by the *COBE* mission in 1992 [809]. These perturbations are the subject of intense experimental study. Hence, density and gravitational wave fluctuations around the FLRW background provide the means to test the predictions of theories of the early universe. Soon after the introduction of the idea of inflation in cosmology, it was realized that in addition to solving the horizon, flatness, and monopole problems, inflation provides the extra advantage of a natural mechanism to generate density perturbations. Such a mechanism is missing in the standard big bang model, where an initial spectrum of perturbations has to be assumed *ad hoc*. Instead, during inflation, quantum fluctuations of the inflaton field provide the required density perturbations. Their physical wavelengths $\lambda_{phys} = a\lambda$ (where λ is the comoving wavelength) scale as the scale factor a, while the horizon size H^{-1} stays almost constant during inflation. As a consequence, perturbations cross outside the horizon during inflation. The spectrum of density perturbations is usually specified by their amplitude at this horizon crossing. While outside the horizon, regions separated by a distance larger than H^{-1} cannot causally communicate with each other and the perturbation is not evolving. After the end of inflation, $H^{-1} \propto a^q$ begins to grow as a power of the scale factor and the perturbations eventually re-enter the horizon during the radiation or the matter era. Density perturbations grow non-linearly during the matter era and begin to form cosmic structures.

The mechanism of generation of structure works in general relativity as well as in scalar-tensor gravity. Many approaches to the study of perturbations of a FLRW background are attempted in the context of BD and of scalar-tensor theories [699, 595, 864, 767, 323, 543, 865, 335, 616, 599, 457, 567, 279, 337, 765, 86, 333, 829, 524, 525, 414, 160, 619, 59, 835, 216, 546, 547, 208, 123, 326, 425]. A general treatment is found in Refs. [499, 500, 502, 503, 507, 670], which we overview in the following sections. It applies to generalized scalar-tensor theories in regions of the phase space that do not contain the critical values of the scalar field discussed in Section 1.12. In Chapter 7, the same computational procedure is applied to the theory of a nonminimally coupled scalar field.

2. Scalar perturbations

It is convenient to write the line element of the perturbed FLRW universe as

$$ds^2 = -(1+2\alpha)\,dt^2 - \chi_{,i}\,dt\,dx^i + a^2(t)\left[(1+2\varphi)\,\delta_{ij} + 2H_T Y_{ij}\right]dx^i\,dx^j\,, \tag{6.1}$$

where α, χ, and φ can be expressed in terms of scalar (Y), vector (Y_i), and tensor (Y_{ij}) spherical harmonics. The scalar field is

$$\phi(t,\vec{x}) = \phi_0(t) + \delta\phi(t,\vec{x})\,, \tag{6.2}$$

where the FLRW background quantity ϕ_0 only depends on the comoving time, while the small perturbations are space- and time-dependent. Motivated by the results of the *Boomerang* [638, 266, 556, 632] and *MAXIMA* [463] experiments, we restrict ourselves here to the consideration of a spatially flat FLRW universe. The background quantities are treated as classical while the perturbations have a quantum nature and are quantized by associating them with suitable quantum operators.

As usual in cosmology, the issue of gauge-dependence of quantities related to the perturbations must be dealt with. In order to obtain gauge-independent results one needs to form gauge-invariant quantities (whose physical interpretation is sometimes rather obscure) and study their evolution. Here the covariant and gauge-invariant formalism of Bardeen, Ellis, Bruni, Hwang and Vishniac [315, 316, 317, 509] is used, following Refs. [499, 502, 503, 507]. We consider generalized scalar-tensor theories described by the action

$$S = \int d^4x\sqrt{-g}\left[\frac{1}{2}f(\phi,R) - \frac{\omega(\phi)}{2}\nabla^c\phi\nabla_c\phi - V(\phi)\right]\,, \tag{6.3}$$

where the matter part of the action is omitted because the inflaton dominates all other forms of matter during inflation. The idea is to write the field equations as effective Einstein equations by moving all the terms different from the Einstein tensor to the right hand side, obtaining

$$G_{ab} = T_{ab}\left[\phi\right],\qquad(6.4)$$

where

$$T_{ab}\left[\phi\right] = \frac{1}{F}\left[\omega\left(\nabla_a\phi\nabla_b\phi - \frac{1}{2}g_{ab}\nabla^c\phi\nabla_c\phi\right) - \frac{1}{2}\left(RF - f + 2V\right)g_{ab}\right.$$
$$\left. + \nabla_a\nabla_b F - g_{ab}\Box F\right]\qquad(6.5)$$

is an effective stress-energy tensor for the scalar ϕ and

$$F(\phi, R) \equiv \frac{\partial f}{\partial R}.\qquad(6.6)$$

In general, T_{ab} has the form of an imperfect fluid

$$T_{ab} = (P + \rho)u_a u_b + P g_{ab} + 2q_{(a}u_{b)} + \pi_{ab},\qquad(6.7)$$

where q^a and $\pi_{ab} = \pi_{ba}$ are the heat current density and the anisotropic stress tensor, respectively, which are purely spatial:

$$q_a u^a = 0,\qquad \pi_{ab}u^a = \pi_{ab}u^b = 0.\qquad(6.8)$$

As already discussed, this effective approach is questionable because the field equations are not true Einstein equations, but it is a very convenient practice in the study of perturbations because it reduces the study of perturbations to the analogous problem in Einstein's theory, which is already solved.

The scalar field ϕ satisfies the equation

$$\Box\phi + \frac{1}{2\omega}\left(\frac{d\omega}{d\phi}\nabla^c\phi\nabla_c\phi + F - 2\frac{dV}{d\phi}\right) = 0.\qquad(6.9)$$

One applies the gauge-invariant formalism for the computation of density and gravitational wave perturbations in Einstein gravity inflation [653, 578] to the effective Einstein theory described by eq. (6.4). It is useful to introduce, in addition to the usual Bardeen variables, the gauge-invariant variable

$$\delta\phi_\varphi \equiv \delta\phi - \frac{\dot\phi}{H}\varphi \equiv -\frac{\dot\phi}{H}\varphi_{\delta\phi},\qquad(6.10)$$

where an overdot denotes differentiation parallel to the fluid four-velocity u^a, i.e., $\dot f \equiv u^c\nabla_c f$ (to first order this operation coincides with differentiation with respect to the comoving time of the background). The evolution of the scalar perturbations can be derived from the second order action [596, 597, 652, 504, 505, 506]

$$S^{(pert)} = \int dt\, d^3\vec{x}\, \mathcal{L}^{(pert)} = \frac{1}{2}\int dt\, d^3\vec{x}\, a^3 Z \left\{ \delta\dot{\phi}_\varphi^2 - \frac{1}{a^2}\delta\phi_{\varphi,}{}^i\, \delta\phi_{\varphi,i} \right.$$
$$\left. + \frac{1}{a^3 Z}\frac{H}{\dot\phi}\left[a^3 Z\left(\frac{\dot\phi}{H}\right)\cdot\right]\cdot \delta\phi_\varphi^2 \right\} \qquad (6.11)$$

which is analogous to the one for inflationary perturbations in general relativity [578], and where

$$Z(t) = \frac{\dfrac{3(\dot F)^2}{2F(\dot\phi)^2} + \omega}{\left(\dfrac{\dot F}{2HF} + 1\right)^2}. \qquad (6.12)$$

In general relativity, by contrast, $F = \kappa^{-1}$, $\omega = 1$ and $Z = 1$. The equation of motion for the perturbations $\delta\phi_\varphi$ obtained by varying the action (6.11) is

$$\delta\ddot\phi_\varphi + \frac{(a^3 Z)^\cdot}{a^3 Z}\delta\dot\phi_\varphi - \left\{\frac{\nabla^2}{a^2} + \frac{1}{a^3 Z}\frac{H}{\dot\phi}\left[a^3 Z\left(\frac{\dot\phi}{H}\right)\cdot\right]\cdot\right\}\delta\phi_\varphi = 0. \qquad (6.13)$$

By introducing the auxiliary variables

$$z(t) \equiv \frac{a\dot\phi}{H}\sqrt{Z}, \qquad (6.14)$$

$$v(t,\vec{x}) \equiv z\frac{H}{\dot\phi}\delta\phi_\varphi = a\sqrt{Z}\,\delta\phi_\varphi, \qquad (6.15)$$

we can write eq. (6.13) as

$$v_{\eta\eta} - \left(\nabla^2 + \frac{z_{\eta\eta}}{z}\right)v = 0, \qquad (6.16)$$

where η is the conformal time.

In order to keep the vacuum state unchanged in time, the perturbations are quantized in the Heisenberg picture by associating the quantum operators $\delta\hat\phi(t,\vec{x})$ and $\hat\varphi$ with the classical variables $\delta\phi(t,\vec{x})$ and φ, with

$$\delta\hat\phi_\varphi = \delta\hat\phi - \frac{\dot\phi}{H}\hat\varphi. \qquad (6.17)$$

The operator $\delta\hat\phi_\varphi$ is decomposed in Fourier modes

$$\delta\hat\phi_\varphi = \frac{1}{(2\pi)^{3/2}}\int d^3\vec{k}\left[\hat a_k\, \delta\phi_{\varphi_k}(t)\, e^{i\vec{k}\cdot\vec{x}} + \hat a_k^\dagger\, \delta\phi_{\varphi_k}^*(t)\, e^{-i\vec{k}\cdot\vec{x}}\right], \qquad (6.18)$$

where the creation and annihilation operators $\hat{a}_k^\dagger$ and $\hat{a}_k$ satisfy the canonical commutation relations

$$[\hat{a}_k, \hat{a}_{k'}] = \left[\hat{a}_k^\dagger, \hat{a}_{k'}^\dagger\right] = 0 , \tag{6.19}$$

$$\left[\hat{a}_k, \hat{a}_{k'}^\dagger\right] = \delta^{(3)}\left(\vec{k} - \vec{k}'\right) . \tag{6.20}$$

The complex Fourier coefficients satisfy the equation

$$\delta\ddot{\phi}_{\varphi k} + \frac{(a^3 Z)^\cdot}{a^3 Z}\delta\dot{\phi}_{\varphi k} + \left\{\frac{k^2}{a^2} - \frac{1}{a^3 Z}\frac{H}{\dot{\phi}}\left[a^3 Z\left(\frac{\dot{\phi}}{H}\right)^\cdot\right]^\cdot\right\}\delta\phi_{\varphi k} = 0 . \tag{6.21}$$

The treatment now mimics that of ordinary inflation in general relativity. The quantity v is Fourier-decomposed, obtaining

$$v(t, \vec{x}) = \frac{1}{(2\pi)^{3/2}}\int d^3\vec{k}\left[v_k(t)\,e^{i\vec{k}\cdot\vec{x}} + v_k^*(t)\,e^{-i\vec{k}\cdot\vec{x}}\right] , \tag{6.22}$$

$$\hat{v} = \frac{zH}{\dot{\phi}}\delta\hat{\phi}_\varphi = a\sqrt{Z}\,\delta\hat{\phi}_\varphi , \tag{6.23}$$

with the components $v_k(t)$ satisfying the equation

$$(v_k)_{\eta\eta} + \left(k^2 - \frac{z_{\eta\eta}}{z}\right)v_k = 0 . \tag{6.24}$$

The momentum conjugated to $\delta\phi_\varphi$ is

$$\delta\pi_\varphi(t, \vec{x}) = \frac{\partial\mathcal{L}^{(\text{pert})}}{\partial(\delta\dot{\phi}_\varphi)} = a^3 Z\,\delta\dot{\phi}_\varphi(t, \vec{x}) , \tag{6.25}$$

and the associated quantum operator is $\delta\hat{\pi}_\varphi$. The operators $\delta\hat{\phi}_\varphi$ and $\delta\hat{\pi}_\varphi$ satisfy the equal time commutation relations

$$\left[\delta\hat{\phi}_\varphi(t, \vec{x}), \delta\hat{\phi}_\varphi(t, \vec{x}')\right] = [\delta\hat{\pi}_\varphi(t, \vec{x}), \delta\hat{\pi}_\varphi(t, \vec{x}')] = 0 , \tag{6.26}$$

$$\left[\delta\hat{\phi}_\varphi(t, \vec{x}), \delta\hat{\pi}_\varphi(t, \vec{x}')\right] = \frac{i}{a^3 Z}\delta^{(3)}\left(\vec{x} - \vec{x}'\right) , \tag{6.27}$$

while the $\delta\phi_{\varphi k}(t)$ satisfy the Wronskian condition

$$\delta\phi_{\varphi k}\,\delta\dot{\phi}_{\varphi k}^* - \delta\phi_{\varphi k}^*\,\delta\dot{\phi}_{\varphi k} = \frac{i}{a^3 Z} . \tag{6.28}$$

In order to proceed it is necessary to use the relation familiar from Einstein gravity inflation

$$\frac{z_{\eta\eta}}{z} = \frac{m}{\eta^2}, \qquad (6.29)$$

where m is a constant. Eq. (6.29) is also valid in the slow-roll inflationary regime of scalar-tensor gravity, which has de Sitter solutions as attractors in phase space. Such a regime exists provided that the functions f and ω satisfy suitable conditions. This statement is proven and the conditions on the coupling functions are made explicit for the non-minimally coupled scalar field theory in Chapter 7. The statement can be generalized to arbitrary scalar-tensor theories of the form (6.3). The regions of the phase space in which slow-roll inflation does not occur are not relevant to the present discussion.

By using eq. (6.29), the evolution equation (6.24) for the Fourier components v_k is reduced to

$$(v_k)_{\eta\eta} + \left[k^2 - \frac{(\nu^2 - 1/4)}{\eta^2} \right] v_k = 0, \qquad (6.30)$$

where

$$\nu = \left(m + \frac{1}{4} \right)^{1/2}. \qquad (6.31)$$

Eq. (6.30) is solved by introducing the quantity

$$s = k\eta \qquad (6.32)$$

which turns v_k into

$$v_k = \sqrt{s}\, J(s), \qquad (6.33)$$

and eq. (6.30) into the Bessel equation

$$\frac{d^2 J}{ds^2} + \frac{1}{s}\frac{dJ}{ds} + \left(1 - \frac{\nu^2}{s^2} \right) J = 0. \qquad (6.34)$$

The solutions are Bessel functions of order ν and

$$v_k(\eta) = \sqrt{k\eta}\, J_\nu(k\eta). \qquad (6.35)$$

Eq. (6.15) yields the solutions for the Fourier coefficients $\delta\phi_{\varphi k}$

$$\delta\phi_{\varphi k}(\eta) = \frac{\dot\phi}{zH} v_k(\eta) = \frac{1}{a\sqrt{Z}} v_k(\eta). \qquad (6.36)$$

The $v_k(\eta)$ can be rewritten by expressing the Bessel functions J_ν in terms of Hankel functions $H_\nu^{(1,2)}$, leading to

$$v_k(\eta) = \frac{\sqrt{\pi |\eta|}}{2} \left[c_1(\vec{k})\, H_\nu^{(1)}(k|\eta|) + c_2(\vec{k})\, H_\nu^{(2)}(k|\eta|) \right] \qquad (6.37)$$

and therefore (cf. eq. (6.36))

$$\delta\phi_{\varphi k}(\eta) = \frac{\sqrt{\pi|\eta|}}{2a\sqrt{Z}}\left[c_1(\vec{k})H_\nu^{(1)}(k|\eta|) + c_2(\vec{k})H_\nu^{(2)}(k|\eta|)\right] . \tag{6.38}$$

The normalization of c_1 and c_2 is such that

$$\left|c_2(\vec{k})\right|^2 - \left|c_1(\vec{k})\right|^2 = 1 , \tag{6.39}$$

which preserves eq. (6.28). Further, in the limit of small scales the vacuum state is required to correspond to positive frequency solutions. This limit is described by

$$\frac{z_{\eta\eta}}{z} \ll k^2 , \tag{6.40}$$

and reduces eq. (6.24) to

$$(v_k)_{\eta\eta} - k^2 v_k = 0 , \tag{6.41}$$

with solutions $v_k \propto e^{\pm ik\eta}$ and

$$\delta\phi_{\varphi k} = \frac{1}{a\sqrt{Z}\sqrt{2k}}\left[c_1\left(\vec{k}\right)e^{ik|\eta|} + c_2\left(\vec{k}\right)e^{-ik|\eta|}\right] . \tag{6.42}$$

One could directly expand eq. (6.38) for $k|\eta| \gg 1$ and obtain the same result. The positive frequency solution in the small scale limit is obtained by setting

$$c_1(\vec{k}) = 0 , \quad c_2(\vec{k}) = 1 , \tag{6.43}$$

which corresponds to selecting the Bunch-Davies vacuum for de Sitter space and yields

$$\delta\phi(\eta,\vec{x}) = \frac{1}{(2\pi)^{3/2}}\int d^3\vec{x}\left[c_2\left(\vec{k}\right)e^{i(\vec{k}\cdot\vec{x}-k\eta)} + c_2^*\left(\vec{k}\right)e^{i(-\vec{k}\cdot\vec{x}+k|\eta|)}\right] . \tag{6.44}$$

The power spectrum of a quantity $g(t,\vec{x})$ is defined by the integral

$$\mathcal{P}(k,t) \equiv \frac{k^3}{2\pi^2}\int d^3\vec{r}\,\langle g(\vec{x}+\vec{r},t)\,g(\vec{x},t)\rangle_{\vec{x}}\,e^{-i\vec{k}\cdot\vec{r}} = \frac{k^3}{2\pi^2}|g_k(t)|^2 , \tag{6.45}$$

where $\langle\ \rangle_{\vec{x}}$ denotes an average over the spatial coordinates $\vec{x}$ and $g_k(t)$ are the Fourier coefficients of

$$g(t,\vec{x}) = \frac{1}{(2\pi)^{3/2}}\int d^3\vec{k}\left[g_k(t)e^{i\vec{k}\cdot\vec{x}} + g_k^*(t)e^{-i\vec{k}\cdot\vec{x}}\right] . \tag{6.46}$$

In particular, by choosing $g(t,\vec{x}) = \langle 0|\delta\hat{\phi}_{\varphi k}|0\rangle$, the power spectrum of the gauge-invariant operator $\delta\hat{\phi}_{\varphi k}$ is obtained

$$\mathcal{P}_{\delta\hat{\phi}_{\varphi}}(k,t) = \frac{k^3}{2\pi^2} \int d^3\vec{r} \langle 0|\delta\hat{\phi}_{\varphi k}(\vec{x}+\vec{r},t)\, \delta\hat{\phi}_{\varphi k}(\vec{x},t)|0\rangle_{\vec{x}}\, e^{-i\vec{k}\cdot\vec{r}}. \quad (6.47)$$

For large-scale perturbations crossing outside the horizon during inflation, one has to take the large scale limit of eq. (6.21), which has the solution

$$\delta\phi_{\varphi}(t,\vec{x}) = -\frac{\dot{\phi}}{H}\left[C(\vec{x}) - D(\vec{x})\int_0^t dt'\, \frac{1}{a^3 Z}\frac{H^2}{\dot{\phi}}\right], \quad (6.48)$$

where $C(\vec{x})$ and $D(\vec{x})$ are the coefficients of a growing component and of a decaying component which we omit from now on, respectively. In the large scale limit $k|\eta| \ll 1$, the solution $\delta\phi_{\varphi k}(\eta)$ for $\nu \neq 0$ becomes

$$\delta\phi_{\varphi k}(\eta) = \frac{i\sqrt{|\eta|}\,\Gamma(\nu)}{2a\sqrt{\pi Z}} \left(\frac{k|\eta|}{2}\right)^{-\nu} [c_2(\eta) - c_1(\eta)], \quad (6.49)$$

where Γ denotes the gamma function. The power spectrum of $\delta\phi_{\varphi k}$ is then given by

$$\mathcal{P}^{1/2}_{\delta\hat{\phi}_{\varphi}}(k,\eta) = \frac{\Gamma(\nu)}{\pi^{3/2} a\,|\eta|\sqrt{Z}} \left(\frac{k|\eta|}{2}\right)^{\frac{3}{2}-\nu} \left|c_2(\vec{k}) - c_1(\vec{k})\right| \quad (6.50)$$

for $\nu \neq 0$. Neglecting the decaying component, eq. (6.48) yields

$$C(\vec{x}) = -\frac{H}{\dot{\phi}}\delta\phi_{\varphi}(t,\vec{x}) \quad (6.51)$$

and, using the definition (6.45) of power spectrum,

$$\mathcal{P}^{1/2}_C(k,t) = \left|\frac{H}{\dot{\phi}}\right| \mathcal{P}^{1/2}_{\delta\hat{\phi}_{\varphi}}(k,t). \quad (6.52)$$

Eqs. (6.10) and (6.51) yield

$$\varphi_{\delta\phi} = -\frac{H}{\dot{\phi}}\delta\phi_{\varphi} = C. \quad (6.53)$$

The variable $C(\vec{x})$ is related to the temperature fluctuations δT in the cosmic microwave background by [502]

$$\frac{\delta T}{T} = \frac{C}{5} \quad (6.54)$$

and the temperature fluctuations spectrum turns out to be

$$\sqrt{\mathcal{P}_{\delta T/T}(k,t)} = \frac{1}{5}\sqrt{\mathcal{P}_C(k,t)}. \quad (6.55)$$

Eq. (6.52) yields

$$\mathcal{P}^{1/2}_{\delta T/T}(k,t) = \frac{1}{5}\left|\frac{H}{\dot\phi}\right|\mathcal{P}^{1/2}_{\delta\phi_\varphi}(k,t) \quad (6.56)$$

with $\mathcal{P}^{1/2}_{\delta\phi_\varphi}$ given by eq. (6.50).

The spectral index of scalar perturbations used in the literature is defined as

$$n_S \equiv 1 + \frac{d\left(\ln \mathcal{P}_{\delta\hat\varphi\delta\phi}\right)}{d(\ln k)}, \quad (6.57)$$

so, from eq. (6.55),

$$n_S = 1 + \frac{d(\ln \mathcal{P}_C)}{d(\ln k)}. \quad (6.58)$$

Eqs. (6.52) and (6.50) yield

$$n_S = 4 - 2\nu \qquad (\nu \neq 0). \quad (6.59)$$

In the slow-roll regime of scalar-tensor inflation it is necessary to introduce the four slow-roll parameters [507, 670]

$$\epsilon_1 = \frac{\dot H}{H^2} = -\epsilon_H, \quad (6.60)$$

$$\epsilon_2 = \frac{\ddot\phi}{H\dot\phi} = -\eta_H, \quad (6.61)$$

$$\epsilon_3 = \frac{\dot F}{2HF}, \quad (6.62)$$

$$\epsilon_4 = \frac{\dot\alpha}{2H\alpha}, \quad (6.63)$$

where

$$\alpha = F\left[\omega + \frac{3\left(\dot F\right)^2}{2F\left(\dot\phi\right)^2}\right]. \quad (6.64)$$

Apart from the sign, ϵ_1 and ϵ_2 coincide with the usual slow-roll parameters ϵ_H and η_H of the Hubble slow-roll approximation to inflation in Einstein gravity, while ϵ_3 and ϵ_4 are new functions. The slow-roll parameters and their time derivatives $\dot\epsilon_i$ are small during slow-roll inflation.

By neglecting the time variation of the ϵ_i one obtains, to first order,

$$\frac{z_{\eta\eta}}{z} = a^2 H^2 \left(2 - 2\epsilon_1 + 3\epsilon_2 - 3\epsilon_3 + 3\epsilon_4\right) . \tag{6.65}$$

By using the standard relation (see Appendix C of Chapter 7)

$$\eta \simeq -\frac{1}{aH}\frac{1}{1+\epsilon_1} , \tag{6.66}$$

one easily obtains

$$m = 2 + 3\left(-2\epsilon_1 + \epsilon_2 - \epsilon_3 + \epsilon_4\right) , \tag{6.67}$$

$$\nu = \frac{3}{2} - 2\epsilon_1 + \epsilon_2 - \epsilon_3 + \epsilon_4 ; \tag{6.68}$$

and finally, from eqs. (6.59) and (6.68), the spectral index of scalar perturbations

$$n_S = 1 + 2\left(2\epsilon_1 - \epsilon_2 + \epsilon_3 - \epsilon_4\right) , \tag{6.69}$$

where the right hand side is computed at the time when the perturbations cross outside the horizon. The spectral index n_S does not deviate much from an exactly scale-invariant Harrison-Zeldovich spectrum with $n_S = 1$.

In the limit to general relativity $f(\phi, R) = R/\kappa$, $\omega = 1$, $\dot{F} = 0$ and $\epsilon_3 = \epsilon_4 = 0$, one recovers the usual result for the spectral index of Einstein gravity inflation with a minimally coupled scalar field [541, 578, 569]

$$n_S = 1 - 4\epsilon_H + 2\eta_H . \tag{6.70}$$

3. Tensor perturbations

Tensor perturbations (gravitational waves) are also generated during inflation and correspond to quantum fluctuations of the metric tensor g_{ab}. They have been calculated for inflation in generalized scalar-tensor gravity in Refs. [506, 670], following the procedure employed in the usual inflationary scenarios of general relativity.

Tensor perturbations correspond to the quantity H_T in eq. (6.1), which can be rewritten using

$$c_{ij}(t, \vec{x}) \equiv H_T(t) Y_{ij}(\vec{x}) , \tag{6.71}$$

where the Y_{ij} are tensor spherical harmonics. The quantities c_{ij} are transverse and traceless:

$$c_i^i = 0 , \qquad \nabla^i c_{ij} = 0 . \tag{6.72}$$

The second order action for the perturbations is [37, 498]

$$S^{(gw)} = \int dt \int d^3\vec{x}\, \frac{a^3 F}{2}\left[\dot{c}_{ij}\dot{c}^{ji} - \frac{1}{a^2}c_{ij,k}\nabla^k c^{ij}\right] \equiv \int dt \int d^3\vec{x}\, \mathcal{L}^{(gw)}, \qquad (6.73)$$

from which one derives the classical evolution equation for the c_{ij}

$$\ddot{c}_{ij} + \left(3H + \frac{\dot{F}}{F}\right)\dot{c}_{ij} - \frac{\nabla^2}{a^2}c_{ij} = 0. \qquad (6.74)$$

By introducing the quantities

$$z_g \equiv a\sqrt{F}, \qquad (6.75)$$

$$v_g(t,\vec{x}) \equiv z_g\, c_{ij}(t,\vec{x}), \qquad (6.76)$$

the evolution equation (6.74) is reduced to

$$(v_g)_{\eta\eta} - \left[\frac{(z_g)_{\eta\eta}}{z} + \nabla^2\right]v_g = 0. \qquad (6.77)$$

In the large scale limit $(z_g)_{\eta\eta}/z_g \gg k^2$, one finds the solution

$$c_{ij}(t,\vec{x}) = C_{ij}(\vec{x}) - D_{ij}(\vec{x})\int_0^t \frac{dt}{a^3 F}, \qquad (6.78)$$

where the term proportional to D_{ij} represents a decaying mode that we omit in the following.

In the limit of small scales the asymptotic solutions for the Fourier coefficients of c_{ij} in terms of conformal time are

$$c_{ij}\left(\eta,\vec{k}\right) = \frac{1}{a\sqrt{F}}\left[c_{ij}^{(1)}\left(\vec{k}\right)e^{ik\eta} + c_{ij}^{(2)}\left(\vec{k}\right)e^{-ik\eta}\right]. \qquad (6.79)$$

One decomposes the classical perturbations c_{ij} into the two possible polarization modes "+" and "×":

$$\begin{aligned}c_{ij}(t,\vec{x}) &= \frac{L^{3/2}}{(2\pi)^3}\int d^3\vec{k} \sum_{l=1}^{l=2} h_{l\vec{k}}(t)\, e_{ij}^{(l)}\left(\vec{k}\right) e^{i\vec{k}\cdot\vec{x}} \\ &\equiv \frac{L^{3/2}}{(2\pi)^3}\int d^3\vec{k}\, \tilde{C}_{ij}\left(t,\vec{x},\vec{k}\right),\end{aligned} \qquad (6.80)$$

where L^3 is a normalization volume which disappears in the final results and $l = +$ or $\times$; and the $e_{ij}^{(l)}$ are polarization tensors for the two polarization states and satisfy

$$e_{ij}^{(l)}\left(\vec{k}\right) e^{(l')\,ij}\left(\vec{k}\right) = 2\delta_{ll'}. \qquad (6.81)$$

By introducing the quantities

$$h_l(t,\vec{x}) \equiv \frac{L^{3/2}}{2(2\pi)^3} \int d^3\vec{k}\, \tilde{C}_{ij}\left(t,\vec{x},\vec{k}\right) e^{(l)\,ij}\left(\vec{k}\right)$$

$$= \frac{L^{3/2}}{(2\pi)^3} \int d^3\vec{k}\, h_{l\vec{k}}(t)\, e^{i\vec{k}\cdot\vec{x}}, \tag{6.82}$$

with the classical power spectrum defined as

$$\mathcal{P}_{c_{ij}}\left(t,\vec{k}\right) = \frac{k^3}{2\pi^2} \int d^3\vec{r}\, \langle c_{ij}(t,\vec{x}+\vec{r})\, c^{ij}(t,\vec{x})\rangle_{\vec{x}}\, e^{-i\vec{k}\cdot\vec{r}}, \tag{6.83}$$

where $\langle\ \rangle_{\vec{x}}$ denotes a spatial average, the following is satisfied

$$\mathcal{P}_{c_{ij}}\left(t,\vec{k}\right) = 2\sum_{l=1}^{2} \mathcal{P}_{h_l}\left(t,\vec{k}\right) = 2\sum_{l=1}^{2} \frac{k^3}{2\pi^2} |h_{l\vec{k}}(t)|^2 \tag{6.84}$$

with $h_{+\vec{k}} = h_{\times\vec{k}} \equiv h_{\vec{k}}$.

Quantization is achieved by associating the quantum operators $\hat{c}_{ij}$ to the classical variables c_{ij}. The $\hat{c}_{ij}$ are Fourier-expanded according to

$$\hat{c}_{ij}(t,\vec{x}) = \frac{1}{(2\pi)^{3/2}} \int d^3\vec{k} \sum_{l=1}^{2} \left[\tilde{h}_{l\vec{k}}(t)\, \hat{a}_{l\vec{k}}\, e_{ij}^{(l)}\left(\vec{k}\right) e^{i\vec{k}\cdot\vec{x}} \right.$$

$$\left. + \tilde{h}_{l\vec{k}}^{*}(t)\, \hat{a}_{l\vec{k}}^{\dagger}\, e_{ij}^{(l)}\left(\vec{k}\right) e^{-i\vec{k}\cdot\vec{x}} \right]$$

$$\equiv \frac{1}{2(2\pi)^{3/2}} \int d^3\vec{k}\, \hat{C}_{ij}\left(t,\vec{x},\vec{k}\right), \tag{6.85}$$

where the $\tilde{h}_{l\vec{k}}(t)$ are mode functions. The creation and annihilation operators $\hat{a}_{l\vec{k}}^{\dagger}$ and $\hat{a}_{l\vec{k}}$ satisfy the canonical commutation relations

$$\left[\hat{a}_{l\vec{k}}, \hat{a}_{l'\vec{k}'}^{\dagger}\right] = \delta_{ll'}\, \delta^{(3)}\left(\vec{k}-\vec{k}'\right), \tag{6.86}$$

$$\left[\hat{a}_{l\vec{k}}, \hat{a}_{l'\vec{k}'}\right] = \left[\hat{a}_{l\vec{k}}^{\dagger}, \hat{a}_{l'\vec{k}'}^{\dagger}\right] = 0. \tag{6.87}$$

The quantum operators

$$\hat{h}_l(t,\vec{x}) \equiv \frac{1}{2(2\pi)^{3/2}} \int d^3\vec{k}\, \hat{C}_{ij}\left(t,\vec{x},\vec{k}\right) e^{(l)\,ij}\left(\vec{k}\right)$$

$$= \frac{1}{(2\pi)^{3/2}} \int d^3\vec{k} \left[\tilde{h}_{l\vec{k}}(t)\, \hat{a}_{l\vec{k}}\, e^{i\vec{k}\cdot\vec{x}} + \tilde{h}_{l\vec{k}}^{*}(t)\, \hat{a}_{l\vec{k}}^{\dagger}\, e^{-i\vec{k}\cdot\vec{x}} \right]$$

$$\tag{6.88}$$

Perturbations

are now introduced and the evolution equation for the $\hat{c}_{ij}$ yields

$$\ddot{\hat{h}}_l + \left(3H + \frac{\dot{F}}{F}\right)\dot{\hat{h}}_l - \frac{\nabla^2 \hat{h}_l}{a^2} = 0 \,. \tag{6.89}$$

The momenta canonically conjugated to the $\hat{h}_l$ are

$$\delta\hat{\pi}_{h_l}(t,\vec{x}) = \frac{\partial \mathcal{L}^{(gw)}}{\partial\left(\dot{\hat{h}}_l\right)} = 2\,a^3\,F\,\dot{\hat{h}}_l \tag{6.90}$$

and satisfy the equal time commutation relation

$$\left[\hat{h}_l(t,\vec{x}),\delta\hat{\pi}_{h_l}(t,\vec{x}')\right] = i\,\delta^{(3)}\left(\vec{x}-\vec{x}'\right)\,, \tag{6.91}$$

while the mode functions $\tilde{h}_{l\vec{k}}(t,\vec{k})$ satisfy the Wronskian condition

$$\tilde{h}_{l\vec{k}}\,\dot{\tilde{h}}^{*}_{l\vec{k}} - \tilde{h}^{*}_{l\vec{k}}\,\dot{\tilde{h}}_{l\vec{k}} = \frac{i}{a^3\,F}\,. \tag{6.92}$$

If the assumption

$$\frac{(z_g)_{\eta\eta}}{z_g} = \frac{m_g}{\eta^2} \tag{6.93}$$

where m_g =const. is verified, one can solve exactly the equation for the mode functions obtaining, in terms of conformal time,

$$\tilde{h}_{l\vec{k}}(\eta) = \frac{\sqrt{\pi|\eta|}}{2a\sqrt{2F}}\left[c_{l1}\left(\vec{k}\right)H^{(1)}_{\nu_g}(k|\eta|) + c_{l2}\left(\vec{k}\right)H^{(2)}_{\nu_g}(k|\eta|)\right]\,, \tag{6.94}$$

where

$$\nu_g = \sqrt{m_g + \frac{1}{4}}\,. \tag{6.95}$$

For each polarization state, the condition

$$|c_{l2}(k)|^2 - |c_{l1}(k)|^2 = 1 \tag{6.96}$$

must be satisfied in order to preserve eq. (6.92). The power spectrum of the tensor fluctuations

$$\mathcal{P}_{\hat{c}_{ij}}(\eta,k) = \frac{k^3}{2\pi^2}\int d^3\vec{r}\,\langle 0|\hat{c}_{ij}(t,\vec{x}+\vec{r})\,\hat{c}^{ij}(t,\vec{x})|0\rangle\,e^{-i\vec{k}\cdot\vec{r}}\,, \tag{6.97}$$

is then shown to be

$$\mathcal{P}_{\hat{c}_{ij}}(\eta,k) = 2\sum_{l=1}^{2}\mathcal{P}_{\hat{h}_l}(\eta,k) = 2\sum_{l=1}^{2}\frac{k^3}{2\pi^2}\left|\tilde{h}_{l\vec{k}}(\eta)\right|^2\,. \tag{6.98}$$

The vacuum state is identified by imposing the condition that quantum field theory in Minkowski space is recovered in the short scale limit, which is equivalent to setting

$$c_1 = 0, \qquad c_2 = 1. \tag{6.99}$$

For $\nu_g \neq 0$ one obtains

$$\sqrt{\mathcal{P}_{\hat{c}_{ij}}(\eta, k)} = \frac{H}{2\pi\sqrt{2F}} \frac{1}{aH|\eta|} \frac{\Gamma(\nu_g)}{\Gamma(3/2)} \left(\frac{k|\eta|}{2}\right)^{\frac{3}{2}-\nu_g}. \tag{6.100}$$

The spectral index of tensor perturbations is defined as

$$n_T \equiv \frac{d\left(\ln \mathcal{P}_{\hat{c}_{ij}}\right)}{d\left(\ln k\right)}, \tag{6.101}$$

yielding

$$n_T = 3 - 2\nu_g. \tag{6.102}$$

In the slow-roll regime, the four slow-roll parameters (6.60)-(6.63) and their derivatives $\dot{\epsilon}_i$ are small and neglecting the $\dot{\epsilon}_i$ one obtains, to first order,

$$\frac{(z_g)_{\eta\eta}}{z_g} = a^2 H^2 \left(2 + \epsilon_1 + 3\epsilon_3\right). \tag{6.103}$$

The relation

$$aH = -\frac{1}{\eta}(1 - \epsilon_1) + O(2) \tag{6.104}$$

— valid for slow-roll inflation with de Sitter attractors — then leads to

$$m_g = 2 - 3\epsilon_1 + 3\epsilon_3, \tag{6.105}$$

$$\nu_g = \frac{3}{2} - (\epsilon_1 - \epsilon_3), \tag{6.106}$$

and the spectral index of tensor perturbations is

$$n_T = 2(\epsilon_1 - \epsilon_3). \tag{6.107}$$

In a *superacceleration* regime during inflation with $\dot{H} > 0$, one has $\epsilon_1 = \dot{H}/H^2 > 0$ and the spectral index of gravitational waves n_T given by eq. (6.107) can be positive (*blue spectrum*), with more power at small wavelengths than the usual inflationary perturbation spectra. This possibility was noted in the context of pole-like inflation in Ref. [594]. Such a feature is very attractive to the gravitational wave community because it enhances the possibility of detection of cosmological gravitational waves. Detailed calculations for specific potentials to assess the resulting gravitational wave amplitudes in the frequency bands covered by the *LIGO*, *VIRGO*, and the other present laser interferometry experiments are not yet available, but it is plausible that blue spectra are meaningful for future space-based interferometers [185]. Blue spectra are impossible with a minimally coupled scalar field in Einstein gravity, for which one obtains $n_T = 4\ddot{H}/H^2 \leq 0$.

The percent temperature fluctuations observed in the cosmic microwave background sky are decomposed in spherical harmonics, removing the monopole term (average temperature) and the dipole term which is caused by the peculiar motion of the Solar System with respect to the comoving reference frame of the unperturbed FLRW universe,

$$\frac{\delta T}{T} = \sum_{l=2}^{+\infty} \sum_{m=-l}^{+l} a_{lm} Y_{lm}(\theta, \varphi) . \qquad (6.108)$$

These temperature fluctuations are due to both scalar and tensor fluctuations,

$$\frac{\delta T}{T} = \left(\frac{\delta T}{T}\right)_S + \left(\frac{\delta T}{T}\right)_T ; \qquad (6.109)$$

by introducing the multipole moments

$$C_l^{(S,T)} \equiv \sum_{m=-l}^{+l} |a_{lm}|^2 \qquad (6.110)$$

for both scalar and tensor modes, the relative contribution of scalar and tensor perturbations to the temperature fluctuations can be quantified by means of the ratio

$$\mathcal{R} \equiv \frac{C_l^{(T)}}{C_l^{(S)}} . \qquad (6.111)$$

For the quadrupole modes one computes [670]

$$\mathcal{R} = 4\pi \left(-\frac{\omega}{8\pi GF} \epsilon_1 + 3 \epsilon_3^2\right) . \qquad (6.112)$$

In Einstein gravity one has $\frac{\omega}{8\pi GF} = 1$, $\epsilon_3 = 0$, and can put eq. (6.112) into the form $\mathcal{R} = -2\pi\, n_T$, which is independent of the inflationary scenario (i.e., of the form of $V(\phi)$) adopted and is regarded as a consistency relation for the idea of inflation [569]. In general, this reduction is not possible for slow-roll inflation in generalized scalar-tensor gravity.

Chapter 7

NONMINIMAL COUPLING

1. Introduction

The theory of a scalar field explicitly coupled to the Ricci curvature of spacetime has received so much attention in the literature in different contexts that it deserves special treatment, although it can formally be reduced to a special case of a scalar-tensor theory. The action is

$$S^{(NMC)} = \int d^4x \sqrt{-g} \left[\left(\frac{1}{2\kappa} - \frac{\xi}{2} \phi^2 \right) R - \frac{1}{2} \nabla^c \phi \nabla_c \phi - V(\phi) \right] + S^{(m)} ,$$
(7.1)

where ξ is a dimensionless coupling constant. Through most of this chapter, the action $S^{(m)}$ describing other forms of matter is omitted, since we want to consider regimes in which the nonminimally coupled scalar dominates the cosmic dynamics, for example during inflation.

Special values of the constant ξ have received particular attention in the literature:

- $\xi = 1/6$ (*conformal coupling*), the value that makes the Klein-Gordon equation (7.3) and the physics of ϕ conformally invariant if $V = 0$ or $V = \lambda \phi^4$ [892],

- $\xi = 0$ (*minimal coupling*),

- *strong coupling* $|\xi| \gg 1$ is also recurrent in the literature [616, 337, 403, 91, 508, 211].

The general case of $\xi \neq 0$ is referred to as *nonminimal coupling*. The nonminimally coupled scalar field was originally introduced in physics in the context of radiation problems [210], and it became well known when it was shown [172] how conformal coupling can be used to renormalize the theory of a scalar field with quartic self-interaction in a curved space.

The field equations obtained from variation of the action (7.1) are

$$\left(1 - \kappa \xi \phi^2\right) G_{ab} = \kappa \left\{ \nabla_a \phi \nabla_b \phi - \frac{1}{2} g_{ab} \nabla^c \phi \nabla_c \phi - V g_{ab} \right.$$
$$\left. + \xi \left[g_{ab} \Box \left(\phi^2\right) - \nabla_a \nabla_b \left(\phi^2\right) \right] \right\}, \quad (7.2)$$

$$\Box \phi - \frac{dV}{d\phi} - \xi R \phi = 0. \quad (7.3)$$

One expects to obtain eq. (7.3) when the Klein-Gordon equation

$$\eta^{cd} \partial_c \partial_d \phi - \frac{dV}{d\phi} = 0 \quad (7.4)$$

valid in Minkowski space is generalized to a curved space and ordinary differentiation is replaced by covariant differentiation. In this case, the non-commutativity of second covariant derivatives leads to the appearance of terms containing the Riemann tensor or its contractions, a situation well known from the correct generalization of the wave equation for the Maxwell four-potential A^a to a curved space [892],

$$\Box A_b - R^c{}_b A_c = -4\pi j_b, \quad (7.5)$$

where j^c is the electromagnetic four-current.

The theory of a nonminimally coupled scalar is formally equivalent to a scalar-tensor theory. In fact, the redefinition of the scalar field

$$\varphi = \frac{1 - 8\pi G \xi \phi^2}{G} \quad (7.6)$$

transforms the action (7.1) into

$$S^{(NMC)} = \int d^4x \frac{\sqrt{-g}}{16\pi} \left[\varphi R - \frac{\omega(\varphi)}{\varphi} \nabla^c \varphi \nabla_c \varphi - U(\varphi) \right] + S^{(m)}, \quad (7.7)$$

where

$$\omega(\varphi) = \frac{G\varphi}{4\xi(1 - G\varphi)} \quad (7.8)$$

and

$$U(\varphi) = 16\pi V \left[\phi(\varphi)\right] = 16\pi V \left(\pm \sqrt{\frac{1 - G\varphi}{8\pi G \xi}} \right). \quad (7.9)$$

In this chapter we generalize the theory of inflation in the early universe to the case of a nonminimally coupled scalar field, while the next chapter discusses the role of this scalar in quintessential scenarios of the present universe. The classic works on inflation did not include the $-\xi R\phi$ term in eq. (7.3), which is equivalent to assuming that $\xi = 0$. Hereafter, this theory is called *ordinary inflation*, as opposed to *generalized inflation*, which corresponds to $\xi \neq 0$. As explained below, the introduction of nonminimal coupling is not an option, but a necessity. Although nonminimal coupling enjoyed a limited popularity in the literature of the 1980s, its necessity is not always known to cosmologists, even though it has profound consequences for the physics of inflation.

1.1 Generalized inflation

Given the unavoidability of nonminimal coupling, one needs to generalize inflation by appropriately including terms corresponding to $\xi \neq 0$ in the relevant equations. This is done for specific inflationary scenarios by a number of authors [71, 333, 409, 546, 91, 405, 547, 559, 767, 336, 335, 337, 508]. However, the approach adopted is largely one in which the coupling constant ξ is regarded as an extra parameter of inflation to be used at will in order to cure pre-existing problems in the inflationary scenario. For example, chaotic inflation with quartic self-interaction $V(\phi) = \lambda \phi^4$ and $\xi = 0$ is fine-tuned: the amplitude of anisotropies of the cosmic microwave background requires the self-coupling constant to satisfy the constraint $\lambda \leq 10^{-12}$. This figure makes the scenario uninteresting from the point of view of particle physics which originally motivated it. The fine-tuning is significantly reduced if one introduces nonminimal coupling with $\xi < 0$ and $|\xi| \simeq 10^4$ [616, 336, 335], however, the trade-off for reducing the fine-tuning of λ is the fine-tuning of ξ. It is hard to agree with the philosophy of this approach because the coupling constant ξ should have, in general, a well-defined value[1] and is not simply an extra free parameter of the theory.

In Section 7.1.3 we review the known prescriptions for the value of the coupling constant ξ and make clear not only that $\xi \neq 0$ in the general case, but also that fine-tuning ξ is not a possibility. We then proceed to analyze the consequences of the inclusion of nonminimal coupling into the equations of inflation. The study proceeds at two levels: consideration of the unperturbed background universe, and the study of scalar and tensor perturbations of the fixed inflationary background universe. The amplitudes and spectra of perturbations are important because they

---

[1]Exceptions are situations in which ξ is a running coupling in grand unified theories.

leave a detectable imprint in the cosmic microwave background. Temperature anisotropies in the sky have been discovered by the *COBE* satellite [809] and their experimental study is one of the primary goals of cosmology. Significant improvements are to be expected from the *WMAP* [913] and *PLANCK* [713] missions.

The goal of this chapter is the formulation of generalized inflation (i.e., including the $\xi \neq 0$ terms) from a general point of view, without committing to a specific inflationary potential. The results for the unperturbed universe are presented in Sections 7.4 and 7.5.

A special role is played by the slow-roll approximation: apart from two exceptions — power-law inflation and the string-inspired, toy model of Ref. [310], see also Ref. [623] — the equations of inflation (both unperturbed and perturbed) cannot be solved exactly, and one needs to resort to the slow-roll approximation. The latter is investigated in great detail for minimal coupling (i.e., $\xi = 0$) [578, 569] and it is also necessary to investigate slow-roll in the case of nonminimal coupling, in which the equations are more complicated. Slow-roll generalized inflation is discussed in Section 7.5. The study of scalar and tensor perturbations with nonminimal coupling is the subject of Section 7.6, where explicit formulas for the observables of inflation are obtained. de Sitter solutions play the role of attractors for the orbits of inflationary solutions in the phase space in generalized, as well as in ordinary, inflation. It is this fact that ultimately makes the slow-roll approximation meaningful.

1.2 Motivations for nonminimal coupling

The nonminimal coupling of the scalar ϕ in eq. (7.3) was introduced by Chernikov and Tagirov [210], although it is better known from the work of Callan, Coleman and Jackiw [172]. Why should one consider $\xi \neq 0$? There are several different answers to this question. Firstly, a nonzero ξ is generated by first loop corrections even if it is absent in the classical action [136, 135, 665, 385, 694, 382]. If one prepares a classical theory with $\xi = 0$, renormalization shifts it to one with $\xi \neq 0$. Even though the shift is small, it can have a drastic effect on an inflationary scenario. The classical example of this effect emerges in the context of chaotic inflation [404] where the shift

$$\xi = 0 \longrightarrow \xi_{renormalized} \simeq 10^{-1} \qquad (7.10)$$

— a typical value predicted by renormalization [21, 512] — is sufficient to ruin the chaotic inflationary scenario with potential $V = \lambda \phi^4$ [404, 406].

Perhaps the main reason to include a $\xi \neq 0$ term in the coupled Einstein-Klein-Gordon equations is that it is introduced by first loop corrections and is required by renormalizability of the theory: this is the motivation for its introduction in Ref. [172]. It is argued that a nonminimal coupling term is expected at high curvatures [385, 382], that classicalization of the universe in quantum cosmology requires $\xi \neq 0$ [682], and that nonminimal coupling can solve potential problems of primordial nucleosynthesis [207, 209]. A pragmatic point of view is that since nonminimal coupling is usually crucial for the success or failure of inflation [1, 404, 406, 341, 344], it is better to take it into account and to decide *a posteriori* whether or not its effects are negligible.

From the minimalist's point of view, one may restrict oneself to the domain of general relativity; then, any value of ξ different from $1/6$ spoils the Einstein equivalence principle and is therefore not admissible [817, 364, 446, 447].

Whichever point of view is adopted — with motivations arising from a broad range of areas spanning quantum field theory in curved spaces, wormholes [462, 248, 245], black holes [481, 877], boson stars [517, 876, 570], specific inflationary scenarios, a pure relativist's approach, or merely a pragmatic one — in general it is wise not to ignore nonminimal coupling and to avoid setting $\xi = 0$, as done in ordinary inflation. Although the inclusion of nonminimal coupling makes the analysis considerably more difficult from the mathematical point of view, and furthermore was neglected in the early days of inflationary theory, it is critical to include nonminimal coupling in scalar field cosmology.

1.3 Which value of ξ?

It is plausible that the value of the coupling constant ξ be fixed by the physics of the problem, and not be left to the choice of the theoretician as a free parameter. For example, particle physicists' answers to the question "what is the value of ξ?" differ according to the theory of the scalar field employed.

If ϕ is a Goldstone boson in a theory with a spontaneously broken global symmetry, then $\xi = 0$ [886]. If the scalar field ϕ is associated with a composite particle, the value of ξ is fixed by the dynamics of its components. In particular, in the large N approximation to the Nambu-Jona-Lasinio model, the value $\xi = 1/6$ is found [478]. In the $O(N)$-symmetric model with quartic self interaction, in which the constituents of the ϕ-particle are themselves bosons, ξ depends on the coupling constants ξ_i of the elementary scalars [735]. In Einstein gravity with the potential

$$V(\phi) = V_0 + \frac{m^2}{2} \phi^2 + \frac{\eta}{3!} \phi^3 + \frac{\lambda}{4!} \phi^4 \qquad (7.11)$$

and back-reaction, the value $\xi = 0$ is found [486, 695]. Higgs fields in the Standard Model have values of ξ in the range $\xi \leq 0$, $\xi \geq 1/6$ [486].

Results are available in the renormalization group approach to quantum field theory in curved spaces. It is shown in Refs. [161, 162, 163, 165, 680, 659, 314, 166, 164] that in asymptotically free grand unified theories ($GUTs$), depending on the gauge group employed ($SU(2)$, $SU(N)$, $SO(N)$, ...) and on the matter content, ξ is a running coupling that converges to $1/6$ (asymptotic conformal invariance), or to a value ξ_0 determined by the initial conditions (usually this occurs for supersymmetric $GUTs$) or formally, $|\xi(\tau)| \to +\infty$. The latter behaviour is often characteristic of large gauge groups such as $SU(10)$, $SO(10)$, ... Here τ is a renormalization group parameter, with $\tau \to +\infty$ corresponding to strong curvature and early universe situations. In Ref. [166] it is also shown that in asymptotically free $GUTs$ with $SU(5)$ as the gauge group, $|\xi(\tau)| \to +\infty$. Finite $GUTs$ — another class — behave similarly to asymptotically free $GUTs$, with $\xi(\tau)$ converging to $1/6$, or to an initial value ξ_0 (e.g., for $N = 4$ supersymmetry), or to infinity. Moreover, for finite $GUTs$ the convergence of $\xi(\tau)$ to its asymptotic value as $\tau \to +\infty$ is much faster than in asymptotically free GUT models, indeed, the convergence is exponentially fast [161, 162, 163, 165, 680, 659, 314, 166, 164]. An exact renormalization group approach to the $\lambda \phi^4$ theory shows that $\xi = 1/6$ is a stable infrared fixed point [141, 142]. Hence, the asymptotic value of ξ in the early universe strongly depends on the choice of the specific GUT and of its gauge group and matter content.

The problem of the value of ξ in this context is not an easy one, as is clear from the example case of the simple $\lambda \phi^4$ theory. The latter is asymptotically free in the infrared limit $\tau \to -\infty$, which does not correspond to high curvature. Nevertheless, it is shown in Refs. [161, 162, 163, 165, 680, 659, 314, 166, 164] that $\xi(\tau) \to 1/6$ as $\tau \to -\infty$. In the limit $\tau \to +\infty$ of strong curvatures, one cannot answer the question of the asymptotic value of $\xi(\tau)$ since the theory is contradictory — not asymptotically free — in this limit. Nevertheless, an exact renormalization group approach to the $\lambda \phi^4$ theory shows that $\xi = 1/6$ is indeed a stable fixed point of the exact renormalization group [695].

Controversies over these results arose for a restricted class of models [141, 142]. The divergence of the running coupling ξ as the energy scale, curvature, and temperature increase going back in time in the history of the universe was introduced in cosmology [478]. It is exploited to make the chaotic inflationary scenario with $\xi < 0$ more plausible in the cases in which $|\xi(\tau)| \to +\infty$ [405]. The divergence of the coupling ξ is also crucial for the success of the so-called *geometric reheating* of the universe after inflation [91, 861, 860], in which particles are created due to the strong coupling of the inflaton to the Ricci curvature R, instead of the usual coupling of ϕ to other fields.

Nonminimal coupling 149

First loop corrections to the classical theory make ξ likely to be a running parameter which depends on the Ricci curvature. In Refs. [385, 695] the effective coupling

$$\xi_{eff} = \xi + \frac{\lambda}{16\pi^2}\left(\xi - \frac{1}{6}\right)\ln s, \qquad (7.12)$$

is found for the self-interaction potential $\lambda\phi^4/4!$, where s is a parameter that tends to zero in the renormalization group approach. In practice, this amounts to having the effective coupling

$$\xi_{eff} \propto \ln(Rl^2), \qquad (7.13)$$

where l^{-1} is a renormalization mass [382]. No other prescriptions for the value of ξ are apparently proposed in the high energy physics literature; instead, a prescription comes from relativity.

In general relativity, and in all other metric theories of gravity in which ϕ is a non-gravitational field, the only value of ξ allowed by the Einstein equivalence principle [909] is the conformal coupling $1/6$. The derivation of this result [817, 364] has nothing to do with conformal invariance, conformal transformations, or conformal flatness of the spacetime metric g_{ab}. Rather, it arises in the study of wave propagation and tails of scalar radiation — violations of the Huygens' principle — in curved spaces. This field of mathematical physics is thought to be somewhat removed from cosmology and *a priori* unlikely to contribute to it, but such is not the case. Principally, one imposes the idea that the structure of tails of ϕ satisfying the wave equation (7.3) becomes closer and closer to the tail structure occurring in Minkowski space, as the curved manifold is progressively approximated by its tangent space. This is the Einstein equivalence principle [909] applied to the physics of ϕ; of course, the rest of the physics also satisfies the Einstein equivalence principle. The requirement that ϕ satisfies it is only a necessary condition for consistency with general relativity. Second, and more importantly, it turns out that $\xi = 1/6$ is necessary in order to avoid the physical pathology of massive fields ϕ propagating along the light cones.

Now let us summarize the derivation of this result, beginning with the physical definition of Huygens' principle due to Hadamard [459]. Assume that a point-like source of (scalar) radiation emits a delta-like pulse at time $t = 0$ in $r = 0$. If at $t = t_1$ radiation is present only on the surface of the sphere with radius $r = ct_1$ and centre $r = 0$, then we say that Huygens' principle is satisfied and that there are no tails. If instead there are waves also at radii $r < ct_1$ (*tails*), Huygens' principle is violated.

Mathematically, the solution for a delta-like pulse is the retarded Green function $G_R(x', x)$ of eq. (7.3) satisfying

$$\left[g^{a'b'}(x') \nabla_{a'} \nabla_{b'} - m^2 - \xi R(x') \right] G_R(x', x) = -\delta^{(4)}(x', x) , \quad (7.14)$$

where $\delta^{(4)}(x', x)$ is the four-dimensional Dirac delta [284] which satisfies the boundary condition $G_R(x', x) = 0$ if x' is in the past of x. For simplicity, we consider the case in which the potential $V(\phi)$ reduces to a mass term — the generalization to arbitrary potentials can be found in Refs. [817, 364]. $G_R(x', x)$ has the general structure [459, 284, 394]

$$G_R(x', x) = \Sigma(x', x) \, \delta_R(\Gamma(x', x)) + W(x', x) \, \Theta_R(-\Gamma(x', x)) , \quad (7.15)$$

where $\Gamma(x', x)$ is the square of the geodesic distance between x' and x. This quantity is well known in quantum field theory in curved spaces [136]. One has $\Gamma = 0$ if x' and x are light-like related, $\Gamma > 0$ if x' and x are space-like related, and $\Gamma < 0$ if x' and x are time-like related. δ_R is the Dirac delta with support in the past of x', and Θ_R is the Heaviside step function with support in the past light cone. The term in $\delta_R(\Gamma)$ describes a contribution to the Green function from ϕ-waves propagating along the light cone ($\Gamma = 0$), while the term $\Theta_R(-\Gamma)$ describes the contribution to G_R from tails of ϕ propagating *inside* the light cone ($\Gamma < 0$). The functions $\Sigma(x', x)$ and $W(x', x)$ are coefficients which are determined once the spacetime metric is fixed [284, 394].

The Einstein equivalence principle is imposed as follows on the physics of the field ϕ. When the spacetime manifold is progressively approximated by its tangent space (i.e., by fixing the point x and considering a small neighborhood of points x' such that $x' \to x$), then the solution $G_R(x', x)$ for a delta-like pulse must reduce to the corresponding one known from Minkowski spacetime, which is

$$G_R^{(Minkowski)}(x', x) = \frac{1}{4\pi} \delta_R(-\Gamma) - \frac{m^2}{8\pi} \Theta_R(\Gamma) . \quad (7.16)$$

By expanding the coefficients Σ and W, one finds [817, 364]

$$\lim_{x' \to x} \Sigma(x', x) = \frac{1}{4\pi} , \quad (7.17)$$

$$\lim_{x' \to x} W(x', x) = -\frac{1}{8\pi} \left[m^2 + \left(\xi - \frac{1}{6} \right) R(x) \right] ; \quad (7.18)$$

hence $G_R \to G_R^{(Minkowski)}$ if and only if

$$\left(\xi - \frac{1}{6} \right) R(x) = 0 . \quad (7.19)$$

In general, this condition is verified only if $\xi = 1/6$. Note that if $\xi \neq 1/6$, a physical pathology may occur: the ϕ-field can have an arbitrarily large mass and still propagate along the light cone at the spacetime points where eq. (7.19) is satisfied. In this situation an arbitrarily massive field would have no tails (see Ref. [282] for the case of anti-de Sitter space with constant curvature). It is even possible to construct an "ultrapathological" de Sitter spacetime in which the value of the constant curvature and of the mass are adjusted in such a way that a scalar field with arbitrarily large mass propagates along the light cone at every point [359].

The result that $\xi = 1/6$ in general relativity is extendable to all metric theories of gravity in which ϕ is not part of the gravitational sector [817, 364]; in fact, in these theories, the Einstein equivalence principle holds [909]. The result that $\xi = 1/6$ in general relativity is also obtainable, of course, if one considers Feynman propagators instead of Green functions [446, 447].

2. Effective energy-momentum tensors

The innocent-looking theory described by the action (7.1) may lead to negative energy fluxes [481, 63, 884, 384], opening the possibility of exotica such as traversable wormholes [63], warp drives [143, 20], and time machines [882]. Note that the system described by eq. (7.1) does not have a quantum nature, such as the quantum systems often considered in the literature for which violations of the weak and null energy conditions are temporary and do not persist on average. The theory (7.1) is, instead, a classical one.

Before discussing the issue of negative energies, it must be noted that there are ambiguities associated with the definition of the stress-energy tensor of a nonminimally coupled scalar field and as a consequence, ambiguity also with the definitions of energy density, pressure, and the equation of state. These ambiguities are discussed in detail below and resemble the situation occurring in BD and scalar-tensor theories. We present three possible and inequivalent ways of writing the field equations, the corresponding ambiguity in the definition of the energy-momentum tensor of the scalar field, and the corresponding conservation laws. In general, in a FLRW universe, the scalar field ϕ only depends on the comoving time, and any effective energy-momentum tensor for ϕ assumes the form corresponding to a perfect fluid without viscosity or heat flux [601, 71].

2.1 Approach à la Callan-Coleman-Jackiw

This procedure for writing the field equations follows the one originally introduced by Callan, Coleman and Jackiw [172] for conformal coupling.

The Einstein equations are written as

$$G_{ab} = \kappa \left(T_{ab}^{(I)} + T_{ab}^{(m)} \right), \quad (7.20)$$

where

$$\begin{aligned} T_{ab}^{(I)} &= \nabla_a \phi \nabla_b \phi - \frac{1}{2} g_{ab} \nabla^c \phi \nabla_c \phi - V(\phi) g_{ab} \\ &+ \xi \left(g_{ab} \Box - \nabla_a \nabla_b \right) (\phi^2) + \xi \phi^2 G_{ab} \end{aligned} \quad (7.21)$$

contains a "geometric" contribution $\xi \phi^2 G_{ab}$ proportional to the Einstein tensor.

First let us discuss the conservation properties of $T_{ab}^{(I)}$: the contracted Bianchi identities $\nabla^b G_{ab} = 0$ yield

$$\nabla^b T_{ab}^{(I)} + \nabla^b T_{ab}^{(m)} = 0. \quad (7.22)$$

Since $T_{ab}^{(m)}$ does not depend on ϕ and $\nabla^b T_{ab}^{(m)}$ vanishes when the action is varied with respect to the matter variables, $T_{ab}^{(m)}$ and $T_{ab}^{(I)}$ are conserved separately,

$$\nabla^b T_{ab}^{(m)} = 0, \quad (7.23)$$

$$\nabla^b T_{ab}^{(I)} = 0. \quad (7.24)$$

The energy density measured by an observer with four-velocity u^c satisfying the normalization $u_c u^c = -1$ is, in this approach,

$$\rho_{(total)}^{(I)} \equiv \left[T_{ab}^{(I)} + T_{ab}^{(m)} \right] u^a u^b = \rho_\phi^{(I)} + \rho^{(m)}, \quad (7.25)$$

where

$$\begin{aligned} \rho_\phi^{(I)} &= T_{ab}^{(I)} u^a u^b = (u^c \nabla_c \phi)^2 + \frac{1}{2} \nabla^c \phi \nabla_c \phi + V(\phi) - \xi \Box (\phi^2) \\ &- \xi u^a u^b \nabla_a \nabla_b (\phi^2) + \xi \phi^2 G_{ab} u^a u^b, \end{aligned} \quad (7.26)$$

and $\rho^{(m)} = T_{ab}^{(m)} u^a u^b$ as usual. The trace of $T_{ab}^{(I)}$ reduces to

$$T^{(I)} = -\partial^c \phi \partial_c \phi - 4V(\phi) + 3\xi \Box (\phi^2) - \xi R \phi^2. \quad (7.27)$$

2.2 Effective coupling

The second way of proceeding is by writing the field equation (7.2) taking the term $\kappa \xi \phi^2 G_{ab}$ to the left hand side,

$$\left(1 - \kappa \xi \phi^2 \right) G_{ab} = \kappa \left(T_{ab}^{(II)} + T_{ab}^{(m)} \right), \quad (7.28)$$

where
$$T_{ab}^{(II)} = \nabla_a\phi\nabla_b\phi - \frac{1}{2}g_{ab}\nabla^c\phi\nabla_c\phi - V(\phi)g_{ab}$$
$$+ \xi(g_{ab}\Box - \nabla_a\nabla_b)(\phi^2) = T_{ab}^{(I)} - \xi\phi^2 G_{ab}; \quad (7.29)$$

Then, one divides eq. (7.28) by the factor $1 - \kappa\xi\phi^2$, with the result
$$G_{ab} = \kappa_{eff}\left(T_{ab}^{(II)} + T_{ab}^{(m)}\right), \quad (7.30)$$
where
$$\kappa_{eff}(\phi) \equiv \frac{\kappa}{1 - \kappa\xi\phi^2} \quad (7.31)$$
is an effective gravitational coupling for both $T_{ab}^{(II)}$ and $T_{ab}^{(m)}$. This procedure is analogous to the familiar identification of the BD scalar field with the inverse of the gravitational coupling in the gravitational sector of the BD action (1.25). The division by the factor $1 - \kappa\xi\phi^2$ leading to eq. (7.31) implies loss of generality for $\xi > 0$. Solutions of eq. (7.2) with scalar field attaining the critical values already considered in Section 1.12
$$\pm\phi_1 \equiv \pm\frac{1}{\sqrt{\kappa\xi}} \quad (\xi > 0), \quad (7.32)$$
are missed when considering eq. (7.30) in the effective coupling approach. For example, solutions in which ϕ is constant and equal to one of the critical values (7.32) appear in models of wormholes [63] and also as special heteroclinics in the phase space of scalar field cosmology [453] (see Appendix A for solutions with constant scalar field). In the effective coupling approach these solutions are missed and can only be recovered by going back to the primitive form (7.2) of the field equations.

Let us now discuss the conservation laws for $T_{ab}^{(II)}$. The contracted Bianchi identities imply that
$$\nabla^b\left(T_{ab}^{(II)} + T_{ab}^{(m)}\right) = \frac{-2\kappa\xi\phi}{1 - \kappa\xi\phi^2}(\nabla^b\phi)\left(T_{ab}^{(II)} + T_{ab}^{(m)}\right). \quad (7.33)$$

By using the argument of the previous case, one obtains $\nabla^b T_{ab}^{(m)} = 0$ and
$$\nabla^b T_{ab}^{(II)} = \frac{-2\kappa\xi\phi}{1 - \kappa\xi\phi^2}(\nabla^b\phi)\left(T_{ab}^{(II)} + T_{ab}^{(m)}\right). \quad (7.34)$$

We get a better idea of the consequences of the conservation law (7.34) by considering ordinary matter in the form of a dust fluid with corresponding energy-momentum tensor $T_{ab}^{(m)} = \rho v_a v_b$, where v^c is the dust four-velocity. Eq. (7.34) then yields

$$\left(\frac{d\rho^{(m)}}{d\lambda} + \rho^{(m)} \nabla^b v_b + \frac{2\kappa\xi\rho^{(m)}\phi}{1-\kappa\xi\phi^2} \frac{d\phi}{d\lambda} \right) v_a + \rho^{(m)} \frac{Dv_a}{D\lambda} = 0 , \quad (7.35)$$

where λ is an affine parameter along the worldlines of fluid particles. Eq. (7.35) is decomposed into the geodesic equation

$$\frac{Dv^a}{D\lambda} \equiv v^b \nabla_b v^a = \frac{d^2 x^a}{d\lambda^2} + \Gamma^a_{bc} \frac{dx^b}{d\lambda} \frac{dx^c}{d\lambda} = 0 \quad (7.36)$$

and the modified conservation equation

$$\frac{d\rho^{(m)}}{d\lambda} + \rho^{(m)} \nabla^b v_b + \frac{2\kappa\xi\phi\,\rho^{(m)}}{1-\kappa\xi\phi^2} \frac{d\phi}{d\lambda} = 0 . \quad (7.37)$$

Test particles move on geodesics, thus verifying the geodesic hypothesis. In the weak field limit, the modified conservation equation (7.37) reduces to

$$\frac{\partial \rho^{(m)}}{\partial t} + \vec{\nabla} \cdot \left(\rho^{(m)} \vec{v} \right) + \frac{2\kappa\xi\phi}{1-\kappa\xi\phi^2} \left(\frac{\partial \phi}{\partial t} + \vec{\nabla}\phi \cdot \vec{v} \right) \rho^{(m)} = 0 , \quad (7.38)$$

where $\vec{v}$ denotes the three-dimensional velocity of the non-relativistic fluid. In the effective coupling approach, the total energy density relative to an observer with four-velocity u^c is

$$\rho^{(II)}_{(total)} \equiv T^{(II)}_{ab} u^a u^b + T^{(m)}_{ab} u^a u^b = \rho^{(II)}_\phi + \rho^{(m)} , \quad (7.39)$$

where

$$\begin{aligned} \rho^{(II)}_\phi &= (u^c \nabla_c \phi)^2 + \frac{1}{2} \nabla^c \phi \nabla_c \phi + V(\phi) - \xi \nabla^c \nabla_c (\phi^2) - \xi u^a u^b \nabla_a \nabla_b (\phi^2) \\ &= \rho^{(I)}_\phi - \xi\phi^2 G_{ab} u^a u^b \end{aligned} \quad (7.40)$$

or, using eq. (7.30),

$$\rho^{(II)}_\phi = \rho^{(I)}_\phi \left(1 - \kappa\xi\phi^2 \right) . \quad (7.41)$$

In the presence of ordinary matter, $\rho^{(II)}_{(total)}$ is a mixture of $\rho^{(I)}_\phi$ and of $\rho^{(m)}$ weighted by the factor $\kappa\xi\phi^2$.

2.3 A mixed approach

The third and last possible way to rewrite the field equation for the nonminimally coupled theory is to bring the term $\xi\phi^2 G_{ab}$ to the left

Nonminimal coupling

hand side but to keep the usual, constant, gravitational coupling, thereby obtaining

$$G_{ab} = \kappa \left(T_{ab}^{(III)} + \frac{T_{ab}^{(m)}}{1 - \kappa \xi \phi^2} \right), \tag{7.42}$$

where

$$T_{ab}^{(III)} = \frac{1}{1 - \kappa \xi \phi^2} \left[\nabla_a \phi \nabla_b \phi - \frac{1}{2} g_{ab} \nabla^c \phi \nabla_c \phi - V g_{ab} \right.$$

$$\left. + \xi \left(g_{ab} \nabla^c \nabla_c - \nabla_a \nabla_b \right) (\phi^2) \right] = \frac{T_{ab}^{(II)}}{1 - \kappa \xi \phi^2}. \tag{7.43}$$

The limitations due to division by the factor $\left(1 - \kappa \xi \phi^2\right)$ occur again in this mixed approach. Note that

$$\kappa T_{ab}^{(III)} = \kappa_{eff} T_{ab}^{(II)} \tag{7.44}$$

and that the total energy density measured by an observer with four-velocity u^c is

$$\rho_{(total)}^{(III)} \equiv T_{ab}^{(II)} u^a u^b + T_{ab}^{(m)} u^a u^b = \rho_\phi^{(II)} + \frac{\rho^{(m)}}{1 - \kappa \xi \phi^2}. \tag{7.45}$$

In the absence of ordinary matter, $\rho_\phi^{(III)} = \rho_\phi^{(II)}$. This time the contracted Bianchi identities $\nabla^b G_{ab} = 0$ yield conservation of the *total* energy-momentum tensor

$$T_{ab}^{(total)} = T_{ab}^{(II)} + \frac{T_{ab}^{(m)}}{1 - \kappa \xi \phi^2}, \tag{7.46}$$

$$\nabla^b T_{ab}^{(total)} = 0, \tag{7.47}$$

but $T_{ab}^{(III)}$ alone is not covariantly conserved:

$$\nabla^b T_{ab}^{(III)} = \frac{-2\kappa \xi \phi}{(1 - \kappa \xi \phi^2)^2} \left(\nabla^b \phi \right) T_{ab}^{(m)}. \tag{7.48}$$

However, in the absence of ordinary matter, $T_{ab}^{(III)}$ is conserved, i.e., $\nabla^b T_{ab}^{(III)} = 0$.

2.4 Discussion

Table 7.1 summarizes the three possible approaches. For $\xi < 0$, the scalar field ϕ does not possess the critical values (7.32) and the three approaches are mathematically equivalent — they only differ by legitimate algebraic manipulations of the field equations (7.2). However, they are physically inequivalent, in fact the identification of the energy density, pressure, and effective equation of state differs in the three approaches. Moreover, the different conservation laws (7.24), (7.34) and (7.48) lead to different physical interpretations. For example, models of quintessence with nonminimal coupling and their observational constraints are often stated in terms of the effective equation of state, but when the definitions of energy density and pressure are fuzzy, the concept of equation of state is not free of ambiguity either.

In order to compare the different scalar field energy densities we assume that ϕ is the only form of matter for the rest of this section. Besides the obvious simplification in eqs. (7.39) and (7.45), this situation is appropriate for the description of inflation and is useful for late quintessence-dominated cosmological scenarios, and also for wormhole models [462, 248, 245, 882, 61, 63]. Results under the assumption are:

- For $\xi < 0$, the effective coupling κ_{eff} is always positive, and $\rho_\phi^{(II)}$ has the same sign as $\rho_\phi^{(III)} = \rho_\phi^{(I)}$.

- For $\xi > 0$, in addition to the physical inequivalence the three approaches are also mathematically inequivalent — as already discussed — and the scalar field possesses the critical values (7.32). If

$$|\phi| < \frac{1}{\sqrt{\kappa \xi}}, \qquad (7.49)$$

then $\kappa_{eff} > 0$ and $\rho_\phi^{(II)}$ has the same sign of $\rho_\phi^{(I)} = \rho_\phi^{(III)}$. If instead $|\phi| > 1/\sqrt{\kappa \xi}$, the effective coupling κ_{eff} becomes negative, a regime called *antigravity* [454, 580, 824, 718]. In this regime

$$\text{sign}\left(\rho_\phi^{(II)}\right) = -\text{sign}\left(\rho_\phi^{(I)}\right) = -\text{sign}\left(\rho_\phi^{(III)}\right). \qquad (7.50)$$

However,

$$\kappa_{eff} \rho_\phi^{(II)} = \kappa \rho_\phi^{(I)} = \kappa \rho_\phi^{(III)}. \qquad (7.51)$$

While it is the product of κ and ρ that ultimately enters the field equations, concerns about the negative energy fluxes associated with nonminimal coupling and leading to time machines are usually expressed using $T_{ab}^{(III)}$ in our notations. Had one proceeded using $T_{ab}^{(II)}$

instead, one would have found positive energy density for $\xi > 0$ and $|\phi| > |\phi_1|$, but the same wormhole solutions.

In conclusion, the definition of energy density and flux must be made explicit in discussions of the energy conditions for a nonminimally coupled scalar. Then, a relevant issue arises as to whether there is a physically preferred stress-energy tensor in the nonminimally coupled theory described by the action (7.1). The answer is to a certain extent a matter of taste depending upon the task pursued; however, the approach à la Callan-Coleman-Jackiw presents certain advantages. First, the corresponding stress-energy tensor $T_{ab}^{(I)}$ is always covariantly conserved, even in the presence of ordinary matter $T_{ab}^{(m)}$. Second, this approach does not lead to loss of generality of the solutions. Third, there are situations in cosmology (described in the next section) in which the energy density $\rho_\phi^{(I)}$ is automatically positive-definite, while $\rho_\phi^{(II)}$ is not.

The difficulty in identifying the "physically correct" stress-energy tensor hints at the problem of the localization of gravitational energy. In fact, we can picture the physical system described by the action (7.1) as two coupled subsystems, the gravitational tensor field g_{ab} and the scalar field ϕ, with the term $-\sqrt{-g}\,\xi\phi^2 R/2$ in the Lagrangian density explicitly displaying this interaction. Currently, no general prescription is available for the energy density of the gravitational field and a satisfactory resolution of the problem of the "physically correct" $T_{ab}[\phi]$ would imply a successful solution of the energy localization problem in general relativity.

2.5 Energy conditions in FLRW cosmology

Once the definitions of effective stress-energy tensor, energy density and pressure are specified, it makes sense to speak of energy conditions. That the energy density of a nonminimally coupled scalar field can become negative has been known for a long time (e.g., [104, 281, 383, 601, 481]), and associated troublesome consequences have been emphasized [61, 63, 884, 384]. There are, however, circumstances in which negative energies do not occur when the definition of energy density $\rho_\phi^{(I)}$ proposed in the previous section is adopted.

Let us consider a FLRW cosmology driven by the nonminimally coupled scalar field. The scale factor $a(t)$ and the scalar field $\phi(t)$ satisfy the Einstein-Friedmann equations

$$\frac{\ddot{a}}{a} = \dot{H} + H^2 = -\frac{\kappa}{6}\left(\rho^{(I)} + 3P^{(I)}\right), \qquad (7.52)$$

Table 7.1. A comparison of the three possible approaches to nonminimal coupling. The field equations are listed in the first line, the T_{ab} and the conservation law it obeys appear in the second and third line, respectively, and the fourth line contains the corresponding energy density.

(I) Approach á la CCJ	(II) Effective Coupling Approach	(III) Mixed Approach
$G_{ab} = \kappa \left(T_{ab}^{(I)} + T_{ab}^{(m)} \right)$	$G_{ab} = \kappa_{eff} \left(T_{ab}^{(II)} + T_{ab}^{(m)} \right)$	$G_{ab} = \kappa \left(T_{ab}^{(III)} + \frac{T_{ab}^{(m)}}{1-\kappa\xi\phi^2} \right)$
$T_{ab}^{(I)} = \partial_a \phi \partial_b \phi - \frac{1}{2} g_{ab} \partial^c \phi \partial_c \phi$ $+ \xi \left(g_{ab} \Box - \nabla_a \nabla_b \right) (\phi^2)$ $- V g_{ab} + \xi \phi^2 G_{ab}$	$T_{ab}^{(II)} = T_{ab}^{(I)} - \xi \phi^2 G_{ab}$	$T_{ab}^{(III)} = \frac{T_{ab}^{(II)}}{1-\kappa\xi\phi^2}$
$\nabla^b T_{ab}^{(I)} = 0$	$\nabla^b T_{ab}^{(II)} = \frac{-2\kappa\xi\phi}{1-\kappa\xi\phi^2} \cdot$ $\left(T_{ab}^{(II)} + T_{ab}^{(m)} \right) \nabla^b \phi$	$\nabla^b T_{ab}^{(III)} = \frac{-2\kappa\xi\phi}{1-\kappa\xi\phi^2} T_{ab}^{(m)} \nabla^b \phi$
$\rho_\phi^{(I)} = (u^c \partial_c \phi)^2 + \frac{1}{2} \partial^c \phi \partial_c \phi$ $- \xi u^a u^b \nabla_a \nabla_b (\phi^2)$ $+ V - \xi \Box (\phi^2) + \xi \phi^2 G_{ab} u^a u^b$	$\rho_\phi^{(II)} = \rho_\phi^{(I)} \left(1 - \kappa\xi\phi^2 \right)$	$\rho_\phi^{(III)} = \rho_\phi^{(II)}$

$$H^2 = \frac{\kappa}{3}\rho^{(I)} - \frac{K}{3a^2}, \tag{7.53}$$

where $P^{(I)}$ is the isotropic pressure obtained from $T_{ab}^{(I)}$ for a time dependent scalar $\phi(t)$. The Hamiltonian constraint (7.53) implies that the energy density $\rho_\phi^{(I)}$ is automatically non-negative for any solutions $(a(t), \phi(t))$ of the Einstein equations for spatially flat ($K = 0$) and for closed ($K = +1$) universes. This conclusion holds in spite of the fact that the expressions for $\rho^{(I)}(t)$ and $P^{(I)}(t)$,

$$\rho^{(I)}(t) = \frac{1}{2}\dot{\phi}^2 + 3\xi H^2 \phi^2 + 6\xi H \phi \dot{\phi} + V(\phi), \tag{7.54}$$

$$P^{(I)}(t) = \left(\frac{1}{2} - 2\xi\right)\dot{\phi}^2 + 2\xi H \phi \dot{\phi} + 2\xi(6\xi - 1)\dot{H}\phi^2$$

$$+ 3\xi(8\xi - 1)H^2 \phi^2 + 2\xi\phi\frac{dV}{d\phi}, \tag{7.55}$$

are complicated, and nothing can *a priori* be concluded by inspecting their form.

2.6 Nonminimal coupling and gravitational waves

Due to the explicit coupling between the gravitational and the scalar fields in the action (7.1), scalar modes are generated when gravitational waves are excited. Let us consider free scalar waves outside material sources, satisfying $V(\phi) = 0$ and $T_{ab}^{(m)} = 0$. The stress-energy tensor (7.21) of ϕ contains the term $\xi(g_{ab}\Box - \nabla_a \nabla_b)(\phi^2)$ linear in the second derivatives of ϕ instead of a canonical term quadratic in its first derivatives $\nabla_c \phi$. This non-canonical structure is analogous to that of the effective stress-energy tensor of the BD scalar considered in Chapter 2, and therefore one could expect to encounter the same problems as discussed for scalar-tensor gravitational waves in BD theory. However this is not the case, at least in the first order perturbation analysis considered here and shown below.

The procedure involves expanding the metric and the scalar field around their flat space values,

$$g_{ab} = \eta_{ab} + h_{ab}, \qquad \phi = \phi_0 + \psi, \tag{7.56}$$

with $\phi_0 =$const. (see Appendix A for solutions with constant scalar field), and $O(h_{ab}) = O(\psi/\phi_0) = O(\epsilon)$. It is easily seen from the field equation (7.2) that the requirement for spacetime to be flat implies ϕ =constant. The offending term in $T_{ab}^{(I)}$ then reduces to

$$\xi \left(g_{ab} \nabla^c \nabla_c - \nabla_a \nabla_b \right) (\phi^2) = -2\xi \phi_0 \, \partial_a \partial_b \psi + O(\epsilon^2) , \qquad (7.57)$$

where we used the equation

$$\Box \psi = O(\epsilon^2) , \qquad (7.58)$$

which follows from the trace of the field equations

$$R = -6\xi \phi_0 \kappa \, \Box \, \psi + O(\epsilon^2) , \qquad (7.59)$$

and from the Klein-Gordon equation. Hence the negative energy problems seem to be present. However, contrary to BD theory, in the theory of a nonminimally coupled scalar the only value of the constant ϕ_0 compatible with a flat background is $\phi_0 = 0$. In fact, a constant nonvanishing scalar field corresponds to Schwarzschild or anti-Schwarzschild solutions [61, 63]. By setting $\phi_0 = 0$ as required for a flat background, the troubles disappear since $T_{ab}^{(I)}$ reduces, to first order, to a canonical form quadratic in the first derivatives of the field, and the energy density of modes of a nonminimally coupled scalar field is positive definite.

3. Conformal transformations

Significant mathematical simplifications occur when conformal transformation techniques are employed to relate a cosmological scenario with a nonminimally coupled scalar field to its conformal cousin with a minimally coupled field, especially in the absence of ordinary matter. The Jordan frame in which the scalar field couples nonminimally to the Ricci curvature is mapped into an Einstein frame in which the transformed scalar $\tilde{\phi}$ is minimally coupled. The two frames are not physically equivalent unless variable units of time, length, and mass are adopted in the Einstein frame.

The conformal transformation is given by eq. (1.141) with conformal factor

$$\Omega = \sqrt{1 - \kappa \xi \phi^2} , \qquad (7.60)$$

and the scalar field is redefined according to

$$d\tilde{\phi} = \frac{\sqrt{1 - \kappa \xi (1 - 6\xi) \phi^2}}{1 - \kappa \xi \phi^2} \, d\phi . \qquad (7.61)$$

The conformal transformation is only defined for $\xi < 0$ and, if $\xi > 0$, for values of ϕ such that $\phi \neq \pm \phi_1 \equiv \pm (\kappa \xi)^{-1/2}$. For large values of

ξ, this may become a serious limitation on the usefulness of conformal transformation techniques. In addition, if $0 < \xi < 1/6$, it must be

$$|\phi| < \phi_2 \equiv \frac{1}{\sqrt{\kappa\xi(1-6\xi)}} \qquad (7.62)$$

where $\phi_1 < \phi_2$. When $\xi > 0$ and the nonminimally coupled scalar field approaches the critical values $\pm\phi_1$, $\tilde{g}_{ab}$ degenerates, $\tilde{\phi}$ diverges and the conformal transformation technique cannot be applied. This happens when $\phi \simeq 0.199\,\xi^{-1/2}m_{pl}$, which induces the constraint $|\phi| < 0.2\,m_{pl}$ if ξ is of order unity. This constraint may be severe in certain scenarios, for example, chaotic inflation requires ϕ larger than about $5\,m_{pl}$. In particular for strong positive coupling $\xi \gg 1$, the critical values $\pm\phi_1$ correspond to very low energies.

The conformal transformation technique cannot provide solutions with ϕ crossing the barriers $\pm\phi_1$. In this sense the conformal technique has the same limitations of the form of the field equations using the effective coupling $\kappa_{eff} = \kappa\left(1 - \kappa\xi\phi^2\right)^{-1}$.

The transformed scalar $\tilde{\phi}$ in the Einstein frame can be explicitly expressed in terms of ϕ by integrating eq. (7.61),

$$\tilde{\phi} = \sqrt{\frac{3}{2\kappa}} \ln\left[\frac{\xi\sqrt{6\kappa\phi^2} + \sqrt{1-\xi(1-6\xi)\kappa\phi^2}}{\xi\sqrt{6\kappa\phi^2} - \sqrt{1-\xi(1-6\xi)\kappa\phi^2}}\right] + f(\phi), \qquad (7.63)$$

where

$$f(\phi) = \left(\frac{1-6\xi}{\kappa\xi}\right)^{1/2} \arcsin\left(\sqrt{\xi(1-6\xi)\kappa\phi^2}\right) \qquad (7.64)$$

for $0 < \xi < 1/6$ and

$$f(\phi) = \left(\frac{6\xi-1}{\kappa\xi}\right)^{1/2} \operatorname{arcsinh}\left(\sqrt{\xi(6\xi-1)\kappa\phi^2}\right) \qquad (7.65)$$

for $\xi > 1/6$. When $\xi = 1/6$, we have

$$\tilde{\phi} = \sqrt{\frac{3}{2\kappa}} \ln\left(\frac{\sqrt{6/\kappa} + \phi}{\sqrt{6/\kappa} - \phi}\right) \qquad \text{if } |\phi| < \sqrt{\frac{6}{\kappa}}, \qquad (7.66)$$

or

$$\tilde{\phi} = \sqrt{\frac{3}{2\kappa}} \ln\left(\frac{\phi - \sqrt{6/\kappa}}{\phi + \sqrt{6/\kappa}}\right) \qquad \text{if } |\phi| > \sqrt{\frac{6}{\kappa}}. \qquad (7.67)$$

It is clear that $\tilde{\phi} \longrightarrow \pm\infty$ in the Einstein frame as $\phi \longrightarrow \pm\phi_1$ in the Jordan frame. Any nonminimally coupled solution ϕ crossing the barriers $\pm\phi_1$ cannot be found by applying the conformal transformation technique. An example of such a solution is the one corresponding to a nonminimally coupled scalar field which is constant and equal to one of the critical values. In this case the field equations yield $R = 6\left(\dot{H} + 2H^2\right) = 0$, $V = 0$, vanishing trace of the energy-momentum tensor $\tilde{T}_{ab}[\phi]$, radiation equation of state $P = \rho/3$, and the solution

$$a = a_0\sqrt{t - t_0}, \qquad \phi = \pm\frac{1}{\sqrt{\kappa\xi}} \qquad (\xi > 0). \tag{7.68}$$

The Einstein frame scalar $\tilde{\phi}$ is minimally coupled and satisfies the Klein-Gordon equation

$$\tilde{\Box}\tilde{\phi} - \frac{d\tilde{V}}{d\tilde{\phi}} = 0, \tag{7.69}$$

where

$$\tilde{V}\left(\tilde{\phi}\right) = \frac{V\left[\phi\left(\tilde{\phi}\right)\right]}{(1 - \kappa\xi\phi^2)^2}, \tag{7.70}$$

with $\phi = \phi\left(\tilde{\phi}\right)$ given by eq. (7.63). The conformal transformation technique is useful in solving the equations of cosmology in the Einstein frame, then mapping the solutions $\left(\tilde{g}_{ab}, \tilde{\phi}\right)$ back into the physical solutions (g_{ab}, ϕ) of the Jordan frame with nonminimal coupling. Although from the mathematical point of view this procedure is convenient (e.g., Ref. [9]), the result is usually not very interesting from the physical point of view. In fact, one starts from a known solution for a potential $\tilde{V}\left(\tilde{\phi}\right)$, which is usually motivated by particle physics in the unphysical Einstein frame, obtaining a solution in the physical Jordan frame which corresponds to a potential $V(\phi)$ with no physical justification. Furthermore if a solution is inflationary in one frame, its conformally transformed counterpart in the other frame is not necessarily inflationary. As an example, consider a conformally coupled scalar field with the self-interaction potential $\tilde{V}\left(\tilde{\phi}\right) = \lambda\tilde{\phi}^4$ in the Einstein frame; one can integrate eq. (7.61) and use eq. (7.70) to obtain

$$V(\phi) = \left(\frac{3}{2\kappa}\right)^2 \lambda \left(1 - \frac{\kappa}{6}\phi^2\right)^2 \ln^4\left[\frac{\sqrt{\kappa/6}\,\phi + 1}{-\sqrt{\kappa/6}\,\phi + 1}\right], \tag{7.71}$$

a potential with scarce physical justification in high energy physics.

There is, however, a situation of physical interest in which an inflationary solution is mapped into another inflationary solution by the conformal transformation: the slow-roll approximation [575, 442, 351]. To begin with, one notices that an exact de Sitter attractor solution is invariant under the conformal transformation (1.141), (7.60), and (7.61).

Nonminimal coupling

In fact, when ϕ is constant, eq. (1.141) reduces to a rescaling of the metric by a constant factor, which can be absorbed into a coordinate rescaling, and the scalar ϕ is mapped into another constant $\tilde{\phi}$. Moreover it will be proven shortly that a de Sitter solution is an attractor point in phase space for suitable values of the coupling constant ξ, with nonminimal coupling as well as with minimal coupling. Hence when the values of ξ lie in a certain range, the conformal transformation maps an attractor of the Jordan frame into an attractor of the Einstein frame. It is therefore meaningful to consider the slow-roll approximation to inflation in both frames.

In the Jordan frame the Hubble parameter is given by

$$a = a_0 \exp[H(t)t] \,, \tag{7.72}$$

$$H(t) = H_0 + \delta H(t) \,, \tag{7.73}$$

where H_0 is constant and $|\delta H| << |H_0|$. In the Einstein frame one has the line element

$$d\tilde{s}^2 = \Omega^2 ds^2 = -d\tilde{t}^2 + \tilde{a}^2 \left(dx^2 + dy^2 + dz^2 \right) \,, \tag{7.74}$$

where $d\tilde{t} = \Omega\, dt$ and $\tilde{a} = \Omega\, a$. The Hubble parameter in the Einstein frame is

$$\tilde{H} \equiv \frac{1}{\tilde{a}} \frac{d\tilde{a}}{d\tilde{t}} = \frac{1}{\Omega}\left(H + \frac{\dot{\Omega}}{\Omega} \right) \,, \tag{7.75}$$

where an overdot denotes differentiation with respect to the Jordan frame comoving time t. For an exact de Sitter solution with constant scalar field, $H =$const. implies $\tilde{H} =$const. and vice-versa. A slow-roll inflationary solution in the Jordan frame satisfies eq. (7.73) and

$$\phi(t) = \phi_0 + \delta\phi(t) \,, \tag{7.76}$$

where ϕ_0 is constant and $|\delta H| << |H_0|, |\delta \phi| << |\phi_0|$. The corresponding Einstein frame quantities are

$$\tilde{H} = \frac{1}{\sqrt{1-\kappa\xi\phi_0^2}} \left(H_0 + \delta H + \frac{\kappa\xi\phi_0 H_0}{1-\kappa\xi\phi_0^2} \delta\phi - \frac{\kappa\xi\phi_0}{1-\kappa\xi\phi_0^2} \delta\dot{\phi} \right)$$

$$= \tilde{H}_0 + \delta\tilde{H} \tag{7.77}$$

and

$$\tilde{\phi} = \tilde{\phi}_0 + \delta\tilde{\phi} \tag{7.78}$$

where, to first order,

$$\tilde{H}_0 = \frac{H_0}{\sqrt{1-\kappa\xi\phi_0^2}} \,, \tag{7.79}$$

$$\frac{\delta \tilde{H}}{\tilde{H}} = \frac{\delta H}{H_0} + \frac{\kappa \xi \phi_0^2}{1 - \kappa \xi \phi_0^2} \left(\frac{\delta \phi}{\phi_0} - \frac{\delta \dot{\phi}}{H_0 \phi_0} \right), \quad (7.80)$$

$$\delta \tilde{\phi} = \frac{\sqrt{1 - \kappa \xi (1 - 6\xi) \phi_0^2}}{1 - \kappa \xi \phi_0^2} \delta \phi. \quad (7.81)$$

If the Jordan frame quantities $\delta H/H_0$, $\delta \phi/\phi_0$, and $\delta \dot{\phi}/(H_0 \phi_0)$ in eq. (7.80) are small, the deviation from a de Sitter solution $\delta \tilde{H}/\tilde{H}$ in the Einstein frame is also small and slow-roll inflation in the Jordan frame implies slow-roll inflation in the Einstein frame. The converse is not true — a counterexample is given in Ref. [9] — and therefore one must be careful when mapping back solutions from the Einstein to the original Jordan frame. These considerations are relevant for the calculation of density and gravitational wave perturbations in generalized inflation.

4. Inflation and $\xi \neq 0$: the unperturbed universe

In this section we summarize the consequences of the appearance of nonminimal coupling in the equations of the unperturbed universe, assuming that the metric is a spatially flat FLRW one.

It is clear from the previous discussion that one cannot arbitrarily set $\xi = 0$ and it has been shown in several works [1, 404, 406, 38, 341, 344] that the value of ξ determines the viability of inflationary scenarios. Several specific inflationary scenarios are analyzed in Ref. [341] and it is shown that the issues are theoretical consistency and fine-tuning.

Regarding theoretical consistency, ask whether any prescription for the value of ξ is applicable. If the answer is affirmative, the consequences for the viability of the specific scenario are examined, i.e., does the value of ξ used correspond to the theoretical prescription? Aspects to be considered include the existence of inflationary solutions and a sufficient amount of inflation.

Fine-tuning is an issue perhaps less fundamental but nevertheless important. The classic example of a situation in which fine-tuning is relevant occurs in the already mentioned chaotic inflationary scenario with $V = \lambda \phi^4$ studied in Ref. [404]. Inflationary solutions turn out to be fine-tuned for $\xi \geq 10^{-3}$, in particular for the value $\xi = 1/6 \simeq 0.16$ predicted by general relativity. These solutions are physically unacceptable since the original purpose of inflation was to eliminate fine-tuning of initial conditions, which is necessary in standard big bang cosmology to solve the horizon and flatness problems.

Usually, positive scalar field potentials are considered in inflation and quintessence scenarios with $\xi = 0$. The reason is that in the slow-rollover approximation $V >> \dot{\phi}^2/2$, one has $\rho \approx V$ and $V > 0$ corresponds to $\rho > 0$ — a minimal requirement on the energy density. However, the approximate relation $\rho \simeq V$ no longer holds in the presence of nonminimal coupling, and ρ is given by a complicated expression. Indeed, negative scalar field potentials are considered in the literature on nonminimal coupling, in cosmology [9, 370] and in other contexts [102, 697, 486, 79, 80, 608], and negative potentials also appear in eleven-dimensional supergravity [155, 470].

Can the positive term $\xi R \phi^2 / 2$ balance a negative potential $V(\phi)$? Sometimes it can. For example, in the toy model of Ref. [98] containing a spatially flat FLRW universe dominated by a conformally coupled scalar field, a negative potential V is balanced by the coupling term $\xi R \phi^2 / 2$ in such a way that inflation is achieved and the energy density of ϕ is always non-negative. From the mathematical point of view the possibility of a negative potential that is inflationary in the presence of nonminimal coupling extends the range of possibilities to explore. However, with a few exceptions [528, 583, 370, 273, 272], a negative scalar field potential $V(\phi)$ is not popular among cosmologists, even though it is often encountered in string or supergravity theories and generalizes the notion of a negative cosmological constant which produces the anti-de Sitter space so popular in M-theory.

4.1 Necessary conditions for generalized inflation

In this section we study necessary conditions for inflation, i.e., for acceleration of the scale factor $\ddot{a} > 0$. Acceleration of the universe — the distinguishing qualitative feature of inflation — is also required at the present epoch of the history of the universe in order to explain the data from high redshift type Ia supernovae [703, 738, 777, 415, 704, 739, 705, 740, 741, 863]. The latter imply that a form of matter with negative pressure (*quintessence*) is beginning to dominate the dynamics of the universe. Scalar fields are proposed as natural models of quintessence [923, 831, 211, 867, 709, 50, 51] and therefore the considerations of this subsection are also relevant for scalar field models of quintessence.

In ordinary inflation driven by a minimally coupled scalar field the Einstein-Friedmann equations

$$\left(\frac{\dot{a}}{a}\right)^2 = \frac{\kappa}{3}\left(\frac{\dot{\phi}^2}{2} + V(\phi)\right), \tag{7.82}$$

$$\frac{\ddot{a}}{a} = -\frac{\kappa}{3}\left(\dot{\phi}^2 - V\right), \tag{7.83}$$

imply that a necessary (but not sufficient) condition for cosmic acceleration is $V \geq 0$. In slow-roll inflation, $\rho \simeq V(\phi) \gg \dot{\phi}^2/2$ and in this case the condition $V \geq 0$ is also necessary to satisfy the weak energy condition.

What is the necessary condition for inflation when $\xi \neq 0$? Manipulation of the equations of inflation with nonminimal coupling yields [351] (see Appendix B)

$$V - \frac{3\xi}{2}\phi \frac{dV}{d\phi} > 0 \qquad (\xi \leq 1/6) \ . \qquad (7.84)$$

This necessary condition is not generalized to values of ξ larger than $1/6$, due to the difficulty of handling the dynamical equations analytically when $\xi \neq 0$. However, the semi-infinite range of values of the coupling constant $\xi \leq 1/6$ covers many of the prescriptions for the value of ξ given in the literature. In the $\xi \to 0$ limit, eq. (7.84) reduces to the well known necessary condition for acceleration $V > 0$.

The necessary condition for inflation (7.84) immediately allows one to reach the following conclusions:

- Consider an even potential $V(\phi) = V(-\phi)$ which is increasing for $\phi > 0$ (e.g., even powers in the MacLaurin approximation of V). For $0 < \xi < 1/6$, one has $\xi \phi \, dV/d\phi > 0$ and it is harder to satisfy the necessary condition (7.84) for inflation than in the minimal coupling case. Hence, for this class of potentials one can say that it is harder to achieve acceleration of the universe and thus inflation. For $\xi < 0$, the necessary condition for cosmic acceleration is more easily satisfied than in the $\xi = 0$ case, but we still cannot conclude that with nonminimal coupling it is easier to achieve inflation, since a necessary but not sufficient condition for acceleration is considered.

- Taking to the extreme the possibility of a balance between the potential $V(\phi)$ and the term $\xi R \phi^2 / 2$ in the action (7.1), one may wonder whether it is possible to obtain inflation with a scalar field that has $V(\phi) = 0$ and zero cosmological constant, caused only by the effect of nonminimal coupling. In particular it is of interest to consider the case of strong coupling $|\xi| \gg 1$ recurrent in the literature [767, 336, 335, 647, 648, 403, 91, 508, 211], in which the effects of nonminimal coupling are enhanced. Here we can immediately conclude that inflation without scalar field potential is not possible for $\xi \leq 1/6$ since by setting $V = 0$ the necessary condition for inflation (7.84) cannot be satisfied.

4.1.1 Specific potentials

As a result of the complication of the coupled Einstein-Klein-Gordon equations when $\xi \neq 0$, general analytical considerations regarding the occurrence of inflation with nonminimally coupled scalar fields are limited. However, the question of whether it is harder or easier to achieve acceleration of the universe with nonminimal coupling for potentials that are known to be inflationary in the minimal coupling case can be answered, at least partially. Since in many situations these potentials are motivated by a high energy physics theory, they are of special interest. In order to appreciate the effect of the inclusion of nonminimal coupling in a given inflationary scenario, we review some exact solutions for popular inflationary potentials, in conjunction with the necessary condition (7.84) for cosmic acceleration. The phase space description of the dynamics of a FLRW universes with a nonminimally coupled scalar field is available for a limited number of specific potentials, with different variables used by the different authors [38, 71, 386, 451, 452, 197, 453, 651].

- $V = 0$

  The qualitative difference between minimal and nonminimal coupling, even for small values of ξ, is exemplified by the solution for $V = 0$, $\xi = 0$, as compared to the corresponding solution for conformal coupling constant. For minimal coupling, one has the stiff equation of state $P = \rho$, and the scale factor[2] $a(t) = a_0 \, t^{1/3}$.

  Instead, in the $\xi = 1/6$ case the Klein-Gordon equation is conformally invariant, corresponding to the vanishing of the trace of $T_{ab}[\phi]$, to the radiation equation of state, and to the scale factor $a(t) = a_0 \, t^{1/2}$. This is in agreement with the fact that there are no accelerated universe solutions for $V = 0$ and any value of ξ, given that the necessary condition (7.84) is not satisfied in this case.

- $V =$ constant

  In the case $V = V_0 =$ const., the necessary condition (7.84) for cosmic acceleration when $\xi \leq 1/6$ coincides with the corresponding condition for minimal coupling $V_0 > 0$.

- $V = m^2 \phi^2/2$

  A mass term is a simple potential for a scalar field, and an example of the class of even potentials for which $\phi \, dV/d\phi > 0$ that is

---
[2] We ignore the trivial Minkowski space solution.

considered above. When $\xi = 0$ the quadratic potential corresponds to chaotic inflation [569] and for $\xi < 0$ it can still generate inflation. For example, the exponentially expanding solution

$$H = H_* = \frac{m}{(12|\xi|)^{1/2}}, \quad \phi = \phi_* = \frac{1}{(\kappa|\xi|)^{1/2}} \quad (7.85)$$

is studied — not in relation to the early universe — but in the search for short periods of unstable exponential expansion of the universe late into the matter dominated era (*late time mild inflation*) [403]. This scenario is somewhat unphysical, since it corresponds to a spacetime in which a *massive* scalar field propagates sharply on the light cone at every point. This phenomenon is due to the cancellation between the usual tail of radiation due to the intrinsic mass m and a second tail term due to backscattering of ϕ waves off the background curvature of spacetime [359].

For $\xi > 0$, the nonminimally coupled scalar field is studied in Refs. [647, 648] with no inflation found, in conjunction with a reported periodicity in the redshift of galaxies [156]. The latter appears to be an artifact of the statistics and not a genuine periodicity, and interest in these scalar field models has not been sustained. However, the models are still interesting from the point of view of the dynamics of scalar fields with nonminimal coupling and they did attract attention when the fictitious periodicity was first reported [647, 648, 763, 342, 764, 724, 437, 436]. The introduction of nonminimal coupling eliminates the acceleration of the universe for large positive values of ξ when $V(\phi) = m^2\phi^2/2$. This is relevant since we are otherwise unable to draw any such conclusions for $\xi > 1/6$.

- $V = \lambda\phi^4$

This potential generates chaotic inflation when $\xi = 0$. As an example for $\xi \neq 0$, consider conformal coupling. For $\xi = 1/6$, the Klein-Gordon equation (7.3) is conformally invariant, corresponding to the radiation equation of state $P = \rho/3$ and to the non-inflationary expansion law $a(t) \propto t^{1/2}$. The introduction of conformal coupling destroys the acceleration occurring in the minimal coupling case for the same potential. However, accelerated solutions can be recovered by breaking the conformal symmetry with the introduction of a mass for the scalar or of a cosmological constant which, in this respect, behaves in the same manner [602]. Exact accelerating and non-accelerating solutions corresponding to integrability of the dynamical system for the potential $V = \frac{\Lambda}{\kappa} + m^2\phi^2/2 + \lambda\phi^4$ are presented in Ref. [742] for special values of the parameters in the potential.

- $V = \lambda \phi^n$

In general, the necessary condition for cosmic acceleration (7.84) depends on the particular solution of the Klein-Gordon equation, which is not known *a priori*. However, this dependence disappears for power-law potentials, which contain the previous cases and also the Ratra-Peebles potential $V \propto \phi^{-|\beta|}$ used in intermediate inflation [75, 658, 84] and in most quintessence models (see Chapter 8). The necessary condition (7.84) for accelerated cosmic expansion is

$$\lambda \left(1 - \frac{3n\xi}{2}\right) > 0 \qquad (7.86)$$

when $\xi \leq 1/6$. Under the assumption $\lambda > 0$ corresponding to a positive scalar field potential, the necessary condition (7.84) fails to be satisfied when $\xi \geq 2/3n$, independently of the solution ϕ and of the initial conditions $(\phi_0, \dot{\phi}_0)$. Also in this case, nonminimal coupling destroys acceleration in the range of values of $\xi \in (2/3n, 1/6]$.

The potential $V = \lambda \phi^n$ with $n > 6$ gives rise to power-law inflation $a = a_0 t^p$ if

$$p = \frac{1 + (n-10)\xi}{(n-4)(n-6)|\xi|} \qquad (7.87)$$

is larger than unity [404, 9]. However, the scenario is fine-tuned for $\xi > 0$ [404]. For $\xi < 0$ the solution is accelerating only if $6 < n < 4 + 2\sqrt{3} \simeq 7.464$.

- $V = V_0 \exp\left(-\sqrt{\frac{2\kappa}{p}}\,\phi\right)$

This exponential potential is associated with power-law inflation $a = a_0 t^p$ when $\xi = 0$ and $\phi > 0$. As seen in Section 1.11.3, an exponential potential is the fingerprint of theories with compactified extra dimensions. The necessary condition for cosmic acceleration is

$$\frac{\phi}{m_{pl}} > \frac{1}{6\xi}\sqrt{\frac{p}{\pi}} \qquad \left(0 < \xi \leq \frac{1}{6}\right), \qquad (7.88)$$

$$\frac{\phi}{m_{pl}} < \frac{1}{6|\xi|}\sqrt{\frac{p}{\pi}} \qquad (\xi < 0). \qquad (7.89)$$

4.2 The effective equation of state with nonminimal coupling

The effective equation of state

$$P = w\rho \qquad (7.90)$$

of the cosmic fluid describing the scalar field has a coefficient w that in general is time-dependent, and it cannot be assigned *a priori* without severely restricting the solutions. For $w =$ constant one has the spatially flat FLRW solutions

$$a(t) = a_0 \, t^{\frac{2}{3(w+1)}} \qquad (7.91)$$

if $w \neq -1$, or

$$a = a_0 \, e^{Ht} \qquad (7.92)$$

if $w = -1$. Solutions for a non-spatially flat universe and arbitrary values of w are presented in Ref. [348]. The function $w(t)$ depends on the particular solution of the equations of motion.

In the case of minimal coupling and for a general potential V, the effective equation of state of the universe is given by

$$\frac{P}{\rho} = \frac{\dot{\phi}^2 - 2V}{\dot{\phi}^2 + 2V} \equiv w(y) , \qquad (7.93)$$

where $y \equiv \dot{\phi}^2/2V$ is the ratio between the kinetic and the potential energy densities of the scalar ϕ. Under the usual assumption $V \geq 0$ which guarantees that the energy density ρ is non-negative when $\dot{\phi} \simeq 0$ during slow-roll inflation, one has $y \geq 0$ and the function

$$w(y) = \frac{y^2 - 1}{y^2 + 1} \qquad (7.94)$$

increases monotonically from its minimum $w_{min} = -1$ attained at $y = 0$ to the horizontal asymptote $w^* = +1$ as $y \to +\infty$. The slow rollover regime corresponds to the region $|y| \ll 1$ and to w near its minimum, where the kinetic energy density of ϕ is negligible in comparison to its potential energy density. As the kinetic energy density $\dot{\phi}^2/2$ increases, the equation of state progressively deviates from $P = -\rho$ and the pressure becomes closer to zero. In this way the system gradually moves away from the slow rollover regime. At the equipartition between the kinetic and the potential energy densities ($y = 1$), one has the dust equation of state $P = 0$. The pressure becomes positive as y increases and, when the kinetic energy density completely dominates the potential energy density ($y \gg 1$), one finally reaches the equation of state of a stiff fluid $P = \rho$.

The limitation $-1 \leq w(y) \leq 1$ for $\xi = 0$ is not valid for $\xi \neq 0$: in the presence of nonminimal coupling the ratio P/ρ is not bounded from below. Examples are the exact solution with $P = -5\rho/3$ obtained in Ref. [742], corresponding to integrability of the equations of motion for a massive conformally coupled scalar field with quartic self-interaction, and the solution (8.82) of Chapter 8. The fact that the range of values $w < -1$ is possible with nonminimal coupling is relevant for quintessence models of today's universe and is discussed in Section 8.3.

4.3 Critical values of the scalar field

Critical values of the scalar field frequently appear in studies of nonminimal coupling and have already been encountered in the discussion of conformal transformations in Section 7.3 and in Section 1.12. Specifically,

- If $\xi \leq 0$, there are no critical values.
- If $0 < \xi < 1/6$ there are four critical values $\{\pm\phi_1, \pm\phi_2\}$ defined by

$$\phi_1 \equiv \frac{1}{\sqrt{\kappa\xi}} \tag{7.95}$$

and

$$\phi_2 \equiv \frac{1}{\sqrt{\kappa\xi(1-6\xi)}}. \tag{7.96}$$

In this case one has $\phi_1 > \sqrt{\frac{3}{4\pi}}\, m_{Pl} \simeq 0.48\, m_{Pl}$ and $\phi_2 \geq \sqrt{\frac{3}{\pi}}\, m_{Pl} \simeq 0.977\, m_{Pl} > \phi_1$.

- If $\xi \geq 1/6$, only the two critical values $\{\pm\phi_1\}$ are present and $\phi_1 < 0.48\, m_{pl}$.

The critical values $\pm\phi_1$ appear when the field equations are divided by $\left(1 - \kappa\xi\phi^2\right)$ to obtain the effective gravitational coupling $G_{eff} = G\left(1 - \kappa\xi\phi^2\right)^{-1}$.

If the scalar field potential $V(\phi)$ is assumed to be non-negative then one can show that in any closed or spatially flat FLRW universe there is no classical solution for $|\phi| > \phi_2$ [404, 38]. In fact, the Hamiltonian constraint

$$\left(1 - \kappa\xi\phi^2\right)\left(H^2 + \frac{K}{a^2}\right) = \frac{\kappa\rho^{(m)}}{3} + \frac{\kappa}{6}\left(\dot\phi\right)^2 + \frac{\kappa V}{3} + 2\kappa\xi H\phi\dot\phi \tag{7.97}$$

can be written as

$$\left[H - \frac{\kappa\xi\phi\dot\phi}{1 - \left(\frac{\phi}{\phi_1}\right)^2}\right]^2 - \frac{\left[1 + (6\xi - 1)\left(\frac{\phi}{\phi_1}\right)^2\right]\kappa\left(\dot\phi\right)^2}{6\left[1 - \left(\frac{\phi}{\phi_1}\right)^2\right]^2} + \frac{K}{a^2}$$

$$-\frac{\kappa V(\phi)}{3\left[1 - \left(\frac{\phi}{\phi_1}\right)^2\right]} = 0. \tag{7.98}$$

If $|\phi| > \phi_2 > \phi_1$ then $(6\xi - 1)\left(\frac{\phi}{\phi_1}\right)^2 < -1$ and we have

$$-\frac{\left[1 + (6\xi - 1)\left(\frac{\phi}{\phi_1}\right)^2\right]\kappa\left(\dot{\phi}\right)^2}{6\left[1 - \left(\frac{\phi}{\phi_1}\right)^2\right]^2} \geq 0. \tag{7.99}$$

Furthermore, one has

$$-\frac{\kappa V(\phi)}{3\left[1 - \left(\frac{\phi}{\phi_1}\right)^2\right]} > 0 \tag{7.100}$$

if $V > 0$ and

$$\left[H - \frac{\kappa \xi \phi \dot{\phi}}{1 - \left(\frac{\phi}{\phi_1}\right)^2}\right]^2 \geq 0, \tag{7.101}$$

and $K/a^2 \geq 0$ for closed or spatially flat universes. Thus, assuming $K = 0$ or $+1$ and $V > 0$, the left hand side of eq. (7.98) is positive, while the right hand side is zero, an absurdity arising from the assumption that $|\phi| > \phi_2$. The solutions for the scalar field are constrained to the range of values $|\phi| < \phi_2$. Since $\phi_1 \to +\infty$ as $\xi \to 0$, for very small values of ξ it is reasonable to assume that $|\phi| < \phi_1 < \phi_2$ at all times; however, this assumption may be unreasonable for large positive values of ξ — the strong coupling condition. Weak coupling situations ($\xi << 1$) occur if the scalar ϕ considered is a quintessence field because of the limits on ξ imposed by present-day Solar System experiments, which constrain the long range force mediated by the ultra-light scalar ϕ. The situation may be very different in this respect during inflation, in which strong couplings are not only allowed but sometimes even sought. In practice, one wants the energy scale of inflation to be well below the Planck scale when scalar and tensor perturbations cross the horizon. Hence it must be

$$V(\phi) << m_{Pl}^4, \tag{7.102}$$

a condition that usually cannot be satisfied without fine-tuning if ϕ assumes values that are too large.

The conformal transformation that reduces the nonminimally coupled theory to that of a minimally coupled field in the Einstein frame is

$$g_{ab} \longrightarrow \tilde{g}_{ab} = \left(1 - \kappa \xi \phi^2\right) g_{ab}, \tag{7.103}$$

Nonminimal coupling 173

$$\phi \longrightarrow \tilde{\phi} = \int d\phi \, \frac{\sqrt{1 - \kappa\xi(1-6\xi)\phi^2}}{1 - \kappa\xi\phi^2} \,, \quad (7.104)$$

and it degenerates when $\phi \to \pm\phi_1$, where the conformal factor vanishes and $\tilde{\phi}$ diverges. Hence, the use of the conformal transformation is limited by the critical points $\pm\phi_1$. Also the perturbation analysis of Section 7.6 loses validity at these points if the unperturbed universe crosses through one of the singularities $\pm\phi_1$. Classical solutions crossing through these critical values do exist, although their physical relevance remains to be established.

5. The slow-roll regime of generalized inflation

The equations of ordinary inflation governing the dynamics of the scale factor and of the scalar field ϕ are solved in the slow-roll approximation. Similarly, the equations for the density and gravitational wave perturbations generated during inflation can, in general, only be solved in the same approximation[3]. We briefly review the slow-roll approximation to ordinary inflation, referring the reader to Refs. [578, 569] for comprehensive discussions.

In the approximation

$$\ddot{\phi} \ll H\dot{\phi} \,, \quad (7.105)$$

$$V(\phi) \approx \rho \gg \frac{\dot{\phi}^2}{2} \,, \quad (7.106)$$

the equations of ordinary inflation (7.82), (7.83) and

$$\ddot{\phi} + 3H\dot{\phi} + \frac{dV}{d\phi} = 0 \,, \quad (7.107)$$

simplify to

$$H^2 \simeq \frac{\kappa}{3} V(\phi) \,, \quad (7.108)$$

$$3H\dot{\phi} + \frac{dV}{d\phi} \simeq 0 \,. \quad (7.109)$$

In this approximation, the equation of state of the cosmic fluid describing the scalar field is close to the vacuum equation of state $P = -\rho$, and the cosmic expansion is almost a de Sitter one,

$$a = a_0 \exp\left[H(t)\, t\right] \,, \quad (7.110)$$

---

[3]Two specific scenarios are known to be exceptions to this statement: power-law inflation and the string-inspired scenario of Ref. [310], a toy model already ruled out by the *COBE* data [310]. Otherwise one may resort to numerical integration of the equations for the perturbations in a specified scenario.

with
$$H(t) = H_0 + H_1 t + \ldots, \quad (7.111)$$

where H_0 is a constant that dominates the small term $H_1 t$ and the next orders in the expansion (7.111) of $H(t)$. The possibility that the kinetic energy density $\dot{\phi}^2/2$ of the inflaton be negligible in comparison to the potential energy density $V(\phi)$ — as expressed by eq. (7.106) — arises if $V(\phi)$ has a shallow section over which ϕ can roll slowly, i.e., with small "speed" $\dot{\phi}$. This is a necessary, but not sufficient, condition for the occurrence of slow-roll inflation: if $V(\phi)$ is too steep, the inflaton will certainly roll fast down the potential, but the latter could have a shallow section and ϕ could still shoot across it with large speed $\dot{\phi}$; it is a matter of initial conditions. Indeed, the slow-roll approximation is an assumption on the solutions of the full equations of inflation (7.82), (7.83) and (7.107). Moreover [569], the reduced equations of slow-roll inflation (7.108) and (7.109) have degree reduced by one in comparison with the full equations of ordinary inflation (7.82), (7.83) and (7.107). Hence, the solution is specified by the reduced set of two initial conditions $(\phi(t_0), a(t_0))$ instead of the full set of four conditions $\left(\phi(t_0), \dot{\phi}(t_0), a(t_0), \dot{a}(t_0)\right)$, with an apparent loss of generality of the solutions. Then, why does the slow-roll approximation work? How is it possible that solutions of slow-roll inflation can be general solutions? If they correspond to a set of zero measure in the set of all initial conditions, they are fine-tuned and unphysical. The answer is that the de Sitter solutions

$$a = a_0 \exp(H_0 t), \quad \phi = 0 \quad (7.112)$$

are attractor points for the orbits of the solutions in phase space [574, 766]. Therefore, the quasi-exponential expansion (7.110) is a general property. Were inflationary de Sitter attractors absent, slow-roll inflation would be an empty theory.

Does an analogous attractor mechanism exist in the case of generalized inflation? Does the attractor property of de Sitter solutions survive when nonminimal coupling is included in the picture? Is a flat section of the potential still a necessary condition for slow-roll inflation? In answering these questions, it is useful to keep in mind that [1, 404, 341] the nonminimal coupling term $\xi R \phi^2 / 2$ in the action (7.1) acts as an effective mass term[4], spoiling the flatness of the potentials that are known to be inflationary for $\xi = 0$. One begins the analysis by writing

---

[4]Although the effect is like that of a mass, the interpretation of curvature as a mass term for the scalar field must not be taken literally even when it is a constant [356].

the equations of generalized inflation as

$$6\left[1 - \xi(1 - 6\xi)\kappa\phi^2\right]\left(\dot{H} + 2H^2\right) - \kappa(6\xi - 1)\dot{\phi}^2 - 4\kappa V$$
$$+ 6\kappa\xi\phi V' = 0, \tag{7.113}$$

$$\frac{\kappa}{2}\dot{\phi}^2 + 6\xi\kappa H\phi\dot{\phi} - 3H^2\left(1 - \kappa\xi\phi^2\right) + \kappa V = 0, \tag{7.114}$$

$$\ddot{\phi} + 3H\dot{\phi} + \xi R\phi + V' = 0. \tag{7.115}$$

Eq. (7.113) corresponds to the trace of the Einstein equations

$$R = -\kappa(\rho - 3P), \tag{7.116}$$

eq. (7.114) is the Hamiltonian constraint $3H^2 = \kappa\rho$, while eq. (7.115) is the well-known Klein-Gordon equation.

Note that in the presence of nonminimal coupling, the stress-energy tensor of the scalar field — and consequently its energy density ρ and pressure P — can be identified in three possible inequivalent ways that correspond to different ways of writing the field equations as explained in Section 7.2. The procedure adopted here is convenient because it is general, i.e., solutions are not lost by manipulating the field equations, and the stress-energy tensor of the scalar field is covariantly conserved, which does not happen for other choices of ρ and P [351, 110].

Explicitly, the effective energy density and pressure of ϕ relative to a comoving observer of the FLRW universe are given by

$$\rho = \frac{\dot{\phi}^2}{2} + 3\xi H^2\phi^2 + 6\xi H\phi\dot{\phi} + V(\phi), \tag{7.117}$$

$$P = \frac{\dot{\phi}^2}{2} - V(\phi) - \xi\left(4H\phi\dot{\phi} + 2\dot{\phi}^2 + 2\phi\ddot{\phi}\right) - \xi\left(2\dot{H} + 3H^2\right)\phi^2. \tag{7.118}$$

Only two equations of the set (7.113)-(7.115) are independent, and the system can be reduced to a two-dimensional phase space manifold with variables (H, ϕ) [38, 386, 451, 452]. It is then straightforward to verify that the solutions

$$(H, \phi) = (H_0, \phi_0), \tag{7.119}$$

where H_0 and ϕ_0 are constants, are all the fixed points of the dynamical system with $\xi \neq 0$, provided that the conditions

$$12\,\xi H_0^2\phi_0 + V_0' = 0, \tag{7.120}$$

$$H_0^2 \left(1 - \kappa\xi\phi_0^2\right) = \frac{\kappa V_0}{3}, \tag{7.121}$$

are satisfied, where $V_0 \equiv V(\phi_0)$ and $V_0' \equiv dV/d\phi|_{\phi_0}$. There are only two such constraints since only two equations in the set (7.113)-(7.115) are independent. The fixed points (7.119) are de Sitter solutions with constant scalar field and generalize the solutions $(H, \phi) = \left(\pm\sqrt{\Lambda/3}, 0\right)$ well known for minimal coupling; $\Lambda > 0$ being the cosmological constant equivalent to the constant potential $V = \Lambda/\kappa$.

In order to assess the stability of the universes (7.119) and to decide whether they are attractors or not, one must perform a stability analysis with respect to perturbations $\delta\phi$ and δH of the scalar field and of the Hubble parameter[5], which are given by

$$\phi(t, \vec{x}) = \phi_0(t) + \delta\phi(t, \vec{x}), \qquad H(t, \vec{x}) = H_0 + \delta H(t, \vec{x}). \tag{7.122}$$

Since general perturbations are space- and time- dependent, there is the recurrent problem of gauge-dependence in cosmology: if the perturbation analysis is performed in a particular gauge (of which many appear in the literature), one cannot be certain that the growing (decaying) modes are genuine perturbations and not pure gauge modes which can be removed by coordinate transformations [541, 581].

To solve the problem of the stability of de Sitter spaces, a gauge-independent analysis is necessary. Here we adopt the covariant and gauge-invariant formalism of Bardeen [65], in the modified formulation of Ellis, Bruni and Hwang [315, 316, 317, 509, 653]. We first present and discuss the results of this analysis [350], postponing the derivation to the next section. One finds that there is stability for expanding de Sitter spaces (7.119) with $H_0 > 0$, and therefore (7.119) is an attractor point, if

$$V_0'' \geq \frac{V_0'}{\phi_0} \frac{1 - 3\xi\kappa\phi_0^2}{1 - \xi\kappa\phi_0^2} \qquad (\phi_0 \neq 0), \tag{7.123}$$

$$V_0'' + 4\xi\kappa V_0 \geq 0 \qquad (\phi_0 = 0). \tag{7.124}$$

By contrast, the contracting fixed points (7.119) with $H_0 < 0$ are always unstable as in the case of minimal coupling. Stability not only depends on the form of the scalar field potential — which is expected

---

[5]In the analysis of the phase space, attention is usually restricted only to time-dependent perturbations (e.g., [461, 657]): however, these perturbations are too special to draw definite conclusions.

— but also on the value of ξ. It is only in particular situations that the ξ-dependence disappears and stability holds irrespective of the value of ξ. For example, this happens:

i) if $V(\phi)$ has a minimum ($V_0' = 0$ and $V_0'' > 0$) at ϕ_0

ii) if $V = \Lambda/\kappa + \lambda \phi^n$ (including the case of a simple mass term $m^2\phi^2/2$) with $\Lambda, \lambda \geq 0$. This space is stable for $n \geq 1 + f(x)$, where

$$x \equiv \kappa \xi \phi_0^2, \qquad f(x) = \frac{1 - 3x}{1 - x} < 1. \qquad (7.125)$$

The stability conditions (7.123) and (7.124) are deduced by assuming that $0 < x < 1$. If $x > 1$ a negative effective gravitational coupling $G_{eff} \equiv G\left(1 - \kappa \xi \phi_0^2\right)^{-1}$ arises [404, 351]. Furthermore, the slow-roll parameter ϵ_3 defined in Section 7.5.2 diverges if the unperturbed solution $\phi(t)$ crosses one of the critical values $\pm \phi_1$ defined for $\xi > 0$, while the slow-roll parameter ϵ_4 diverges if $\phi(t)$ crosses one of the other critical values $\pm \phi_2$ defined for $0 < \xi < 1/6$.

As we have already seen, under the usual assumption that V is nonnegative, the Hamiltonian constraint (7.114) forces $|\phi|$ to be smaller than ϕ_2; we further assume that $|\phi| < \phi_1$. If instead $|\phi| > \phi_1$, the direction of the inequality (7.123) is reversed.

The case $\phi = \pm \phi_1$, not considered so far, corresponds to a class of solutions with constant Ricci curvature containing a de Sitter representative [451, 452]. However, the latter is clearly fine-tuned and is unstable with respect to perturbations $\delta \phi$.

For $\xi = 0$, eq. (7.121) yields $V_0' = 0$ for the fixed point, while the stability condition (7.123) gives $V_0'' > 0$. This happens, for example, when $V(\phi)$ has a minimum Λ/κ in ϕ_0 which intuitively corresponds to stability. A solution starting at any value of ϕ is attracted towards the minimum. If ϕ identically coincides with ϕ_0 and there is no kinetic energy ($\dot{\phi} = 0$), eqs. (7.117) and (7.118) yield the energy density $\rho = 3\xi H_0^2 \phi_0^2 + V_0 = -P$ and the vacuum equation of state corresponding to de Sitter solutions holds. If instead $V_0'' < 0$ and the potential has a maximum $V_0 = \Lambda/\kappa$ in ϕ_0, a solution starting near ϕ_0 will run away from it.

When $\xi \neq 0$ the interpretation of the stability conditions (7.123) and (7.124) is complicated by the balance between $V(\phi)$ and $\xi R \phi^2/2$ in the action (7.1). Eqs. (7.123) and (7.124) describe in exact terms this balance. Qualitative considerations [1, 337, 404, 341] help understanding of these equations: an almost flat section of the potential $V(\phi)$ that allows slow-roll inflation for a minimally coupled scalar is distorted when the nonminimal coupling term $\xi R\phi^2/2$ is introduced in the Lagrangian density. The extra term plays the role of an effective mass term for the inflaton. While R is only constant for a de Sitter space and the effective mass changes with time, the time-dependence is relatively weak during slow roll-inflation when the solution is attracted by a de Sitter space.

In conclusion,

slow-roll inflation only makes sense for $\xi \neq 0$ when the conditions (7.123) or (7.124) are satisfied.

In this case, the expanding de Sitter spaces (7.119) satisfying eqs. (7.120) and (7.121) are attractor points in phase space. Caution is indicated in checking that eqs. (7.123) or (7.124) are satisfied before applying the slow-roll formalism presented in the next section to an inflationary scenario with a nonminimally coupled scalar field. The importance of the inflationary attractors is made clear once again by the example of the contracting spaces (7.119) with $H_0 < 0$, for which the slow-roll approximation is exact, in the sense that the slow-roll parameters ϵ_i defined in the next section exactly vanish. However, this bears no relationship to the actual inflationary solutions because the contracting spaces (7.119) with $H_0 < 0$ are not attractors.

5.1 Derivation of the stability conditions

The derivation of eqs. (7.123) and (7.124) proceeds as follows. The metric perturbations are identified by the quantities A, B, H_L and H_T in the spacetime metric

$$ds^2 = a^2(t)\left\{-(1+2AY)dt^2 - 2B\,Y_i dt\,dx^i + [\delta_{ij}(1+2H_L) + 2H_T Y_{ij}]\,dx^i dx^j\right\}, \quad (7.126)$$

where the Y are scalar spherical harmonics satisfying

$$\nabla^2 Y = \left(\frac{\partial^2}{\partial x^2} + \frac{\partial^2}{\partial y^2} + \frac{\partial^2}{\partial z^2}\right) Y = -k^2 Y, \quad (7.127)$$

Y_i and Y_{ij} are related to the derivatives of the spherical harmonics Y by

$$Y_i = \frac{1}{k^2} \partial_i Y, \quad (7.128)$$

$$Y_{ij} = \frac{1}{k^2} \partial_i \partial_j Y + \frac{1}{3}\delta_{ij} Y, \quad (7.129)$$

respectively [65], and k is the eigenvalue defined by eq. (7.127). Here we shall use Bardeen's gauge-invariant potentials

$$\Phi_H = H_L + \frac{\dot{a}}{k}\left(B - \frac{a}{k}\dot{H}_T\right), \quad (7.130)$$

$$\Phi_A = A + \frac{\dot{a}}{k}\left(B - \frac{a}{k}\dot{H}_T\right) + \frac{a}{k}\left[\dot{B} - \frac{1}{k}\left(a\dot{H}_T\right)^{\cdot}\right],\qquad (7.131)$$

and the Ellis-Bruni-Hwang [315, 316, 317] variables

$$\Delta\phi(t,\vec{x}) = \delta\phi + \frac{a}{k}\dot{\phi}\left(B - \frac{a}{k}\dot{H}_T\right),\qquad (7.132)$$

$$\Delta R(t,\vec{x}) = \delta R + \frac{a}{k}\dot{R}\left(B - \frac{a}{k}\dot{H}_T\right).\qquad (7.133)$$

The evolution equations for the gauge-invariant variables $\Phi_{H,A}$ and $\Delta\phi$ are [498]

$$\dot{\Phi}_H + \left(\frac{\xi\kappa\phi\dot{\phi}}{1-\kappa\xi\phi^2} - H\right)\Phi_A - \frac{\kappa}{1-\kappa\xi\phi^2}\left\{\xi\phi\Delta\dot{\phi}\right.$$
$$\left. + \left[\xi\phi\left(\frac{\dot{\phi}}{\phi} - H\right) - \frac{\dot{\phi}}{2}\right]\Delta\phi\right\} = 0,\qquad (7.134)$$

$$\left(\frac{k}{a}\right)^2 \Phi_H + \frac{1}{1-\kappa\xi\phi^2}\left(\frac{3\xi^2\kappa\phi^2}{1-\kappa\xi\phi^2} + \frac{1}{2}\right)\kappa\dot{\phi}^2\,\Phi_A$$
$$- \frac{1}{1-\kappa\xi\phi^2}\left\{\left(\frac{3\xi^2\kappa\phi^2}{1-\kappa\xi\phi^2} + \frac{1}{2}\right)\kappa\dot{\phi}\Delta\dot{\phi}\right.$$
$$\left. + \left[\left(\frac{k}{a}\right)^2\xi\phi - \ddot{\phi}\left(\frac{3\xi^2\kappa\phi^2}{1-\kappa\xi\phi^2} + \frac{1}{2}\right)\right]\kappa\Delta\phi\right\} = 0,\quad (7.135)$$

$$\Phi_A + \Phi_H - \frac{2\xi\kappa\phi\Delta\phi}{1-\kappa\xi\phi^2} = 0,\qquad (7.136)$$

$$\ddot{\Phi}_H + H\dot{\Phi}_H + \left(H - \frac{\xi\kappa\phi\dot{\phi}}{1-\kappa\xi\phi^2}\right)\left(2\dot{\Phi}_H - \dot{\Phi}_A\right) - \frac{\kappa V}{1-\kappa\xi\phi^2}\Phi_A$$
$$+ \frac{\kappa}{1-\kappa\xi\phi^2}\left\{-\xi\phi\Delta\ddot{\phi} + \left[\frac{\dot{\phi}}{2} - 2\xi\left(\dot{\phi} + H\phi\right)\right]\Delta\dot{\phi}\right.$$
$$\left. + \left[\xi\phi\left(\kappa p_H - \frac{\ddot{\phi}}{\phi} - \frac{2H\dot{\phi}}{\phi}\right) - \frac{V'}{2\kappa}\right]\kappa\Delta\phi\right\} = 0,\qquad (7.137)$$

$$\Delta\ddot{\phi} + 3H\Delta\dot{\phi} + \left(\frac{k^2}{a^2} + \xi R + V''\right)\Delta\phi + \dot{\phi}\left(3\dot{\Phi}_H - \dot{\Phi}_A\right)$$

$$+ 2\left(V' + \xi R\phi\right)\Phi_A + \xi\phi\Delta R = 0, \qquad (7.138)$$

where

$$p_H = \frac{1}{1 - \kappa\xi\phi^2}\left[\frac{\dot{\phi}^2}{2} - V - 2\xi\phi\left(\ddot{\phi} + 3H\dot{\phi} + \frac{\dot{\phi}^2}{\phi}\right)\right]. \qquad (7.139)$$

An overdot denotes differentiation with respect to the comoving time of the unperturbed background, and the subscript zero denotes unperturbed quantities. Only first order calculations in the perturbations are presented.

Considerable simplifications occur in eqs. (7.133)-(7.139) for the case of a de Sitter space with constant scalar field (7.119) as the background universe. To first order, one obtains

$$\Phi_H = \Phi_A = \frac{\xi\kappa\phi_0}{1 - \kappa\xi\phi_0^2}\Delta\phi, \qquad (7.140)$$

$$\Delta\ddot{\phi} + 3H_0\Delta\dot{\phi} + \left[\frac{k^2}{a^2} + V_0'' + \frac{\xi R_0(1 + \kappa\xi\phi_0^2) + 2V_0'\kappa\xi\phi_0}{1 - \kappa\xi\phi_0^2}\right]\Delta\phi + \xi\phi_0\Delta R = 0 \qquad (7.141)$$

while eq. (7.136) reduces to the constraint

$$-\frac{V_0\xi\phi_0}{1 - \kappa\xi\phi_0^2} = V_0' - p_H\xi\phi_0 \qquad (7.142)$$

which, using eq. (7.139), is written as

$$\frac{V_0'}{V_0} = -\frac{4\kappa\xi\phi_0}{1 - \kappa\xi\phi_0^2}; \qquad (7.143)$$

eq. (7.143) can also be obtained by division of eq. (7.120) by eq. (7.121). Using the expression of the Ricci scalar $R = 6(\dot{H} + 2H^2)$ and eq. (7.120) one obtains

$$\Delta\ddot{\phi} + 3H_0\Delta\dot{\phi} + \left(V_0'' + 12\xi H_0^2 + \frac{k^2}{a^2}\right)\Delta\phi + \xi\phi_0\Delta R = 0. \qquad (7.144)$$

For a de Sitter background (7.119) the gauge-invariant variables $\Delta\phi$ and ΔR coincide with the scalar field and curvature perturbations $\delta\phi$ and δR respectively, to first order,

$$\Delta\phi = \delta\phi, \qquad \Delta R = \delta R = 6\left(\delta\dot{H} + 4H_0\delta H\right), \qquad (7.145)$$

Nonminimal coupling

and therefore

$$\Delta R = \delta R = \frac{-6\xi\kappa\phi_0 \left[V_0'' + 4\left(1 + 3\xi\right) H_0^2\right]}{1 - \xi\left(1 - 6\xi\right)\kappa\phi_0^2} \Delta\phi . \tag{7.146}$$

One can now substitute eq. (7.146) into eq. (7.144) for $\Delta\phi$; the use of eq. (7.120) then yields

$$\Delta\ddot{\phi} + 3H_0\Delta\dot{\phi} + \left(\frac{k^2}{a^2} + \alpha\right)\Delta\phi = 0 , \tag{7.147}$$

where

$$\alpha = \frac{V_0''\phi_0\left(1 - \kappa\xi\phi_0^2\right) - V_0'\left(1 - 3\kappa\xi\phi_0^2\right)}{\phi_0\left[1 - \xi\left(1 - 6\xi\right)\kappa\phi_0^2\right]} \tag{7.148}$$

and $a = a_0 \exp(H_0 t)$.

Let us consider the expanding ($H_0 > 0$) de Sitter spaces (7.119). At late times $t \to +\infty$ one can neglect the term proportional to $(k/a)^2 \propto e^{-2H_0 t}$ in eq. (7.147), and look for solutions of the form

$$\Delta\phi(t, \vec{x}) = \frac{1}{(2\pi)^{3/2}} \int d^3\vec{l}\, \Delta\phi_l(t)\, e^{i\vec{l}\cdot\vec{x}} , \qquad \Delta\phi_l(t) = \epsilon_l\, e^{\beta_l t} . \tag{7.149}$$

Note that the Fourier expansion (7.149) is well defined due to the fact that the universe has flat spatial sections. The constants β_l must satisfy the algebraic equation

$$\beta_l^2 + 3H_0\beta_l + \alpha = 0 , \tag{7.150}$$

with roots

$$\beta_l^{(\pm)} = \frac{3H_0}{2}\left(-1 \pm \sqrt{1 - \frac{4\alpha}{9H_0^2}}\right) . \tag{7.151}$$

While $Re(\beta_l^{(-)}) < 0$, the sign of $Re(\beta_l^{(+)})$ depends on α: $Re(\beta_l^{(+)}) > 0$ if $\alpha < 0$ and $Re(\beta_l^{(+)}) \leq 0$ if $\alpha \geq 0$. Hence one has stability for $\alpha \geq 0$ which, for $\phi_0 \neq 0$ translates into the advertised result (7.123). If instead $\alpha < 0$, the gauge-invariant perturbations $\Delta\phi$ and ΔR are proportional to $\Delta\phi$ (cf. eq. 7.146)), they grow without bound and there is instability.

Now let us discuss the $\phi_0 = 0$ case. Eqs. (7.133)-(7.137) yield

$$\Phi_H = \Phi_A = 0 , \tag{7.152}$$

$$\Delta\ddot{\phi} + 3H_0\Delta\dot{\phi} + \left(\frac{k^2}{a^2} + \alpha_1\right)\Delta\phi = 0 , \tag{7.153}$$

where $\Delta R = 0$ and $\alpha_1 = V_0'' + 4\xi\kappa V_0$. Hence for $\phi_0 = 0$, there is stability if eq. (7.124) is satisfied and instability otherwise.

Finally, consider the contracting ($H_0 < 0$) fixed points (7.119). In this case it is convenient to use conformal time η and the auxiliary variable $u \equiv a\Delta\phi$. Eq. (7.147) becomes, in these variables,

$$\frac{d^2 u}{d\eta^2} + \left[k^2 - U(\eta)\right] u = 0 , \tag{7.154}$$

where

$$U(\eta) = \left(4 - \frac{\alpha_1}{H_0^2}\right) \frac{1}{\eta^2} + \frac{2}{H_0 \eta^3} , \tag{7.155}$$

and we used the relation

$$\eta = -\frac{1}{aH_0} \tag{7.156}$$

valid in the background (7.119) (see Appendix C). Formally, eq. (7.154) is a one-dimensional Schrödinger equation for a quantum particle of unit mass in the potential $U(\eta)$. Its asymptotic solutions at large η — corresponding to $t \to +\infty$ for a contracting de Sitter background — are free waves $u \simeq e^{\pm ik\eta}$, and $\Delta\phi \propto H_0\eta$ diverges. The solutions (7.119) with $H_0 < 0$ are unstable, as in the $\xi = 0$ case.

5.2 Slow-roll parameters

The Hubble slow-roll approximation known for ordinary inflation [541, 578, 569] is characterized by two slow-roll parameters

$$\epsilon_H = -\dot{H}/H^2 , \qquad \eta_H = -\frac{\ddot{\phi}}{H\dot{\phi}} \tag{7.157}$$

which stay small during slow-roll inflation. When ϵ_H and η_H increase, the kinetic energy of the inflaton increases and, when ϵ_H and η_H become of order unity, the slow-roll approximation breaks down and inflation ends.

Slow-roll parameters are also identified for generalized inflation [498, 524, 525, 507] and correspond to eqs. (6.60)-(6.63) for the special choice $F = \kappa^{-1} - \xi\phi^2$. In this case one has to satisfy four slow-roll necessary conditions instead of two. The slow-roll parameters are the dimensionless quantities[6]

$$\epsilon_1 = \frac{\dot{H}}{H^2} = -\epsilon_H , \tag{7.158}$$

---

[6] For a different definition of the slow-roll parameters see Ref. [649].

$$\epsilon_2 = \frac{\ddot{\phi}}{H\dot{\phi}} = -\eta_H \,, \qquad (7.159)$$

$$\epsilon_3 = -\frac{\xi\kappa\phi\dot{\phi}}{H\left[1-\left(\frac{\phi}{\phi_1}\right)^2\right]} \,, \qquad (7.160)$$

$$\epsilon_4 = -\frac{\xi(1-6\xi)\kappa\phi\dot{\phi}}{H\left[1-\left(\frac{\phi}{\phi_2}\right)^2\right]} \,. \qquad (7.161)$$

The parameters ϵ_3 and ϵ_4 vanish in the limit $\xi \to 0$ of ordinary inflation; ϵ_4 also vanishes for conformal coupling ($\xi = 1/6$). One has $|\epsilon_i| << 1$ for every solution attracted by the expanding de Sitter spaces (7.119) at sufficiently large times, when the latter are attractor points. Moreover, $\epsilon_i = 0$ exactly for the de Sitter attractor solutions.

6. Inflation and $\xi \neq 0$: perturbations

The quantum fluctuations of the inflaton field which unavoidably occur during inflation generate density (scalar) perturbations that act as seeds for the formation of the structures observed in the universe today, from galaxies to superclusters [541, 581]. Similarly, quantum fluctuations δg_{ab} of the metric tensor are generated during inflation, corresponding to gravitational waves [541, 581, 569]. Both scalar and tensor perturbations leave an imprint in the cosmic microwave background by generating temperature fluctuations. The latter have been detected by the *COBE* [809] satellite and by other experiments and are/are intended to be, the subject of more accurate study by the *WMAP* [913] and the *PLANCK* [713] missions.

In order to confront itself with the present and future observations, the theory must predict observables such as the amplitudes and spectra of perturbations. For ordinary inflation, these are well known[7]. The generalization to arbitrary scalar-tensor theories is presented in Chapter 6. The case of a nonminimally coupled scalar field is recovered by setting

$$f(\phi, R) = \frac{R}{\kappa} - \xi R \phi^2 \,, \qquad \omega = 1 \,. \qquad (7.162)$$

The original calculation is covariant and gauge-invariant, and builds upon the formalism developed by Bardeen [65], Ellis, Bruni and Hwang

---

[7] The calculation of perturbations took a long time to be completed, beginning with efforts in the early 1980s. See Ref. [578] for a review and Ref. [456] for a historical perspective.

[315, 316, 317] and Hwang and Vishniac [509], and considers a FLRW universe with arbitrary curvature index. Motivated by inflation, we restrict ourselves to the spatially flat case.

6.1 Density perturbations

The metric is written as

$$ds^2 = -(1 + 2\alpha)\, dt^2 - \chi_{,i}\, dt\, dx^i + a^2(t)\, \delta_{ij}\, (1 + 2\varphi)\, dx^i\, dx^j\,, \quad (7.163)$$

while the scalar field is given by eq. (7.122). One introduces the additional gauge-invariant variable

$$\delta\phi_\varphi = \delta\phi - \frac{\dot\phi}{H}\varphi \equiv -\frac{\dot\phi}{H}\varphi_{\delta\phi}\,. \quad (7.164)$$

The second order action for the perturbations (analogous to the one for perturbations in ordinary inflation [578]) is [504, 505, 506]

$$S^{(pert)} = \int dt\, d^3\vec{x}\, \mathcal{L}^{(pert)} = \frac{1}{2}\int dt\, d^3\vec{x}\, a^3 Z \left\{ \delta\dot\phi_\varphi^2 - \frac{1}{a^2}\delta\phi_{\varphi,}{}^i\, \delta\phi_{\varphi,i} \right.$$

$$\left. + \frac{1}{a^3 Z}\frac{H}{\dot\phi}\left[a^3 Z\left(\frac{\dot\phi}{H}\right)\right]^{\cdot}\delta\phi_\varphi^2 \right\}\,, \quad (7.165)$$

where

$$Z(t) = \frac{H^2\left[1 - \kappa\xi(1 - 6\xi)\phi^2\right](1 - \kappa\xi\phi^2)}{\left[H(1 - \kappa\xi\phi^2) - \xi\kappa\phi\dot\phi\right]^2}\,. \quad (7.166)$$

The action (7.165) yields the evolution equation for the perturbations $\delta\phi_\varphi$

$$\delta\ddot\phi_\varphi + \frac{(a^3 Z)^{\cdot}}{a^3 Z}\delta\dot\phi_\varphi - \left\{\frac{\nabla^2}{a^2} + \frac{1}{a^3 Z}\frac{H}{\dot\phi}\left[a^3 Z\left(\frac{\dot\phi}{H}\right)\right]^{\cdot}\right\}\delta\phi_\varphi = 0\,. \quad (7.167)$$

By using the auxiliary variables[8]

$$z(t) = \frac{a\dot\phi}{H}\sqrt{Z}\,, \quad (7.168)$$

$$v(t, \vec{x}) = z\frac{H}{\dot\phi}\delta\phi_\varphi = a\sqrt{Z}\,\delta\phi_\varphi\,, \quad (7.169)$$

---

[8]The variable z of eq. (7.168) agrees with the z-variable of Ref. [653] and with the z of Ref. [578] multiplied by the factor $\sqrt{Z}$ (note that $Z = 1$ corresponds to ordinary inflation).

eq. (7.167) is reduced to

$$v_{\eta\eta} - \left(\nabla^2 + \frac{z_{\eta\eta}}{z}\right)v = 0, \qquad (7.170)$$

where η denotes conformal time.

Quantization is achieved by assuming that the background is classical while the perturbations have quantum nature. A Heisenberg picture is used in which quantum operators change in time while the state vectors remain constant. The vacuum state of the system is identified with the adiabatic vacuum [136], and one wants this vacuum state to remain unchanged in time. The fluctuations $\delta\phi(t, \vec{x})$ of the scalar field are associated with a quantum operator $\delta\hat{\phi}(t, \vec{x})$; similarly $\varphi \to \hat{\varphi}$, and the gauge-invariant variable (7.164) is associated with the quantum operator

$$\delta\hat{\phi}_\varphi = \delta\hat{\phi} - \frac{\dot{\phi}}{H}\hat{\varphi} \qquad (7.171)$$

(in the following, carets denote quantum operators). The unperturbed quantities are regarded as classical.

Since the three-dimensional spatial sections are flat, it is meaningful to perform a Fourier decomposition of the operator $\delta\hat{\phi}_\varphi$,

$$\delta\hat{\phi}_\varphi = \frac{1}{(2\pi)^{3/2}} \int d^3\vec{k} \left[\hat{a}_k\, \delta\phi_{\varphi k}(t)\, e^{i\vec{k}\cdot\vec{x}} + \hat{a}_k^\dagger\, \delta\phi_{\varphi k}^*(t)\, e^{-i\vec{k}\cdot\vec{x}}\right], \qquad (7.172)$$

where the annihilation and creation operators $\hat{a}_k$ and $\hat{a}_k^\dagger$ satisfy canonical commutation relations

$$[\hat{a}_k, \hat{a}_{k'}] = \left[\hat{a}_k^\dagger, \hat{a}_{k'}^\dagger\right] = 0, \qquad (7.173)$$

$$\left[\hat{a}_k, \hat{a}_{k'}^\dagger\right] = \delta^{(3)}\left(\vec{k} - \vec{k}'\right), \qquad (7.174)$$

and the mode functions $\delta\phi_{\varphi k}(t)$ are complex Fourier coefficients satisfying the classical equations obtained from eq. (7.162)

$$\delta\ddot{\phi}_{\varphi k} + \frac{(a^3 Z)^{\cdot}}{a^3 Z}\delta\dot{\phi}_{\varphi k} + \left\{\frac{k^2}{a^2} - \frac{1}{a^3 Z}\frac{H}{\dot{\phi}}\left[a^3 Z \left(\frac{\dot{\phi}}{H}\right)\right]^{\cdot}\right\}\delta\phi_{\varphi k} = 0. \quad (7.175)$$

Similarly,

$$v(t, \vec{x}) = \frac{1}{(2\pi)^{3/2}} \int d^3\vec{k} \left[v_k(t)\, e^{i\vec{k}\cdot\vec{x}} + v_k^*(t)\, e^{-i\vec{k}\cdot\vec{x}}\right], \qquad (7.176)$$

$$\hat{v} = \frac{zH}{\dot{\phi}}\delta\hat{\phi}_\varphi = a\sqrt{Z}\,\delta\hat{\phi}_\varphi, \qquad (7.177)$$

and the $v_k(t)$ satisfy the equation

$$(v_k)_{\eta\eta} + \left(k^2 - \frac{z_{\eta\eta}}{z}\right) v_k = 0 . \tag{7.178}$$

The momentum conjugated to $\delta\phi_\varphi$ is

$$\delta\pi_\phi(t,\vec{x}) = \frac{\partial \mathcal{L}^{(\text{pert})}}{\partial(\delta\dot\phi_\varphi)} = a^3 Z \, \delta\dot\phi_\varphi(t,\vec{x}) , \tag{7.179}$$

and the associated quantum operator is $\delta\hat\pi_\varphi$. The operators $\delta\hat\phi_\varphi$ and $\delta\hat\pi_\varphi$ satisfy the equal time commutation relations

$$\left[\delta\hat\phi_\varphi(t,\vec{x}), \delta\hat\phi_\varphi(t,\vec{x}')\right] = \left[\delta\hat\pi_\varphi(t,\vec{x}), \delta\hat\pi_\varphi(t,\vec{x}')\right] = 0 , \tag{7.180}$$

$$\left[\delta\hat\phi_\varphi(t,\vec{x}), \delta\hat\pi_\varphi(t,\vec{x}')\right] = \frac{i}{a^3 Z} \delta^{(3)}\left(\vec{x} - \vec{x}'\right) . \tag{7.181}$$

The $\delta\phi_{\varphi k}(t)$ satisfy the Wronskian condition

$$\delta\phi_{\varphi k} \, \delta\dot\phi_{\varphi k}^* - \delta\phi_{\varphi k}^* \, \delta\dot\phi_{\varphi k} = \frac{i}{a^3 Z} . \tag{7.182}$$

It is usually assumed [499, 502, 503, 507] that

$$\frac{z_{\eta\eta}}{z} = \frac{m}{\eta^2} , \tag{7.183}$$

where m is a constant. We comment later on the validity of this assumption. Under the assumption (7.183), eq. (7.178) for the Fourier modes v_k reduces to

$$(v_k)_{\eta\eta} + \left[k^2 - \frac{(\nu^2 - 1/4)}{\eta^2}\right] v_k = 0 , \tag{7.184}$$

where

$$\nu = \left(m + \frac{1}{4}\right)^{1/2} . \tag{7.185}$$

By making the substitutions

$$s = k\eta , \quad v_k = \sqrt{s} \, J(s) , \tag{7.186}$$

eq. (7.184) is reduced to the Bessel equation

$$\frac{d^2 J}{ds^2} + \frac{1}{s}\frac{dJ}{ds} + \left(1 - \frac{\nu^2}{s^2}\right) J = 0 ; \tag{7.187}$$

therefore the solutions $v_k(\eta)$ can be expressed as

$$v_k(\eta) = \sqrt{k\eta}\, J_\nu(k\eta) , \qquad (7.188)$$

where $J_\nu(s)$ are Bessel functions of order ν. Eq. (7.169) yields the solutions for the Fourier coefficients $\delta\phi_{\varphi k}$

$$\delta\phi_{\varphi k}(\eta) = \frac{\dot\phi}{zH} v_k(\eta) = \frac{1}{a\sqrt{Z}} v_k(\eta) . \qquad (7.189)$$

The $v_k(\eta)$ which solve the Bessel equation (7.184) are expressed in terms of Hankel functions $H_\nu^{(1,2)}$, leading to (see Appendix D)

$$v_k(\eta) = \frac{\sqrt{\pi|\eta|}}{2} \left[c_1(\vec{k})\, H_\nu^{(1)}(k|\eta|) + c_2(\vec{k})\, H_\nu^{(2)}(k|\eta|) \right] \qquad (7.190)$$

and, by using eq. (7.189), to

$$\delta\phi_{\varphi k}(\eta) = \frac{\sqrt{\pi|\eta|}}{2a\sqrt{Z}} \left[c_1(\vec{k}) H_\nu^{(1)}(k|\eta|) + c_2(\vec{k}) H_\nu^{(2)}(k|\eta|) \right] . \qquad (7.191)$$

The normalization is chosen in such a way that the relation

$$\left|c_2(\vec{k})\right|^2 - \left|c_1(\vec{k})\right|^2 = 1 \qquad (7.192)$$

holds, in order to preserve eq. (7.182). Furthermore, the coefficients are completely fixed by requiring that, in the limit of small scales, the vacuum corresponds to positive frequency solutions. In fact, field theory in Minkowski space must be recovered in this limit. The small scale (large wavenumber) limit corresponds to

$$\frac{z_{\eta\eta}}{z} \ll k^2 , \qquad (7.193)$$

and eq. (7.178) reduces to

$$(v_k)_{\eta\eta} - k^2 v_k = 0 \qquad (7.194)$$

in this limit, with solutions $v_k \propto e^{\pm ik\eta}$. Eq. (7.189) yields

$$\delta\phi_{\varphi k} = \frac{1}{a\sqrt{Z}\sqrt{2k}} \left[c_1(\vec{k})\, e^{ik|\eta|} + c_2(\vec{k})\, e^{-ik|\eta|} \right] , \qquad (7.195)$$

which can also be obtained by expansion of the solutions (7.191) for $k|\eta| \gg 1$. Obviously, the positive frequency solution at small scales is obtained by setting

$$c_1(\vec{k}) = 0 , \qquad c_2(\vec{k}) = 1 \qquad (7.196)$$

so that

$$\delta\phi(\eta, \vec{x}) = \frac{1}{(2\pi)^{3/2}} \int d^3\vec{x} \left[c_2(\vec{k}) \, e^{i(\vec{k}\cdot\vec{x}-k\eta)} + c_2^*(\vec{k}) \, e^{i(-\vec{k}\cdot\vec{x}+k|\eta|)} \right]. \tag{7.197}$$

The power spectrum of a quantity $f(t, \vec{x})$ is defined as

$$\mathcal{P}(k,t) \equiv \frac{k^3}{2\pi^2} \int d^3\vec{r} \, \langle f(\vec{x}+\vec{r},t) f(\vec{x},t) \rangle_{\vec{x}} \, e^{-i\vec{k}\cdot\vec{r}} = \frac{k^3}{2\pi^2} |f_k(t)|^2, \tag{7.198}$$

where $\langle \, \rangle_{\vec{x}}$ denotes an average over the spatial coordinates $\vec{x}$ and $f_k(t)$ are the coefficients of the Fourier expansion

$$f(t,\vec{x}) = \frac{1}{(2\pi)^{3/2}} \int d^3\vec{k} \left[f_k(t) \, e^{i\vec{k}\cdot\vec{x}} + f_k^*(t) \, e^{-i\vec{k}\cdot\vec{x}} \right]. \tag{7.199}$$

The power spectrum of the gauge-invariant operator $\delta\hat{\phi}_\varphi$ is

$$\mathcal{P}_{\delta\hat{\phi}_\varphi}(k,t) = \frac{k^3}{2\pi^2} \int d^3\vec{r} \, \langle 0 | \delta\hat{\phi}_{\varphi k}(\vec{x}+\vec{r},t) \, \delta\hat{\phi}_{\varphi k}(\vec{x},t) | 0 \rangle_{\vec{x}} \, e^{-i\vec{k}\cdot\vec{r}}, \tag{7.200}$$

where $\langle 0 | \hat{A} | 0 \rangle$ denotes the expectation value of the operator $\hat{A}$ on the vacuum state. One is interested in computing the power spectrum for large-scale perturbations, i.e., perturbations that cross outside the horizon during inflation, subsequently remain "frozen" while outside the horizon, and only after the end of inflation — during the radiation- or the matter-dominated era — re-enter the horizon to seed the formation of structures. In the large scale limit the solution of eq. (7.175) is

$$\delta\phi_\varphi(t,\vec{x}) = -\frac{\dot\phi}{H} \left[C(\vec{x}) - D(\vec{x}) \int^t dt' \, \frac{1}{a^3 Z} \frac{H^2}{\dot\phi} \right], \tag{7.201}$$

where $C(\vec{x})$ and $D(\vec{x})$ are the coefficients of a growing component and of a decaying component respectively; the latter is neglected in the following. Accordingly, the solution $\delta\phi_{\varphi k}(\eta)$ in the large scale limit $k|\eta| \ll 1$ is

$$\delta\phi_{\varphi k}(\eta) = \frac{i\sqrt{|\eta|}\,\Gamma(\nu)}{2a\sqrt{\pi Z}} \left(\frac{k|\eta|}{2} \right)^{-\nu} [c_2(\eta) - c_1(\eta)] \tag{7.202}$$

for $\nu \neq 0$, where Γ denotes the gamma function. The power spectrum (7.200) therefore is given by

$$\mathcal{P}_{\delta\hat{\phi}_\varphi}^{1/2}(k,\eta) = \frac{\Gamma(\nu)}{\pi^{3/2} a |\eta| \sqrt{Z}} \left(\frac{k|\eta|}{2} \right)^{3/2-\nu} \left| c_2(\vec{k}) - c_1(\vec{k}) \right| \tag{7.203}$$

for $\nu \neq 0$, while one obtains [499, 502, 503, 507]

$$\mathcal{P}^{1/2}_{\delta\hat{\phi}_\varphi}(k,\eta) = \frac{2\sqrt{|\eta|}}{a\sqrt{Z}}\left(\frac{k}{2\pi}\right)^{3/2} \ln(k|\eta|) \left|c_2(\vec{k}) - c_1(\vec{k})\right| \qquad (7.204)$$

for $\nu = 0$. Now, eq. (7.201) yields (neglecting the decaying component),

$$C(\vec{x}) = -\frac{H}{\dot{\phi}}\delta\phi_\varphi(t,\vec{x}) \qquad (7.205)$$

and therefore, using eq. (7.198),

$$\mathcal{P}^{1/2}_C(k,t) = \left|\frac{H}{\dot{\phi}}\right| \mathcal{P}^{1/2}_{\delta\hat{\phi}_\varphi}(k,t) . \qquad (7.206)$$

By combining eqs. (7.161) and (7.205) one obtains

$$\varphi_{\delta\phi} = -\frac{H}{\dot{\phi}}\delta\phi_\varphi = C . \qquad (7.207)$$

The relation between temperature fluctuations of the cosmic microwave background and the variable $C(\vec{x})$ is [502]

$$\frac{\delta T}{T} = \frac{C}{5} \qquad (7.208)$$

and therefore the spectrum of temperature fluctuations is

$$\mathcal{P}^{1/2}_{\delta T/T}(k,t) = \frac{1}{5}\mathcal{P}^{1/2}_C(k,t) \qquad (7.209)$$

and eq. (7.206) yields

$$\mathcal{P}^{1/2}_{\delta T/T}(k,t) = \frac{1}{5}\left|\frac{H}{\dot{\phi}}\right|\mathcal{P}^{1/2}_{\delta\hat{\phi}_\varphi}(k,t) \qquad (7.210)$$

with $\mathcal{P}^{1/2}_{\delta\hat{\phi}_\varphi}$ given by eqs. (7.203) and (7.204).

The spectral index of scalar perturbations is defined as

$$n_S \equiv 1 + \frac{d\left(\ln \mathcal{P}_{\delta\hat{\varphi}_{\delta\phi}}\right)}{d\left(\ln k\right)} , \qquad (7.211)$$

and eq. (7.209) immediately yields

$$n_S = 1 + \frac{d\left(\ln \mathcal{P}_C\right)}{d\left(\ln k\right)} . \qquad (7.212)$$

By using eqs. (7.206) and (7.203) one obtains

$$n_S = 4 - 2\nu \qquad (\nu \neq 0) . \tag{7.213}$$

Note that, in the following, we do not need the expression for $\nu = 0$. To proceed, however, we do need extra input which is connected to the validity of the assumption (7.183) now discussed. Eq. (7.183) is satisfied in slow-roll inflation, to first order. This is interesting because we know from Section 7.5 that for suitable values of ξ there is a de Sitter attractor for nonminimal coupling, and therefore that it is sensible to consider the slow-roll approximation. Eq. (7.183) is also satisfied in pole-like inflation $a(t) \propto (t - t_0)^{-q}$ with $q > 0$ [499, 502, 503, 507].

Most models of ordinary inflation are set and solved in the context of the slow-roll approximation. In generalized inflation as well, the fact that the slow-roll regime satisfies eq. (7.183) allows one to solve eq. (7.178) for the perturbations.

Let us have a deeper look at eq. (7.183) and at the value of m for the slow-roll approximation. The quantity $z_{\eta\eta}/z$ in eq. (7.178) is computed exactly in Refs. [499, 502, 503, 507] in terms of the slow-roll coefficients (7.158)-(7.161). Upon assuming that the ϵ_i are small and that their derivatives $\dot\epsilon_i$ can be neglected ($i = 1, ..., 4$), in a situation mimicking the usual one of ordinary inflation [541, 578, 569], one obtains to lowest order

$$\frac{z_{\eta\eta}}{z} = a^2 H^2 \left(2 - 2\epsilon_1 + 3\epsilon_2 - 3\epsilon_3 + 3\epsilon_4 \right) . \tag{7.214}$$

The standard relation

$$\eta \simeq -\frac{1}{aH}\frac{1}{1+\epsilon_1} , \tag{7.215}$$

yields eq. (7.183) with

$$m = 2 + 3 \left(-2\epsilon_1 + \epsilon_2 - \epsilon_3 + \epsilon_4 \right) \tag{7.216}$$

and

$$\nu = \frac{3}{2} - 2\epsilon_1 + \epsilon_2 - \epsilon_3 + \epsilon_4 . \tag{7.217}$$

Eqs. (7.213) and (7.217) yield the spectral index of scalar perturbations for generalized slow-roll inflation,

$$n_S = 1 + 2 \left(2\epsilon_1 - \epsilon_2 + \epsilon_3 - \epsilon_4 \right) , \tag{7.218}$$

where the right hand side is computed at the time when the perturbations cross outside the horizon during inflation. The deviations of n_S

Nonminimal coupling 191

from unity (i.e., from an exactly scale-invariant Harrison-Zeldovich spectrum) are small during generalized slow-rolling. For $\xi = 0$ one recovers the well known formula for the spectral index of ordinary inflation [541, 578, 569]

$$n_S = 1 - 4\epsilon_H + 2\eta_H \qquad (\xi = 0) \ . \qquad (7.219)$$

6.2 Tensor perturbations

Tensor perturbations (gravitational waves) are generated as quantum fluctuations of the metric tensor g_{ab} in nonminimally coupled inflation and are calculated in Ref. [506] with a procedure similar to the one used for scalar perturbations.

It is convenient to introduce tensor perturbations as the trace-free and transverse quantities c_{ij} in the metric

$$ds^2 = -dt^2 + a^2(t)\left(\delta_{ij} + 2c_{ij}\right) dx^i dx^j \ , \qquad (7.220)$$

with

$$c^i{}_i = 0 \ , \qquad c_{ij}{}^{,j} = 0 \ . \qquad (7.221)$$

The power spectrum is

$$\mathcal{P}_{c_{ij}}(k,t) = \frac{k^3}{2\pi^2} \int d^3\vec{r}\, \langle c_{ij}\left(\vec{x}+\vec{r},t\right) c_{ij}\left(\vec{x},t\right)\rangle_{\vec{x}}\, e^{-i\vec{k}\cdot\vec{r}} \qquad (7.222)$$

and the spectral index of tensor perturbations is

$$n_T = \frac{d\left(\ln \mathcal{P}_{c_{ij}}\right)}{d\left(\ln k\right)} \ . \qquad (7.223)$$

One obtains

$$\mathcal{P}^{1/2}_{c_{ij}}(k,\eta) = \frac{8\pi G H}{\sqrt{2\pi}} \frac{1}{\sqrt{1-\xi\kappa^2\phi^2}} \frac{\Gamma(\nu)}{\Gamma(3/2)} \left(\frac{k|\eta|}{2}\right)^{3/2-\nu_g}$$

$$\cdot \sqrt{\frac{1}{2}\Sigma_l \left|c_{l1}\left(\vec{k}\right) - c_{l2}\left(\vec{k}\right)\right|^2} \ , \qquad (7.224)$$

where the summation Σ_l is intended over the two polarization states $\times$ and $+$ of gravitational waves. In the slow-roll approximation one has

$$\frac{z_{\eta\eta}}{z} = \frac{m_g}{\eta^2} \ , \qquad (7.225)$$

where m_g is the linear combination of the slow-roll parameters

$$m_g = 2 - 3(\epsilon_1 - \epsilon_3) \qquad (7.226)$$

and $\nu_g = (m_g + 1/4)^{1/2}$, as usual. Hence, $\nu_g \simeq 3/2 - \epsilon_1 + \epsilon_3$ and, to first order,

$$n_T = 2(\epsilon_1 - \epsilon_3) \ . \qquad (7.227)$$

Eq. (2.106) reduces to the well known spectral index of tensor perturbations [541, 578, 569] of ordinary inflation when $\xi \to 0$. Note that n_T is very small for slow-roll inflation, similar to slow-roll ordinary inflation.

As already noted in Section 6.3, in a superacceleration regime during inflation with $\dot{H} > 0$, one has $\epsilon_1 = \dot{H}/H^2 > 0$ and the spectral index of gravitational waves n_T given by eq. (7.227) can be positive, giving the possibility of a blue spectrum, which is instead ruled out with minimal coupling, for which $n_T = 4\dot{H}/H^2 \leq 0$.

7. Conclusion

The inclusion of nonminimal coupling in the equations of inflation seems to be indicated in most inflationary theories, leading to important consequences. Inflationary solutions are changed into noninflationary ones, and fine-tuning problems appear. The much needed slow-roll approximation to inflation is meaningful only when particular relations between the scalar field potential, its derivative, and the value of ξ are satisfied. The programme of rethinking inflationary theory by including the nonminimal coupling of the scalar field — a generally unavoidable, crucial element too often forgotten — is not exhausted by the results presented here. We outline open problems.

It is useful to remark that, in addition to inflation, nonminimal coupling changes the description and the results of the dynamical systems approach to cosmology (e.g., [451, 452]), quantum cosmology [90, 327, 533, 89, 261, 682], classical and quantum wormholes [462, 248, 245], and constitutes an avenue of approach to the cosmological constant problem by attempting to cancel the cosmological constant with a negative nonminimal coupling term [298, 299, 486, 382, 834, 402]. Unfortunately, these attempts fail because the cancellation of the cosmological constant is accompanied by strong variations in the gravitational coupling that are forbidden by the available experimental data.

Cosmic no-hair theorems [541] in the presence of nonminimal coupling are not known: one would like to know whether inflation is still a generic phenomenon when $\xi \neq 0$. In other words, starting with an anisotropic Bianchi model, does inflation occur and lead the universe toward a highly spatially homogeneous and isotropic FLRW state, with flat spatial sections? This question is unanswered. Preliminary results [823, 406] show that the convergence to the $K = 0$ FLRW universe can disappear by going from $\xi = 0$ to $\xi \neq 0$. The perturbation analysis of

Section 7.6 shows that the de Sitter solutions are still inflationary attractors in the phase space when the deviations from homogeneity and isotropy are small, but an analysis for large deviations from a FLRW space in the presence of a general potential $V(\phi)$ is needed. Cosmic no-hair theorems are not expected to be as general as in Einstein's theory, on the basis of the close analogy between nonminimal coupling and BD theory for which many $K = \pm 1$ solutions do not approach spatially flat universes at late times [545, 482].

The reconstruction of the inflationary potential from cosmological observations is a task that has been undertaken for $\xi = 0$. Given the unavoidability of nonminimal coupling in the general case, a similar formalism for $\xi \neq 0$ would be useful. The reconstruction of the potential in the context of quintessence models of the present-day universe is studied in Ref. [138], while Doppler peaks in the cosmic microwave background are computed for the specific case of inverse power law potentials [709, 50, 51].

APPENDIX 7.A: Solutions with constant scalar field

Let us adopt approach I and assume that $T_{ab}^{(m)} = 0$; then the field equations are

$$G_{ab} = \kappa T_{ab}^{(I)} = \kappa \left[\partial_a \phi \, \partial_b \phi - \frac{1}{2} g_{ab} \, \partial^c \phi \, \partial_c \phi - V(\phi) g_{ab} + \xi \left(g_{ab} \Box - \nabla_a \nabla_b \right) (\phi^2) \right.$$

$$\left. - \xi \phi^2 G_{ab} \right] , \tag{7.A.1}$$

$$\Box \phi - \xi R \phi - \frac{dV}{d\phi} = 0 . \tag{7.A.2}$$

By assuming that $\phi = \text{const.} \equiv \phi_0$, one has

$$G_{ab} = -\kappa V_0 g_{ab} + \kappa \xi \phi_0^2 G_{ab} , \tag{7.A.3}$$

$$\xi R \phi_0 + V_0' = 0 . \tag{7.A.4}$$

In the non-trivial case $\phi_0 \neq 0$, one has

$$R = -\frac{V_0'}{\xi \phi_0} = \text{const.} , \tag{7.A.5}$$

i.e., all these solutions have constant Ricci curvature. Eq. (7.A.3) then implies that

$$R = 4\kappa V_0 + \kappa \xi \phi_0^2 R \tag{7.A.6}$$

and, if $\phi_0^2 \neq 1/\kappa\xi$,

$$R = \frac{4\kappa V_0}{1 - \kappa \xi \phi_0^2} . \tag{7.A.7}$$

By comparing eqs. (7.A.5) and (7.A.7) one obtains the constraint between V, ϕ_0, and ξ

$$\phi_0 V_0 = \frac{-V_0'}{4\kappa\xi \left(1 - \kappa\xi\phi_0^2\right)} . \tag{7.A.8}$$

If instead, $\phi^2 = \phi_1^2 \equiv 1/(\kappa\xi)$ with $\xi > 0$, then it must be $V_0 = 0$, $R = 0$, and $V_0' = 0$; the potential $V(\phi)$ must have a zero and horizontal tangent in $\pm\phi_1$. These critical scalar field values have been studied in scalar field cosmology for the potential

$$V(\phi) = \frac{\alpha}{2}\phi^2 \left(\frac{6}{\kappa} - \phi^2\right) \tag{7.A.9}$$

and for $\xi = 1/6$; in this case all the solutions with constant Ricci curvature are classified [451]. For Lorentzian wormholes without potential V, all the solutions corresponding to the critical scalar field values are also classified [63].

APPENDIX 7.B: Necessary condition for cosmic acceleration

Cosmic acceleration corresponds to the condition $\rho + 3P < 0$; the expressions for the energy density and pressure of a nonminimally coupled scalar field then yield

$$(1 - 3\xi)\dot{\phi}^2 - V - 3\xi\phi\left(\ddot{\phi} + H\dot{\phi}\right) < 0. \tag{7.B.1}$$

The Klein-Gordon equation (7.3) is used to eliminate $\ddot{\phi}$, obtaining

$$(1 - 3\xi)\dot{\phi}^2 - V + 3\xi^2 R\phi^2 + 6\xi H\phi\dot{\phi} + 3\xi\phi\frac{dV}{d\phi} < 0. \tag{7.B.2}$$

The expression of the energy density is used again to rewrite the condition for cosmic acceleration $\ddot{a} > 0$ as

$$x \equiv \left(1 - \kappa\xi\phi^2\right)\rho - 2V + \dot{\phi}^2\left(\frac{1}{2} - 3\xi\right) + 3\xi^2 R\phi^2 + 3\xi\phi\frac{dV}{d\phi} < 0. \tag{7.B.3}$$

The Hamiltonian constraint guarantees that, for a solution of the field equations in a spatially flat universe, the energy density of ϕ is non-negative, $\rho \geq 0$. If $\xi \leq 1/6$, one has $-2V + 3\xi\phi dV/d\phi \leq x < 0$ and a necessary condition for cosmic acceleration to occur is

$$V - \frac{3\xi}{2}\phi\frac{dV}{d\phi} > 0. \tag{7.B.4}$$

In the limit $\xi \to 0$, eq. (7.84) reduces to the well known necessary condition for acceleration $V > 0$.

If V or ϕ, or both, are negative and $0 \leq \xi \leq 1/6$, the inequality (7.84) can be rewritten as

$$\frac{d}{d\phi}\left\{\ln\left[\frac{V}{V_0}\left(\frac{\phi_0}{\phi}\right)^{\frac{2}{3\xi}}\right]\right\} < 0, \tag{7.B.5}$$

where V_0 and ϕ_0 are arbitrary (positive) constants. As a result of the fact that the logarithm is a monotonically increasing function of its argument, the necessary condition for cosmic acceleration (7.84) requires that the potential $V(\phi)$ grow with ϕ more slowly than the power-law potential $V_{crit}(\phi) \equiv V_0 \left(\phi/\phi_0\right)^{\frac{2}{3\xi}}$. If instead $\xi < 0$, (7.84) requires that V grow faster than $V_{crit}(\phi)$ as ϕ increases.

APPENDIX 7.C: Proof of eq. (7.156)

In de Sitter space, the dependence of the scale factor on the comoving time t

$$a = a_0 \exp(H_0 t) \tag{7.C.1}$$

and the definition of conformal time

$$\eta = \int^t \frac{dt'}{a(t')}, \tag{7.C.2}$$

yield the relation

$$\eta = -\frac{1}{aH} = -\frac{e^{-H_0 t}}{a_0 H_0}. \tag{7.C.3}$$

For expanding de Sitter spaces ($H_0 > 0$), $t \to +\infty$ corresponds to $\eta \to 0$, while for contracting ($H_0 < 0$) de Sitter spaces, $t \to +\infty$ corresponds to $\eta \to +\infty$. During slow-roll inflation, eq. (7.C.3) is corrected to

$$\eta = -\frac{1}{aH}\frac{1}{1+\epsilon_1}, \tag{7.C.4}$$

which holds in the approximation that the derivatives of the slow-roll parameters ϵ_i can be neglected [504, 505, 506].

APPENDIX 7.D: Proof of eq. (7.190)

The Bessel function $J_\nu(s)$ can be expressed as

$$J_\nu(s) = \frac{H^{(1)}_\nu(s) + H^{(2)}_\nu(s)}{2}, \tag{7.D.1}$$

and therefore

$$v_k(\eta) = \sqrt{k\eta}\, J_\nu(k\eta) = \frac{\sqrt{k\eta}}{2}\left[H^{(1)}_\nu(k\eta) + H^{(2)}_\nu(k\eta)\right]. \tag{7.D.2}$$

In addition, the property

$$J_p(z) = \sum_{k=0}^{\infty} \frac{(-1)^k}{k!\,\Gamma(p+k+1)} \left(\frac{z}{2}\right)^{p+2k} \tag{7.D.3}$$

yields

$$J_p(-z) = (-1)^p J_p(z). \tag{7.D.4}$$

Then, $J_\nu(k\eta) = (-1)^\nu J_\nu(k|\eta|)$ if $\eta < 0$.

Chapter 8

THE PRESENT UNIVERSE

1. Present acceleration of the universe and quintessence

In 1998 it was discovered that, contrary to the popular belief at that time, the expansion of the universe at the present cosmological epoch is *accelerated*, i.e., $d^2a/dt^2 > 0$. This conclusion is based on observations of distant type Ia supernovae [738, 703], which have a characteristic light curve and constitute reasonable standard candles for cosmological studies [149, 438]. The systematic search for type Ia supernovae followed two decades of progress in the study of such objects and improved understanding of their light curves. The observations revealed that these supernovae are fainter than expected, an effect attributed to the structure of the universe. Then, at fixed redshift, the supernovae must be more distant than one expects in a decelerating universe, according to the expression (valid for moderate redshifts z)

$$H_0 D_L = z + \frac{1}{2}(1 - q_0) z^2 + \dots, \qquad (8.1)$$

where $q \equiv -\ddot{a}a/(\dot{a})^2$ is the deceleration parameter and D_L is the luminosity distance. For an accelerating universe ($q < 0$), D_L is larger than for a decelerating one with $q > 0$. If the universe expanded more slowly in the past than today, it is actually older than previously thought and the apparent faintness of the supernovae is explained. Cosmic acceleration also helps reconcile the age of the universe with that of globular clusters, a problem that resurfaces now and again in cosmology.

The observational evidence from the *Boomerang* [638, 266, 556, 632] and *MAXIMA* [463] data is now rather compelling for a spatially flat universe with total energy density $\Omega = \rho/\rho_c$, in units of the critical density ρ_c, given by

$$\Omega = \Omega^{(m)} + \Omega^{(q)} = 1 \tag{8.2}$$

where $\Omega^{(m)} \simeq 0.3$ is the matter energy density, of which 5% is baryonic, and the rest is dark matter, and $\Omega^{(q)} \simeq 0.7$ describes an unknown form of energy called *dark energy* or *quintessence*, which remains unclustered at all scales [738, 703, 777, 376, 739, 740, 741, 311]. The Einstein-Friedmann equation

$$\frac{\ddot{a}}{a} = -\frac{4\pi G}{3}(\rho + 3P) \tag{8.3}$$

allows an accelerated expansion if and only if the dominant form of energy in the universe is distributed homogeneously and is described by an effective equation of state satisfying $P < -\rho/3$. This fact is in contrast to the properties of dark matter, which has non-negative pressure, and points to an unknown form of energy with exotic properties.

It is not yet clear how negative the pressure of dark energy can be. Observational reports often quote the constraints on the present value of the function $w_q \equiv P^{(q)}/\rho^{(q)}$ as $-1 \leq w_q \leq -1/3$, but in reality values $w_q < -1$ are also compatible with the available observations, and it is even argued that the range of values $w_q \leq -1$ is favoured [168, 918, 919, 390]. This seems to be the case, in particular, when radio galaxies are used as standard candles instead of type Ia supernovae [253, 252, 251].

An obvious candidate for dark energy — and one that has been advocated many times before — is the cosmological constant Λ, which is characterized by the constant equation of state $P_\Lambda = -\rho_\Lambda$ [757, 134, 187, 755]. However the case for a cosmological constant is not very strong, since this explanation runs into severe problems due to the very fact that the equation of state, as well as P_Λ and ρ_Λ, is constant. Λ-based models express even more acutely the conundrum of the cosmological constant problem [899, 900]: the value of the vacuum energy density ρ_Λ is about 120 orders of magnitude smaller than its natural value originating at the Planck scale ($\Lambda \sim c^3/(\hbar G)$), or 40 orders of magnitude smaller than the value predicted by a cutoff at the QCD scale. Exact cancellation of the cosmological constant is preferred to such an extreme fine-tuning, although a plausible mechanism to achieve it is unknown. A further problem is posed by the fact that the dark energy just begins to dominate the expansion of the universe at redshifts $z \sim 1$, after the radiation- and matter-dominated eras. In order for this to happen around the present time, the energy density of the cosmological constant must be fine-tuned. In particular, the dark energy density must be subdominant at the time of primordial nucleosynthesis, in order not to affect the expansion rate of the universe (i.e., the speedup factor (5.12)) and alter the observed abundance of light elements. Moreover, the vacuum energy density must stay dormant during the matter-dominated era to allow for the growth of density perturbations. The problem of why the dark energy is only now beginning to dominate the expansion of the universe while it was negligible in the past, is called the *cosmic coincidence problem*. In other words, the matter and the dark energy densities were different in the past, and will be different in the future. This begs the question: why is it that the time at which the matter and dark energy densities are equal coincides with the epoch in the history of the universe in which we live? The cosmological constant explanation is currently rejected by most authors in favour of different models in which the energy density of quintessence is time-dependent and begins to manifest itself during the matter era. Various models of quintessence are proposed, including

- Minimally coupled scalar fields in general relativity

- Scalar fields coupled nonminimally to the Ricci curvature, or BD-like fields in scalar-tensor gravity

- A dissipative cosmological fluid with bulk viscosity, or a kinetic theory generalization of it [217, 218, 792, 53]. Semiclassical particle production also leads to viscosity

- Domain walls which yield $w_q \simeq -2/3$, or a frustrated cosmic strings network which produces $w_q \simeq -1/3$ [167, 97]. A network of non-intercommuting topological defects produces an effective equation of state parameter $w_q = -n/3$, where n is the dimension of the defect [880, 819]

- Scalar field models with non-canonical — possibly nonlinear — kinetic terms in the Lagrangian and zero potential energy (*kinetically driven quintessence* or *k-essence*) [750, 471, 215, 46]. These are described by the action

$$S^{(k)} = \int d^4x \sqrt{-g} \left[\frac{R}{2\kappa} + p(\phi, \nabla_c \phi) \right] + S^{(m)}, \quad (8.4)$$

where the dependence of the function p on the gradient of ϕ is often chosen to be of the form $p = p(\phi, X)$ with $X = \nabla_c \phi \nabla^c \phi$

- Particles with variable masses composing dark energy [260, 484, 195, 407, 40, 407]

- Brane-world models [600, 238, 496, 267, 756, 759]

- Models with an explicit coupling between quintessence and dark matter or other matter fields[1] [186, 28, 30, 34]

- A rolling tachyon field in string theories, which generates power-law accelerated expansion, and is characterized by energy density and pressure given by [788, 787, 428, 429, 654, 689, 397, 369]

$$\rho^{(q)} = \frac{V(\phi)}{\sqrt{1 - \left(\dot{\phi}/\dot{\phi}_0\right)^2}}, \qquad (8.5)$$

$$P^{(q)} = -V(\phi)\sqrt{1 - \left(\frac{\dot{\phi}}{\dot{\phi}_0}\right)^2} \qquad (8.6)$$

- A Chaplygin gas characterized by the equation of state

$$P = -\frac{A}{\rho}, \qquad (8.7)$$

where A is a positive constant. This rather exotic form of matter is found in higher-dimensional gravity and in Born-Infeld theories [534, 129, 118, 119, 439, 283, 113, 232, 324, 617, 100, 19, 233, 803]. A generalized Chaplygin gas with equation of state

$$P = -\frac{A}{\rho^\alpha}, \qquad (8.8)$$

with $\alpha > 0$ is also studied in the literature

- New gravitational physics described by terms proportional to $1/R^n$ ($n > 0$) in the gravitational Lagrangian, and no scalar fields [189, 181, 180, 885, 289, 213, 671, 380, 634, 818], or scalar field models in which the kinetic term evolves to zero, stabilizing the scalar field to an asymptotic value ϕ_0. The resulting $V(\phi_0)$ cancels a bare negative cosmological constant [656, 655].

The possibility that quintessence couples differently to dark and ordinary matter is also considered [260, 28, 867, 99]. The most popular models by far are simple scalar-field based ones. After all, it is well

---

[1]The direct coupling of quintessence to ordinary matter is constrained by tests of the equivalence principle and of the time variation of the fundamental constants [186].

known from inflationary theories that a scalar field slowly rolling in a flat potential can fuel an accelerated expansion of the universe and is equivalent to a fluid. Furthermore, the fact that the scalar field evolves in time causes the effective equation of state to be time-dependent — a necessary feature to solve the cosmic coincidence problem and provide a dynamical vacuum energy.

Many scenarios are presented in which quintessence is modeled by a minimally coupled scalar field ϕ in Einstein gravity, and attempts are made to identify the quintessence scalar with a field known from particle physics (e.g., [228]). There are also several scenarios in which the quintessence field either couples nonminimally to the Ricci curvature, or is a BD-like field in the context of a true scalar-tensor theory of gravity.

We first discuss the simplest models based on a single minimally coupled scalar field. If the potential $V(\phi)$ satisfies the condition [831]

$$\Gamma \equiv V \frac{\frac{d^2 V}{d\phi^2}}{\left(\frac{dV}{d\phi}\right)^2} \geq 1 , \qquad (8.9)$$

then the phase space of the solutions of the field equations contains late time attractors with a very large attraction basin. For a given potential, one can test for the presence of these attractors by checking the condition $\Gamma > 1$ without solving the field equations. These attractor solutions are time-dependent, hence they are not fixed points in phase space. They are present in a wide class of potentials used in the literature and are called *tracking solutions* [831, 923, 571]. Such attractors have been known since before the discovery of cosmic acceleration, as mechanisms to implement an effective time-dependent cosmological "constant" [730, 903, 396, 374, 373], which otherwise would have had to be assumed *ad hoc* by assigning Λ as a function of the cosmic time t or of the scale factor a. Indeed, the idea of a time-dependent vacuum energy has a long history (see Ref. [686] for a review), and was studied in order to reconcile the low value of the matter density $\Omega^{(m)}$ observed in the 1990s with the inflationary prediction $\Omega^{(m)} + \Omega^{(\Lambda)} = 1$.

For an enormous range of initial conditions spanning 150 orders of magnitude, the solution converges to the attractor before the present era and begins to dominate the expansion of the universe after the matter era, with equipartition $\Omega^{(q)} = \Omega^{(m)}$ occurring at redshift $z \sim 1$ (there is direct evidence of a decelerated epoch of the universe from the supernova SN1997ff at $z = 1.7$ [741, 432] and two other supernovae at $z = 1.2$

[850, 863]). The cosmic coincidence problem is then resolved, contrary to what happens with a fixed cosmological constant energy density.

Many scalar field potentials are used in the literature, including the following.

- The Ratra-Peebles potential ([697, 730] — see also [896, 904])

$$V(\phi) = \frac{M^4}{(\phi/M)^\alpha}, \qquad (8.10)$$

where $\alpha > 0$ and $M \leq H_0 \sim 10^{-33}$ eV, is associated with an extremely light scalar [730, 239, 169, 923, 831, 571, 709, 277, 276]. Inverse power-law potentials appear in supersymmetric QCD theories [842, 16, 17, 624] and supersymmetry breaking models [318, 912, 667, 133]. The values of α recurring most often in the literature are 1 and 2.

While the parameter α is not constrained very strongly, the scale M is fixed by the requirement that $\Omega^{(q)} \simeq 0.7$ today; this implies that

$$\lg_{10}\left(\frac{M}{1\,\text{GeV}}\right) \simeq \frac{19\alpha - 47}{4(\alpha + 1)}. \qquad (8.11)$$

- The potential

$$V(\phi) = \frac{M^4}{(\phi/M)^\alpha} e^{\kappa \phi^2/2} \qquad (8.12)$$

is motivated by supergravity theories [153, 154], which introduce the exponential factor as a correction to the Ratra-Peebles potential (8.10). This correction is unimportant at early times and does not modify the scaling properties of the tracking solutions during the matter- and radiation-dominated eras, but it affects the dynamics later by pushing the w_q factor in the effective equation of state of quintessence from values around -0.7 obtained for the Ratra-Peebles potential closer to the value -1 which fits the observations better [153, 154].

- Another choice is the potential [923, 831, 211]

$$V(\phi) = M^4 \left[\exp\left(\frac{1}{\sqrt{8\pi G}\,\phi}\right) - 1\right]. \qquad (8.13)$$

- The cosine potential

$$V(\phi) = M^4 \left[\cos\left(\frac{\phi}{f}\right) + 1\right] \qquad (8.14)$$

is sometimes considered, where $V_0 = M^4$ is a constant and ϕ is an ultra-light pseudo Nambu-Goldstone boson; the mass scale M may be related to a neutrino mass and is naturally small [230, 535, 212]. This potential does not satisfy the condition (8.9) and does not have tracking solutions. The model only works if ϕ is tuned near the maximum of the potential [831], a situation somehow ameliorated by the inclusion of nonminimal coupling with $\xi > 0$ [212].

- The exponential potential

$$V(\phi) = V_0 \left[\exp(\alpha \phi) + \exp(\beta \phi)\right] \tag{8.15}$$

(with both signs of α and β possible) is invoked, or obtained as the asymptotic form of other potentials [903, 374, 373, 879, 239, 131, 571, 69, 867, 212, 99, 96, 791, 615], and more often is considered with a single exponential term.

- The potential [758, 518]

$$V(\phi) = V_0 \left[\cosh(\lambda\phi) - 1\right]^p \tag{8.16}$$

has the asymptotic forms

$$V(\phi) \simeq \tilde{V}_0 \exp(-p\lambda\phi) \tag{8.17}$$

for $|\lambda\phi| \gg 1$ and $\phi < 0$, or

$$V(\phi) \simeq \tilde{V}_0 (\lambda\phi)^{2p} \tag{8.18}$$

for $|\lambda\phi| \ll 1$, where $\tilde{V}_0 = V_0/2^p$.

- The potential [866, 757, 518]

$$V(\phi) = V_0 \left[\sinh(\alpha\phi + \beta)\right]^{-\gamma} \qquad (\gamma > 0) \tag{8.19}$$

has an exponential asymptotic form and is encountered in the literature on field theory, statistical mechanics, and condensed matter physics. The potential

$$V(\phi) = V_0 \sinh^2(\alpha\phi) \tag{8.20}$$

is also considered [866, 276].

- Another possibility is the potential [789]

$$V(\phi) = V_0 \, \phi^q \, (\ln \phi)^p \,, \tag{8.21}$$

- or the potential [18, 808]

$$V(\phi) = V_0 \left[(\phi - A)^\alpha + B \right] \exp(-\lambda \phi) , \tag{8.22}$$

where A, B, α and λ are constants.

- Also the double exponential potential

$$V(\phi) = V_0 \exp\left(-A\, e^{\sqrt{2\kappa}\,\phi}\right) \tag{8.23}$$

with $A > 0$ is studied [133, 271, 666]. This potential is inspired by supersymmetry breaking via gaugino condensation [275, 292] or by T-duality in string theories [271], but it does not satisfy the condition (8.9). In fact, it exhibits a late time attractor that is decelerating and matter-dominated [133, 271, 666] and can only describe quintessence for a transient period.

- Another potential that does not satisfy the condition (8.9) is

$$V(\phi) = \lambda \phi^4 , \tag{8.24}$$

which can be used for toy models without the essential feature of tracking solutions [614].

Perhaps the potential studied most often is the Ratra-Peebles potential (8.10), which is used to illustrate many features common to scalar-field based quintessence scenarios. For a minimally coupled quintessence field ϕ and ordinary matter described by the equation of state $P^{(m)} = w_m \rho^{(m)}$, one finds the attractor scaling solution [923, 831, 211]

$$H = H_0\, a^{-\frac{3(1+w_m)}{2}} , \tag{8.25}$$

$$\phi = \phi_0\, a^{\frac{3(1+w_m)}{\alpha+2}} , \tag{8.26}$$

where

$$\phi_0 = \left\{ \frac{2\alpha\,(\alpha+2)^2\, M^{\alpha+4}}{9 H_0^2\,(1+w_m)\,[4+(1-w_m)\,\alpha]} \right\}^{\frac{1}{\alpha+2}} . \tag{8.27}$$

The effective equation of state of the quintessence fluid is $P^{(q)} = w_q \rho^{(q)}$ with

$$w_q = \frac{\alpha\, w_m - 2}{\alpha + 2} , \tag{8.28}$$

which is always negative if $w_m = 0$. With the exception of this case and of a few others [785, 48], the parameter w_q obtained in the literature changes with time in order to avoid severe constraints on the potential [297]). However, these constraints are much looser if the cosmic fluid is a dissipative — instead of a perfect fluid — and there is a quintessence field [217, 218].

The quintessence energy density redshifts more slowly than the energy density of ordinary matter and radiation and eventually comes to dominate late into the matter era, even if its numerical value is initially negligible as the ratio $\rho^{(m)}/\rho^{(q)}$ decreases during the evolution of the universe, until it becomes less than unity and $\rho^{(q)}$ starts to dominate. The equipartition time is determined by the mass scale M in the quintessence potential, which is fixed by the requirement that $\Omega_0 \simeq 1$ today. During the radiation- and matter-dominated eras, quintessence tracks the dominant radiation or matter energy density and emerges only later on. Remember that the radiation energy density $\rho^{(r)} \propto a^{-4}$ initially dominates, but it redshifts faster than the matter energy density $\rho^{(m)} \propto a^{-3}$; similarly, the quintessence energy density $\rho^{(q)}$ is initially much smaller than $\rho^{(m)}$ but redshifts more slowly and eventually comes to dominate.

In certain scenarios, the scalar field that drives inflation is the same as the quintessence scalar dominating the cosmic expansion today. Models with this interesting feature are called *quintessential inflationary* scenarios [393, 523, 698, 371, 700, 433, 434, 747] and are sometimes studied in the context of the brane-world [238, 496, 759], which is not discussed here. Quintessential inflation is regarded with suspicion by some authors because of the unwanted outcome of isocurvature fluctuations and dangerous relics in these models [371]. Indeed, the identification of the quintessence field with the inflaton may be asking too much of nature, although it has aesthetic appeal. Other authors regard these models with favour due to the large production of gravitational waves at the end of inflation, which gives $\Omega_{gw} \simeq 10^{-6}$, about eight orders of magnitude larger than in ordinary inflation [433, 434, 736] and is perhaps interesting for the Laser Interferometer Space Antenna (*LISA* [514]).

As far as the nature of the quintessence field is concerned, nothing is definitive in the early 21st century. Attempts have been made to relate this field to a known particle physics theory [228] and to identify quintessence with a supergravity field [153, 154, 240, 748], an axion, the string dilaton, or moduli fields in string theories [536, 227, 226, 666, 101, 473, 418, 858, 423, 477, 389, 388]. However, these identifications are completely speculative.

1.1 Coupled quintessence

In standard models of quintessence, $\Omega^{(q)}$ begins to dominate at redshifts $z \sim 1$; it cannot begin dominating earlier without compromising the growth of structures in the universe [374, 301]. Indeed, a non-negligible, subdominant amount of quintessence during the epoch of structure formation would leave an imprint in the cosmic microwave background that could lead to its detection [170]. The anthropic principle can be invoked to argue that $\Omega^{(q)}$ cannot possibly be much larger than $\Omega^{(m)}$ at the present epoch [881]. However the situation is different in less conventional models in which the quintessence field couples directly to matter [30, 31, 34, 35, 32]. In these models, density perturbations can grow even when $\Omega^{(q)}$ is dominating and the cosmic acceleration can start early, even as early as $z \sim 5$ [32]. A direct coupling between quintessence and ordinary matter is present in the Einstein frame version of scalar-tensor theories, but it must be suppressed to a large extent because it creates a long range fifth force that is constrained by present experiments. Models in which this problem is avoided by coupling quintessence to dark matter — but not to baryons — are studied in the literature [30, 29, 34, 35, 848, 33]. This preferential coupling of dark energy to dark matter looks like "conceptual fine tuning," but given our complete ignorance of the nature and physical properties of both dark energy and dark matter at the time of writing, it seems to be worth exploring.

A direct coupling of quintessence to the fields of the Standard Model arises rather naturally in particle physics, and can be of the form

$$\beta \frac{\phi}{M} \mathcal{L} , \qquad (8.29)$$

where M is a mass scale, $\mathcal{L}$ is the covariant and gauge-invariant Lagrangian density describing a certain matter field, and β describes the strength of the coupling, with different β-parameters for different fields in the Standard Model. *A priori*, one expects values of β of order unity, but there are observational constraints on these coupling strength parameters [186]. It is possible that an unbroken (exact or approximate) global symmetry suppresses these couplings. Such a symmetry (studied in Ref. [186]), allows for a coupling of the quintessence field to the Maxwell field through the term in the Lagrangian

$$\beta \frac{\phi}{M} F_{ab} \tilde{F}^{ab} , \qquad (8.30)$$

where F_{ab} is the Maxwell tensor, $\tilde{F}^{ab} = \epsilon^{abcd} F_{cd}$ is its dual, and $F_{ab} \tilde{F}^{ab}$ is the pseudoscalar constructed from them. Couplings of this kind are not mandatory in particle physics, but are present only in certain models, and they have been studied for a long time because they lead to potentially observable effects. If ϕ is a homogeneous cosmological field that varies with time, the coupling (8.30) induces a rotation of polarization in the electromagnetic radiation from distant radio sources through a dependence of the dispersion relation on the polarization [190, 191]. One has

$$\omega(k) = \sqrt{k^2 \pm \frac{\beta}{M}\dot{\phi}k}, \qquad (8.31)$$

with the upper (lower) sign corresponding to right- (left-) handed polarization in the mode of angular frequency ω and wave vector $\vec{k}$. If θ is the angle between a fiducial direction in the sky and $\vec{k}$, the resulting rotation is

$$\delta\theta = \frac{\beta}{M}\delta\phi \qquad (8.32)$$

in the Wentzel-Kramers-Brillouin limit [190, 191]. The rotation of polarization effect was well constrained by observations before the discovery of cosmic acceleration, in the context of a debate on cosmological birifringence [669, 313, 188, 586, 558], but the coupling (8.30) is not yet excluded and it could even lead to the direct detection of quintessence if this coupling is actually present [186].

1.2 Multiple field quintessence

Quintessence can be modeled by two or more scalar fields — instead of a single one — on the lines of what is done in models of hybrid inflation [582, 115], or in models of reheating after inflation [538, 539]. Models with two quintessence scalar fields minimally coupled to the Ricci curvature yet mutually coupled are studied in Refs. [117, 116, 219]. An analogy with particle dynamics developed for a single scalar turns out to be useful in dealing with the case of a large number of quintessence fields [219]. However, it is to be hoped that quintessence is simpler than this situation.

1.3 Falsifying quintessence models

Proposals to verify (or falsify) the idea and scenarios of quintessence exist, suggesting the reconstruction of the effective potential $V(\phi)$ and of the effective equation of state of quintessence from the luminosity distance-redshift relation $D_L(z)$ of type Ia supernovae [827, 214, 661, 497, 138, 320, 760, 518]. A minimally coupled scalar field and its potential can be expressed in terms of the variable $x \equiv 1 + z$ as [760]

$$\frac{V(x)}{\rho_c^{(0)}} = \frac{H^2}{H_0^2} - \frac{x}{6H_0^2}\frac{d(H^2)}{dx} - \frac{\Omega^{(m)}}{2}x^3, \qquad (8.33)$$

$$\frac{1}{\rho_c^{(0)}} \left(\frac{d\phi}{dx}\right)^2 = \frac{2}{3H_0^2 x} \frac{d(\ln H)}{dx} - \frac{\Omega^{(m)}}{H^2} x, \qquad (8.34)$$

and the equation of state parameter is

$$w_q(x) = \frac{2x \frac{d(\ln H)}{dx} - 3}{3\left[1 - \left(\frac{H_0}{H}\right)^2 \Omega^{(m)} x^3\right]}, \qquad (8.35)$$

where $\rho_c^{(0)} = 3H_0^2/(8\pi G)$ is the present value of the critical density. The Hubble parameter as a function of redshift z can be derived from the kinematic relation with the luminosity distance $D_L(z)$ in a $K=0$ FLRW universe

$$H = \left\{ \frac{d}{dz} \left[\frac{D_L(z)}{1+z} \right] \right\}^{-1}. \qquad (8.36)$$

If observations determine the luminosity distance $D_L(z)$, then eq. (8.36) allows reconstruction of the potential V and the parameter w_q [138, 760]. Another proposal consists of testing the effective equation of state of the universe by using weak gravitational lensing [112]. Finally, for a minimally coupled scalar in general relativity, the relation

$$\frac{d^2 H(z)}{dz^2} \geq 3\Omega_0^{(m)} H_0 (1+z)^2 \qquad (8.37)$$

must be satisfied [757]. If observations show a violation of this inequality, we would have evidence for non-Einsteinian gravity.

2. Quintessence with nonminimal coupling

The same motivations that lead one to inclusion of the nonminimal coupling of the scalar field to the Ricci curvature in inflationary theory induce consideration of models in which the quintessence scalar is nonminimally coupled. Quintessence modeled by a nonminimally coupled scalar field, or by a BD-like field [211, 867, 709, 29, 277, 351, 207, 209, 320, 790, 50, 51, 400, 401, 123, 87, 30, 138, 31, 117, 212, 736, 321, 56, 354, 855, 707, 737] is called *extended quintessence*; *superquintessence* or *phantom energy* if associated with values of the w_q-parameter $w_q < -1$.

A further reason to consider nonminimally coupled models of quintessence is that the range of values for the equation of state parameter w_q indicated by observations include $w_q < -1$, and there are claims that these values fit the observations better than $w_q \geq -1$. Although there is some theoretical prejudice against the possibility $w_q < -1$ (see Ref. [630] for a discussion), this range of values should not be discarded *a priori*, at least if w_q is not constant. An effective equation of state with $w_q < -1$

cannot be obtained in the context of a minimally coupled scalar field model in Einstein gravity, and to explain it one has to resort to new physics such as nonminimally coupled models [736, 693, 351, 354, 737], or models in which the kinetic energy of the quintessence scalar field appears with the wrong sign [785], or models with "stringy" matter [635, 389, 388, 92] or decaying cold dark matter [918, 919]. Even the use of negative potentials leading to negative energy density for a minimally coupled scalar field is contemplated [273, 272, 527], with little theoretical motivation.

Specific models [211, 867, 709, 29, 277, 351, 207, 209, 50, 51, 87, 30, 351, 693, 736, 212, 354, 855, 707, 708, 706, 737] using a nonminimally coupled scalar field as quintessence include particular choices of the potential $V(\phi)$. As a general feature, attractor (tracking) solutions are still present when nonminimal coupling terms are introduced into the field equations.

2.1 Models using the Ratra-Peebles potential

Detailed models depend on specific choices of the quintessence potential $V(\phi)$: we focus on scenarios using the Ratra-Peebles potential (8.10). When the Ratra-Peebles potential is used with $\xi \neq 0$, the tracking solution (8.25)-(8.27) is still valid and its stability does not depend on the value of ξ; the parameter w_q is given again by eq. (8.28)) and is independent of ξ [867, 211, 736, 737]. The stability analysis performed for $\xi = 0$ [571] can be generalized to the $\xi \neq 0$ case, with the result that the value of ξ determines only the nature of the fixed point in phase space (stable spiral or node) [867]. The value of ξ however, cannot be arbitrary — since the quintessence field is ultra-light, it mediates a long range ($\sim H_0^{-1}$) force which is constrained by Solar System experiments [211, 736, 737]. The case of weak coupling ($|\xi| << 1$) is the easiest to study in an approach which treats it as a perturbation of the $\xi = 0$ case [211]. Nonminimal coupling corresponds to the effective BD coupling

$$\omega_0 \simeq \frac{3}{4\alpha\left(\alpha+2\right)\xi^2} \quad (8.38)$$

and yields the constraint [211]

$$|\xi| < \frac{3.9 \cdot 10^{-2}}{\sqrt{\alpha\left(\alpha+2\right)}}. \quad (8.39)$$

If one uses the time varying effective coupling $G_{eff} = G\left(1 - 8\pi G\xi\phi^2\right)^{-1}$, one obtains [211]

$$\left.\frac{\dot{G}_{eff}}{G_{eff}}\right|_0 = \frac{16\pi G\xi\phi_0\dot{\phi}_0}{1 - 8\pi G\xi\phi_0^2} \simeq 2\xi\alpha H_0 \; ; \tag{8.40}$$

the current limits on the variation of the gravitational coupling then yield the constraints on ξ

$$-10^{-2} \lesssim \xi \lesssim 10^{-1} \; , \tag{8.41}$$

where the upper bound is lowered to a value of the order of 10^{-2} if more stringent limits on $\dot{G}/G$ [847] are employed. Remember that these limits on ξ are specific to the particular choice of the Ratra-Peebles potential.

The case $\alpha = 1$, $V(\phi) = M^5/\phi$, is studied in detail in Ref. [709], where the scale M is fixed so that $\Omega^{(m)} = 0.6$ today. Both the unperturbed background universe and perturbations (in the synchronous gauge) are analyzed. The quintessence scalar field is an additional source of fluctuations, which can in principle cause new and observable effects in the cosmic microwave background and in the formation of large scale structures [879]. These fluctuations are present during the radiation- and matter-dominated eras and superpose to the primordial density fluctuations generated during inflation. A minimally coupled quintessence field does not clump on subhorizon scales the way dark matter does [706, 708].

The inclusion of nonminimal coupling transforms the term $dV/d\phi$ in the Klein-Gordon equation into

$$\frac{dV_{eff}}{d\phi} = \frac{dV}{d\phi} - \xi R\phi \; ; \tag{8.42}$$

the correction induces a time variation of the gravitational potential between the last scattering surface and the present time and enhances the integrated Sachs-Wolfe or Rees-Sciama [733] effect. This seems to be more a signature of nonminimal coupling than a feature of the specific Ratra-Peebles potential. If, in addition, the quintessence field is also responsible for inflation earlier on — quintessential inflation — then there is an overproduction of gravitational waves leading to $\Omega^{(gw)} \simeq 10^{-6}$, about 10^8 times the density of gravitational waves produced during ordinary inflation [433, 434]. These extra gravitational waves contribute to the polarization of the cosmic microwave background at the last scattering, and to temperature fluctuations $\delta T/T$ in the microwave background due to the Rees-Sciama effect computed between the last scattering and today. The details of these extra contributions in comparison to ordinary inflation are studied in Refs. [49, 736].

Nonminimally coupled quintessence has yet another effect. On subhorizon scales, perturbations of the energy density of the quintessence field follow the evolution of the dark matter energy density: an effect called *gravitational dragging*. As a result, clumps of nonminimally coupled quintessence could form at these scales, perhaps with considerable effects on the formation of galaxies and clusters when the perturbations go nonlinear [708, 706].

2.2 Necessary conditions for accelerated expansion

Whenever possible it is interesting to derive properties of quintessence models that are independent of the specific potential used. An example is the derivation of a general necessary condition for accelerated expansion of the universe with a nonminimally coupled quintessence field.

In quintessence models based on a nonminimally coupled scalar field, the energy density of the latter is just beginning to dominate the dynamics of the universe (equipartition $\Omega^{(q)} = \Omega^{(m)}$ occurring at $z \sim 1$ in most models), and the effect of matter with energy density $\rho^{(m)} \propto a^{-3}$ and vanishing pressure $P^{(m)} = 0$ must be taken into account in the study of the dynamics. The total energy density of matter plus quintessence can be written in the effective coupling approach as

$$\rho = \left(1 - \kappa\xi\phi^2\right)^{-1} \left[\rho^{(m)} + \frac{(\dot\phi)^2}{2} + V(\phi) + 6\xi H \phi \dot\phi\right] = \frac{\rho^{(m)}}{1 - \kappa\xi\phi^2} + \rho^{(q)}, \tag{8.43}$$

where $V(\phi)$ is the quintessence potential. The necessary and sufficient condition for the acceleration of the universe $\rho + 3P < 0$ is written as

$$y \equiv \frac{\rho^{(m)}}{2} + (1 - 3\xi) \dot\phi^2 - V + 6\xi H \phi \dot\phi + 3\xi^2 R \phi^2 + 3\xi\phi \frac{dV}{d\phi} < 0, \tag{8.44}$$

where the Klein-Gordon equation has been used to eliminate $\ddot\phi$. One obtains a necessary condition for the accelerated expansion of the universe in this approach by rewriting the quantity y as

$$y = \frac{\rho^{(m)}}{2} + \left(1 - \kappa\xi\phi^2\right)\rho^{(q)} - 2V + \dot\phi^2 \left(\frac{1}{2} - 3\xi\right) + 3\xi^2 R \phi^2 + 3\xi\phi \frac{dV}{d\phi} < 0, \tag{8.45}$$

and by assuming that $\rho^{(m)}$ and $\rho^{(q)}$ are non-negative. In this way eq. (7.84) of the previous chapter is reobtained. The analysis is now refined by noting that, when quintessence dominates,

$$\rho^{(q)} \approx \frac{V}{1 - \kappa\xi\phi^2}. \tag{8.46}$$

By measuring matter and scalar field energy densities in units of the critical density ρ_c with $\Omega^{(m)} \equiv \rho^{(m)}/\rho_c$ and $\Omega^{(q)} \equiv \rho^{(q)}/\rho_c$), one has

$$\rho^{(m)} \simeq V \frac{\Omega^{(m)}}{\Omega^{(q)}} \tag{8.47}$$

and

$$y = \left(-1 + \frac{\Omega^{(m)}}{2\Omega^{(q)}} - \kappa\xi\phi^2 \right) V + \left(\frac{1}{2} - 3\xi \right) \dot{\phi}^2 + 3\xi^2 R\phi^2 + 3\xi\phi \frac{dV}{d\phi} < 0 . \tag{8.48}$$

For $\xi \leq 1/6$ one has

$$\left(-1 + \frac{\Omega^{(m)}}{2\Omega^{(q)}} \bigg|_0 - \kappa\xi\phi^2 \right) V + 3\xi\phi \frac{dV}{d\phi} \leq y < 0 , \tag{8.49}$$

where the ratio $\Omega^{(m)}/\Omega^{(q)}$ is approximated by its present value — a good approximation for moderate redshifts. By assuming, for simplicity, that V and ϕ are positive, the necessary condition for cosmic acceleration if $\xi \leq 1/6$ is

$$\frac{d}{d\phi} \left\{ \ln \left[\frac{V}{V_0} \left(\frac{\phi_0}{\phi} \right)^\alpha \exp\left(-\frac{\kappa}{6}\phi^2 \right) \right] \right\} < 0 , \tag{8.50}$$

where V_0 and ϕ_0 are constants and

$$\alpha = \left(1 - \frac{\Omega^{(m)}}{2\Omega^{(q)}} \bigg|_0 \right) \frac{1}{3\xi} \simeq \frac{0.26}{\xi} \tag{8.51}$$

by assuming $\Omega^{(m)} = 0.3$ and $\Omega^{(q)} = 0.7$ today. Thus, to have quintessential expansion with $0 < \xi \leq 1/6$, one needs a potential $V(\phi)$ that does not grow with ϕ faster than the function

$$C(\phi) \equiv V_0 \left(\frac{\phi}{\phi_0} \right)^\alpha \exp\left(\frac{\kappa}{6}\phi^2 \right) . \tag{8.52}$$

If instead $\xi < 0$, then $V(\phi)$ must grow faster than $C(\phi)$. In this case α is negative and, upon rescaling of ϕ, the function (8.52) reduces to the supergravity potential (8.12).

2.3 Doppler peaks with nonminimal coupling

The acoustic oscillations or *Doppler peaks* in the cosmic microwave background that are well known from minimal coupling [569] are modified by nonminimal coupling. A detailed calculation of the Doppler peaks predicted by a certain theory requires the specification of the potential

$V(\phi)$, the value of the coupling constant ξ, details about the end of inflation and reheating, *et cetera* Again, we focus on models using the Ratra-Peebles potential (8.10) for $\alpha = 1$ and 2, which are investigated in detail in Refs. [709, 50, 51]. When nonminimal coupling is added to the quintessence scenario with this potential, the acoustic peaks and the spectrum turnover are displaced with respect to the $\xi = 0$ case. Specifically, the effects of nonminimal coupling in this model are:

- An enhancement of the large scale ($\theta = 180°/l$), low multipoles l region of the Doppler peaks, due to an enhancement of the integrated Sachs-Wolfe/Rees-Sciama [733] effect. This occurs because of the *R-boost* — the fact that the coupling term $\xi R \phi^2/2$ term in the action affects the dynamics of ϕ together with $V(\phi)$.

- The oscillating region of the Doppler peaks is attenuated due to the fact that the Hubble length H^{-1} is smaller in the past in this model than in a $\xi = 0$ model. As a consequence a given wavelength crosses back into the horizon later, there is less matter at decoupling, and the radiation density is higher. This situation makes the amplitude of the acoustic oscillations slightly smaller.

- The locations of the Doppler peaks are shifted toward higher multipoles because, at decoupling, the Hubble length H^{-1} subtends a smaller angle $\theta \simeq 180°/l$ in the sky, corresponding to larger multipole l.

These features are derived by direct integration of the equations for the perturbations with a modified *CMBFAST* code [229]. Similar qualitative effects appear in models based on induced gravity and models with negative kinetic energy terms [785], and are interpreted as the signature of a broad class of scalar-tensor gravity theories in the cosmic microwave background [709, 50].

3. Superquintessence

The available constraints on the quintessence equation of state are in agreement with values of the parameter $w_q < -1$ even for very negative w_q: the limits reported are (see Refs. [464, 633, 252, 784], respectively)

$$-2.68 < w_q < -0.78, \tag{8.53}$$

$$-1.62 < w_q < -0.74, \tag{8.54}$$

$$-2.6 < w_q < -0.25, \tag{8.55}$$

or

$$-1.30 < w_q < -0.65. \tag{8.56}$$

It is often argued that there is evidence that the pressure to density ratio is strictly less than -1 [168, 630, 192], and the confidence level curves in the (Ω, w) plane from observational limits allow this interpretation (e.g., [251]-[253]). The possibility that $w_q < -1$ is investigated by several authors [168, 215, 918, 919, 785, 633, 391, 684, 564, 563] using models based on a scalar field with the "wrong" sign of the kinetic energy term of the scalar, or other modifications of the Lagrangian; it is also argued that stringy matter requires $w_q < -1$ [389]. Dark energy with the property $w_q < -1$ is often called *phantom energy*.

The search for a model of phantom energy outside of Einstein gravity with minimally coupled fields is justified by the following considerations. The inequality $P < -\rho$ is equivalent to $\dot{H} > 0$, and the Hubble parameter satisfies the equations

$$H^2 = \frac{\kappa}{3}\rho, \qquad (8.57)$$

$$\frac{\ddot{a}}{a} = \dot{H} + H^2 = -\frac{\kappa}{6}\left(\rho + 3P\right). \qquad (8.58)$$

Assuming for simplicity that a single quintessence fluid dominates the cosmic expansion, eqs. (8.57) and (8.58) imply that

$$\dot{H} = -\frac{\kappa}{2}\left(\rho^{(q)} + P^{(q)}\right) \qquad (8.59)$$

and therefore

$$P^{(q)} < -\rho^{(q)} \quad \Leftrightarrow \quad w_q < -1 \quad \Leftrightarrow \quad \dot{H} > 0. \qquad (8.60)$$

If the matter component has a non-negligible energy density, the pressure of quintessence must be even more negative.

A regime with $\dot{H} > 0$ was originally investigated in the context of inflationary models of the early universe and was called *superinflation* [594], a name later adopted by string cosmology. A more accurate name in the present context is *superacceleration*, which stresses the possibility that $\dot{H} > 0$ in a quintessence-dominated universe today, well after the end of inflation [736, 351, 354].

The simplest models of quintessence and inflation based on a single scalar field minimally coupled to gravity cannot contain a superacceleration regime because, in such models, the scalar field $\phi(t)$ behaves as a perfect fluid with energy density and pressure

$$\rho = \frac{\dot{\phi}^2}{2} + V(\phi), \qquad (8.61)$$

The present universe

$$P = \frac{\dot{\phi}^2}{2} - V(\phi), \tag{8.62}$$

respectively, where $V(\phi)$ is the scalar field potential. Then, eq. (8.59) becomes

$$\dot{H} = -\frac{\kappa}{2}\dot{\phi}^2 \tag{8.63}$$

and $\dot{H} \leq 0$ in any minimally coupled model. The case $\dot{H} = 0$ corresponds to a de Sitter solution with scale factor $a(t) = a_0\, e^{Ht}$. For any potential $V(\phi)$, the effective equation of state of a FLRW universe dominated by a minimally coupled scalar field is

$$\frac{P}{\rho} = \frac{\dot{\phi}^2 - 2V}{\dot{\phi}^2 + 2V} \equiv w(x), \tag{8.64}$$

where $x \equiv \dot{\phi}^2/2V$ is the ratio between the kinetic and the potential energy densities of the scalar. Assuming that $V \geq 0$, which guarantees positivity of the energy density of ϕ,

$$w(x) = \frac{x-1}{x+1} \tag{8.65}$$

monotonically increases from its minimum $w_{min} = -1$ attained at $x = 0$ to the horizontal asymptote $+1$ as $x \to +\infty$, corresponding to $V = 0$. At most, the effective equation of state parameter of a minimally coupled scalar field can span the range of equations of state

$$-1 \leq w \leq 1. \tag{8.66}$$

Observational data producing $w < -1$ are not explained by this "canonical" scalar field model for quintessence, unless one is willing to accept a negative potential $V(\phi) < -\dot{\phi}^2/2$ which makes the energy density (8.61) of the scalar negative, a situation that is *a priori* disturbing and is forbidden by accelerated dynamics (eq. (8.58)). Thus we do not further consider this possibility. The superacceleration regimes known in the pre-1998 literature consist of pole-like inflation with scale factor

$$a(t) = \frac{a_0}{t - t_0}, \tag{8.67}$$

a special form of superacceleration studied in early inflationary theories [594, 721], in pre-big bang cosmology [579], and in BD theory [247].

A phenomenology with negative pressures or even $P < -\rho$ is also present in higher derivative theories of gravity [720], in induced gravity [721], or due to semiclassical particle production resulting in nonzero bulk viscosity [916, 493, 74, 920], dissipative fluids in the presence of a quintessence field [217, 218] or, in general, with quantum fields that violate the weak energy condition [6, 684]. Rather unconventional, supergravity-inspired models with kinetic energy density of the scalar field $-\dot{\phi}^2/2$ instead of $\dot{\phi}^2/2$, or nonlinear kinetic terms are investigated [168, 215, 918, 919, 785, 633] (see also [693]). When the scalar field potential V is absent, this form of matter is called *kinetically driven quintessence* or, in short, *k-essence*. It is also possible that a time-dependent and positive parameter w_q, when interpreted as a constant, give an effective $w_q^{(eff)} < -1$, or vice-versa that the assumptions w_q =const. and $w_q \geq -1$ in a likelihood analysis lead to gross errors disguising an equation of state $w_q < -1$ [620].

Here we consider what is perhaps the simplest model of superquintessence, a scalar field ϕ nonminimally coupled to gravity, assuming that it has already begun to dominate the dynamics of the universe. The nonminimally coupled scalar has pressure given by

$$P = \frac{\dot{\phi}^2}{2} - V(\phi) - \xi \left[4H\phi\dot{\phi} + 2\dot{\phi}^2 + 2\phi\ddot{\phi} + \left(2\dot{H} + 3H^2\right)\phi^2 \right] \quad (8.68)$$

and obeys the equations

$$6\left[1 - \xi(1 - 6\xi)\kappa\phi^2\right]\left(\dot{H} + 2H^2\right) - \kappa(6\xi - 1)\dot{\phi}^2 - 4\kappa V + 6\kappa\xi\phi V' = 0, \quad (8.69)$$

$$\frac{\kappa}{2}\dot{\phi}^2 + 6\xi\kappa H\phi\dot{\phi} - 3H^2\left(1 - \kappa\xi\phi^2\right) + \kappa V = 0, \quad (8.70)$$

$$\ddot{\phi} + 3H\dot{\phi} + \xi R\phi + V' = 0. \quad (8.71)$$

In this approach the scalar field energy density and pressure satisfy the usual conservation equation

$$\dot{\rho} + 3H(\rho + P) = 0, \quad (8.72)$$

which, using the effective equation of state $P = w\rho$, yields

$$\frac{\dot{\rho}}{\rho} = -3(w+1)H. \quad (8.73)$$

Hence, a peculiar feature of a superacceleration regime with $w < -1$ is that the energy density of the scalar field — or of any fluid satisfying the conservation equation (8.72) — grows with time instead of redshifting as in ordinary acceleration, or staying constant as for a true cosmological constant. After equipartition $\Omega^{(q)} = \Omega^{(m)}$ the energy density of such a field becomes dominant very quickly.

The fact that a nonminimally coupled scalar field can generate a superacceleration regime has long been known in the context of pole-like inflation [594] and in studies of scalar field dynamics with a dynamical system approach [452, 453]. Exact and numerical solutions with $\dot{H} > 0$ are known and inspire the speculation that a superacceleration regime could play a role in the genesis of the universe from Minkowski space. Prior to the conception of inflationary and quintessence scenarios, this idea was implemented in a phenomenological approach by imposing the required superaccelerated behaviour of the scale factor and by deriving the time-dependent equation of state deemed necessary for the cosmic fluid [144, 902]. Similar solutions originating from, and ending in, Minkowski space also appear in string theories [324] and are possible due to the nonminimal coupling of the dilaton to gravity. Superacceleration in scalar field dynamics manifests itself in the dynamics of a simple model containing a conformally coupled scalar in the potential commonly encountered in scalar field cosmology

$$V(\phi) = \frac{m^2}{2} \phi^2 + \lambda \phi^4 , \qquad (8.74)$$

keeping in mind that the value $\xi = 1/6$ of the coupling constant is an infrared fixed point of the renormalization group and is required by the Einstein equivalence principle (see Section 7.1). The dynamics of this system turn out to be much richer than in the minimally coupled case [38, 71, 386, 451, 453, 452]. One can have spontaneous bouncing, i.e., transition from a contracting ($H < 0$) regime to an expanding ($H > 0$) regime which requires superacceleration ($\dot{H} > 0$). Although the attention of the cosmological community to superacceleration is motivated by the observed behaviour of the universe, the superacceleration phenomenon is well known in nonminimally coupled scalar field theories. Furthermore, superacceleration is not unique to nonminimal coupling: BD theory is also known to have bouncing and superaccelerating solutions, which are present, in particular, in the $\omega = -1$ case constituting the low-energy limit of bosonic string theory [454, 545, 235, 324, 45].

3.1 An exact superaccelerating solution

In the following a time-dependent parameter w_q is allowed, a general potential $V(\phi)$ is considered, and we present an exact superaccelerating solution [451, 354] as a toy model.

The assumption of a constant equation of state is very restrictive. For both minimal and nonminimal coupling, by imposing $w =$ constant in a spatially flat FLRW universe one can only obtain power-law and de Sitter solutions, or a third exact solution[2] [352].

A superaccelerating exact solution of the field equations (8.69)-(8.71) is obtained for $\xi > 0$ by setting ϕ equal to one of the critical scalar field values $\pm\phi_1 = \pm 1/\sqrt{\kappa\xi}$. The trace of the field equations

$$R = 6\left(\dot{H} + 2H^2\right) = \kappa\left(\rho - 3P\right) \tag{8.75}$$

then yields

$$\dot{H} + 2H^2 = C, \tag{8.76}$$

where C is a constant. If $C > 0$ one has

$$\frac{\dot{H}}{1 - 2H^2/C} = C, \tag{8.77}$$

an equation which is immediately reduced to a quadrature,

$$\int \frac{dx}{1 - x^2} = \sqrt{2C}\,(t - t_0), \tag{8.78}$$

where $x \equiv H\sqrt{2/C}$. One has

$$\operatorname{arctanh} x = \sqrt{2C}\,(t - t_0) \tag{8.79}$$

if $x^2 > 1$ and

$$\ln\left[\sqrt{\frac{1+x}{1-x}}\right] = \sqrt{2C}\,(t - t_0) \tag{8.80}$$

if $x^2 < 1$, and hence

$$H = \sqrt{\frac{C}{2}}\tanh\left[\sqrt{2C}\,(t - t_0)\right] \tag{8.81}$$

in both cases $H > \sqrt{C/2}$ and $H < \sqrt{C/2}$. The remaining cases $H = \pm\sqrt{C/2}$ (or $x^2 = 1$) correspond to trivial de Sitter solutions, while $C < 0$ yields no superacceleration. Eq. (8.81) is in contradiction with the limit $H > \sqrt{C/2}$ and hence (8.81) is only a solution for $H < \sqrt{C/2}$. This case corresponds to the scale factor

$$a(t) = a_0 \cosh^{1/2}\left[\sqrt{2C}\,(t - t_0)\right], \tag{8.82}$$

---

[2]Solutions for spatially curved FLRW universes and arbitrary values of $w =$const. are summarized in Ref. [348].

The present universe 219

which represents an asymptotic contracting de Sitter space as $t \longrightarrow -\infty$, reaching a minimum size at $t = t_0$, and then bouncing, expanding, and superaccelerating with $\dot{H} > 0$ and $w < -1$. At late times $t \longrightarrow +\infty$ this solution approaches an expanding de Sitter space. The effective equation of state parameter

$$w(t) = -\frac{A + \xi\left(2\dot{H} + 3H^2\right)}{3H^2 + A} = C\left\{2 - \frac{1}{2}\tanh^2\left[\sqrt{2C}\,(t - t_0)\right]\right\}, \tag{8.83}$$

where $A = \kappa V(\pm\phi_1)$, is time-dependent. While this exact bouncing solution is clearly fine-tuned in the scalar field value, it solves the field equations for any potential and explicitly illustrates the superacceleration phenomenon, the latter being a generic feature of the dynamics of a nonminimally coupled scalar field [736, 351, 354].

3.2 Big Smash singularities

A nonminimally coupled scalar field can act as quintessence and exhibits peculiar features of interest in connection with the present acceleration of the universe which are not obtainable in models with a minimally coupled scalar. These features include the spontaneous exit of the universe from a superaccelerated regime to enter an ordinary accelerated one, followed again by spontaneous exit and entry into a decelerated epoch. These features are absent in BD and string-based models that keep accelerating forever [473], or in inflationary models where inflation is terminated by an *ad hoc* modification of the equations or of the shape of $V(\phi)$, and provide for the possibility of the universe ending the superacceleration era and returning to a decelerated epoch.

There is however a trade-off for superacceleration. The condition $w < -1$ violates the weak and dominant energy conditions and, as such, it is often regarded with diffidence as it opens the door to violations of causality[3]. However, this is not the case here, at least if the w-parameter is time dependent [629, 630, 274]. In fact the speed of sound is not larger than the speed of light because $P = w\rho$ describes an *effective* equation of state, not a true one. Acoustic waves are space-dependent perturbations of the exact model and they do not obey the equation of state satisfied by the unperturbed P and ρ, which instead depend only on time.

Another potential result with superacceleration is that if the superacceleration epoch continues, the universe runs the risk of ending its history

---
[3]The violation of the dominant energy condition is not unique to nonminimally coupled quintessence, but also occurs in k-essence models [46] and in scalar-tensor theories.

in a finite time, i.e., the scale factor $a(t) \to +\infty$ as $t \to t_*$, with t_* finite. The energy density and pressure also diverge as $t \to t_*$ (remember that the energy density grows in a superacceleration regime). This possibility, which poses problems of interpretation, is called a *big smash* or *big rip* [168, 828, 629, 630, 391, 171, 355] and from the mathematical point of view, is commonly encountered in the study of ordinary differential equations in which the growth of the solutions is too fast, and the solution cannot be maximally extended to an infinite domain (see Appendix A for an example). While for the Friedmann equations a big smash is unavoidable when the equation of state is constant [828, 391, 630] and the scale factor is given by

$$a(t) = \frac{a_0}{|t-t_0|^{\frac{2}{3|1+w|}}} \qquad (8.84)$$

(this being perhaps one more reason to reject the unphysical assumption w_q =constant), it does not need to occur in general. Indeed there are examples of universes with nonconstant $w < -1$ which extend to the infinite future. One of them is the bouncing solution (8.82) and other examples are known[4] [453, 452, 629, 804].

Finally, we remark that the $\dot{H} > 0$ regime achieved with nonminimal coupling is a true superacceleration regime, contrary to the case of pre-big bang cosmology and of BD theory. As explained in Section 5.4, in these theories while the universe expands, the Planck length also grows and the ratio of any physical length to the Planck length — which is a true measure of the inflation of a universe starting from Planck size — actually decreases [247]. In string-inspired pre-big bang cosmology this phenomenon is due to the fact that the gravitational coupling is Ge^{Φ}, where $\Phi(t)$ is the string dilaton [247], while in BD theory the effective coupling is ϕ^{-1}.

For a nonminimally coupled scalar field, using the approach selected here and the corresponding T_{ab} for the scalar field discussed in Section 7.2, the gravitational coupling and the Planck length are true constants and superacceleration is genuine. A different approach using the effective gravitational coupling $G_{eff} = G\left(1 - 8\pi G \xi \phi^2\right)^{-1}$ suffers the same problem of pre-big bang cosmology for $\xi > 0$ and cannot produce the exact solution (8.82) since G_{eff} diverges at the critical scalar field values $\pm\phi_1$ corresponding to the solution (8.82).

If the inequality $w_q < -1$ is supported by the future reconstruction of the equation of state of quintessence, the more conventional quintess-

---

[4]The example of Ref. [629] is a slight generalization of the exact bouncing solution (8.82) in the context of string theory.

ence models based on minimally coupled scalars have to be abandoned in favour of alternative ones. Among these, the nonminimally coupled theory described here appears as the simplest and most promising, although different models also produce $w_q < -1$ and superacceleration.

When the parameters of scalar-tensor models are subjected to the available constraints, one finds that the conventional scalar field potentials used in the literature — some of which are motivated by high energy physics — produce only a very small amount of superacceleration, as shown by numerical solution of the field equations [855]. The deviation of the w_q parameter from -1 is so small as to be probably undetectable by present technology and, should future measurements definitely provide a value $w_q < -1$, the current models should either be abandoned or significantly modified. However, even a very small amount of superquintessence leads to a fast (on a cosmological timescale) growth of $\rho^{(q)}$ and may lead to a big smash, thus enlarging the possibilities for the future evolution of the universe from expansion in an infinite time to expansion into a singularity in a finite time. In addition, the quintessence potential is unknown, and potentials different from the ones used in the literature may provide more superacceleration at the present epoch and still pass the tests imposed by the available observational constraints. Finally, even though the unperturbed scenarios are indistinguishable from minimally coupled models at present, density perturbations developing in the presence of dark energy may well have the potential to discriminate between general relativity and scalar-tensor gravity [855].

4. Quintessence in scalar-tensor gravity

Scalar-tensor theories are discussed in the context of attempts to find a mechanism to fix the cosmological constant problem [298, 299, 382, 834, 402], because they may contain a negative decaying effective energy which could in principle cancel Λ. However, this mechanism does not work because it induces unrealistic variations in the gravitational coupling. Although these attempts to solve the cosmological constant problem are unsuccessful [899, 900], they show how scalar-tensor theories lend themselves to the role of natural candidates for quintessence models in the absence of a true cosmological constant — which is to be removed by a different, as yet unknown, mechanism.

The BD-like scalar playing the role of quintessence in scenarios based on scalar-tensor gravity is called *extended quintessence*, analogous to the extended inflationary scenario based on BD theory. We consider the scalar-tensor action (1.84); in the presence of a matter fluid with density $\rho^{(m)}$ and pressure $P^{(m)}$, the field equations in an Einstein-de

Sitter universe are given by eqs. (1.86)-(1.88). As for minimally coupled quintessence, tracking solutions are sought with energy density that is negligible and tracks the radiation or matter energy density during the radiation and matter eras, and only later starts to grow with respect to $\rho^{(m)}$, eventually coming to dominate the dynamics. These tracking solutions must be attractors with a large attraction basin to avoid fine-tuning problems. When considering the effective equation of state parameter w_q, one must pay attention to the fact that different possible definitions of effective energy-momentum tensors for the BD-like scalar exist in scalar-tensor theories, and the effective energy density, pressure, and equation of state are unavoidably linked to one of these definitions, as explained in detail in the previous chapters (see also Refs. [110] for non-minimal coupling and [855] for BD theory). Of course, an unambiguous definition of superacceleration is still given by $\dot{H} > 0$.

Let us begin by considering simple BD theory. One quickly realizes that a massless, non self-interacting BD scalar cannot be a form of quintessence. In fact, consider the Nariai solution (4.28)-(4.31) in the presence of a matter fluid with $\gamma = 1$, which coincides with the Brans-Dicke dust solution (4.22)-(4.25). This power-law solution $a(t) \propto t^q$ is accelerating ($q > 1$) only if $-2 < \omega < -4/3$, a range of values of the BD parameter that is in violent contrast with the observational limit $|\omega| > 3300$ [910]. Similarly, a pole-like solution is obtained with a non self-interacting BD scalar only in the unrealistic range $-4/3 < \omega < -1$ of the BD parameter. Hence, a self-interaction potential $V(\phi)$ must be included in any scenario of BD quintessence, the alternative being to resort to a scalar-tensor theory with variable $\omega(\phi)$. Tracking solutions do exist when a suitable potential $V(\phi)$ is included in the action. However, scalar-tensor quintessence scenarios must satisfy the experimental upper bounds on the variation of the gravitational coupling and the lower bounds on the present value of ω.

For a massive scalar, $V(\phi) = m^2\phi^2/2$ and for $\omega < 0$, accelerating power-law solutions exist [123] with

$$a(t) = a_0 \left(\frac{t}{t_0}\right)^{4/3}, \qquad (8.85)$$

$$\phi(t) = \phi_0 \left(\frac{t_0}{t}\right)^2, \qquad (8.86)$$

and with deceleration parameter $q = -1/4$. The variation of the gravitational coupling in this model is

$$\frac{\dot{G}}{G} = \frac{3}{2} H_0, \qquad (8.87)$$

which is compatible with the present limits, but the age of the universe turns out to be

$$t_0 = \sqrt{\frac{4\,|\omega + 2|}{3\,\Omega^{(m)}}}\,H_0^{-1} \geq 79\,H_0^{-1} \qquad (8.88)$$

if the limit $|\omega| > 3300$ is used. This has the effect of making the universe much older than is currently accepted. Nevertheless, this model shows how close a minimalist BD scenario can get to producing acceptable quintessence solutions. A perturbation analysis [123] shows that density perturbations can grow in the matter era if $\omega < 0$, consistent with the assumptions of the model on the sign of ω.

As usual in relativistic theories of gravity, instead of prescribing an effective equation of state or a potential one can assume a particular solution $(a(t), \phi(t))$ of the field equations and run them backward to determine a form of matter that generates the specified solution. This is the approach taken, for example[5], in Refs. [790, 789, 792, 286]. A dissipative fluid in the context of relativistic thermodynamics [513] can generate cosmic acceleration with tracking attractors in general relativity [217, 218] and is considered in BD theory as well [792].

Exact solutions for BD quintessence with a quartic potential are found in Ref. [614], but only for values of the BD parameter ω of order unity. In Ref. [56] a self-interacting scalar field ψ distinct from the BD scalar ϕ and minimally coupled to gravity is considered: power-law solutions for $a(t)$ and $\phi(t)$ are assumed, leading to two classes of solutions:

- Accelerating spatially flat universes for negative values of ω, with a scalar $\psi \propto t^{-\alpha}$ ($\alpha > 0$). Here the potential $V(\psi)$ for ψ is forced to be a quartic one.

- Non-spatially flat coasting universes — both positively and negatively curved — with $a(t) = a_0 t$ and $\phi(t) = \phi_0 t^{-2}$. This is inconsistent with the observed acceleration.

Occam's razor dictates that a single quintessence field be used in the context of general relativity instead of resorting to alternative gravity plus an extra field. However, if gravity is really described by a scalar-tensor theory, quintessence could be the gravitational scalar field — in which case quintessence is severely constrained by the time variability of the effective gravitational coupling — or a different kind of model is mandatory.

---

[5]Beware of a sign error in the wave equation for the scalar field in Refs. [790, 789, 792].

Power-law attractor solutions of BD theory are found for the power-law potential

$$V(\phi) = V_0 \, \phi^{2n} , \qquad (8.89)$$

for which a detailed phase space analysis is available [483]. This example shows that the necessary condition (8.9) for the existence of tracking solutions in minimally coupled gravity does not hold in scalar-tensor theories.

A special class of exact solutions of BD theory which are actually superaccelerating — although only very slightly at the present epoch — is obtained by changing from comoving time t and scalar field ϕ to new variables τ and φ defined by $dt = e^{\varphi/2} d\tau$, $\phi = G^{-1} e^{\varphi}$ and by assuming the special relation $a = \phi^{\alpha}$ between scale factor and BD scalar [45]. This class of exact solutions corresponds to large values of the BD parameter ω for suitable values of the parameter α.

In general, the problem of small values of ω seems to plague most attempts to describe quintessence in BD theory. It is more promising to search for models in scalar-tensor gravity, in which ω is allowed to vary with time. An attempt to model the present universe in a scalar-tensor theory with variable BD parameter given by

$$2\omega(\phi) + 3 = (G\phi - 1)^2 \qquad (8.90)$$

is made in Ref. [57]. Again, this form leads to values of ω of order unity at late times, in conflict with Solar System limits.

Formal solutions of the theory (1.84) are found [28, 36] in the strong coupling limit and assuming the relation between the coupling function $f(\phi)$ and the potential

$$V(\phi) = V_0 \, [f(\phi)]^M . \qquad (8.91)$$

However, these solutions are ruled out by the present constraints on the variability of the gravitational coupling, and they only allow for an energy density $\Omega^{(q)} \simeq 0.04$ in the form of quintessence.

A viable scenario in scalar-tensor gravity is obtained when the constraints on the BD parameter and the variability of the gravitational coupling are satisfied by the attractor mechanism of the scalar-tensor theory toward general relativity discussed in the previous chapters, while attraction toward the tracker solution makes the scalar field operate as quintessence [87]. In this scenario, by defining

$$\alpha^2(\phi) = \frac{1}{2\omega(\phi) + 3} , \qquad (8.92)$$

and by choosing the coupling function as

$$\alpha(\phi) = -B \, e^{-\beta \phi} , \qquad (8.93)$$

The present universe 225

and the Ratra-Peebles potential $V(\phi) = V_0 \phi^{-m}$ (with B, V_0 and m constants) for the scalar field, one obtains the approximate solution during the radiation- and matter-dominated eras

$$\phi(a) = \frac{1}{\beta} \ln \left[\beta B \ln \left(\frac{2}{3} + \frac{a}{a_{eq}} \right) + C \right], \qquad (8.94)$$

where a_{eq} is the value of the scale factor at the equivalence epoch and C is another constant. This solution is different from the usual power-law tracker of most quintessence models, is an attractor in phase space, and behaves as a subdominant contribution to the energy density that starts increasing during the matter era and eventually comes to dominate in the present era [87]. The Doppler peaks are shifted toward higher multipoles and their height is changed with respect to minimally coupled models, providing a signature to look for in future microwave background experiments.

Consider a scalar-tensor theory described by the action

$$S = \int d^4x \sqrt{-g} \left[\frac{f(\phi)}{2} R - \frac{1}{2} \nabla^c \phi \nabla_c \phi - V(\phi) \right] + S^{(m)} ; \qquad (8.95)$$

observations provide the means to reconstruct the coupling function $f(\phi)$ and the scalar field potential $V(\phi)$. While in general relativity only the scalar field potential needs to be reconstructed, in scalar-tensor gravity there are two functions to determine, hence two independent types of observations are needed to reconstruct the two independent functions. The reconstruction is achieved by determining observationally the luminosity distance $D_L(z)$ from the study of distant type Ia supernovae and the matter perturbations $\delta_m(z)$ from studies of clustering and of weak gravitational lensing [138, 320, 321].

The steps of the reconstruction are as follows [138, 320, 321]. One begins by determining the Hubble parameter $H(z)$ as a function of redshift by using measurements of $D_L(z)$ and the kinematic relation (8.36), as in general relativity. From the knowledge of the present value of the matter density $\Omega_0^{(m)}$ the present value of the effective equation of state parameter is also determined

$$w_q = \frac{\frac{2}{3} \left[\frac{d}{dz} (\ln H) \right]_0 - 1}{1 - \Omega_0^{(m)}}. \qquad (8.96)$$

The study of matter density perturbations δ_m yields [138, 320, 321].

$$H^2 \frac{d^2(\delta_m)}{dz^2} + \left[\frac{1}{2} \frac{d(H^2)}{dz} - \frac{H^2}{1+z} \right] \frac{d(\delta_m)}{dz}$$

$$-\frac{3}{2}(1+z)H_0^2 \frac{G_{eff}(z)}{G}\Omega_0^{(m)}\delta_m = 0.\qquad(8.97)$$

where G_{eff} is the effective gravitational coupling of the scalar-tensor theory given by

$$G_{eff} = \frac{1}{8\pi f}\frac{2f + 4(df/d\phi)^2}{2f + 3(df/d\phi)^2}.\qquad(8.98)$$

Once $H(z)$ is determined, one can then determine the matter density perturbations $\delta_m(z)$ and reconstruct $G_{eff}(z)$. Since the effective BD parameter

$$\omega = \frac{f(\phi)}{(df/d\phi)^2}\qquad(8.99)$$

is constrained by Solar System experiments to assume a large value ω_0 at the present time, an expansion for large ω_0 is meaningful and leads to

$$f(z) = \frac{1}{8\pi G_{eff}(z)}.\qquad(8.100)$$

The potential $V(z)$ as a function of redshift is determined by using the equation for the BD-like scalar rewritten in terms of z instead of comoving time,

$$\frac{d^2 f}{dz^2} + \left[\frac{d(\ln H)}{dz} - \frac{4}{1+z}\right]\frac{df}{dz} + \left[\frac{6}{(1+z)^2} - \frac{2}{1+z}\frac{d(\ln H)}{dz}\right]f$$
$$-\frac{2V(z)}{(1+z)^2 H^2} - 3(1+z)\left(\frac{H_0}{H}\right)^2 f_0\Omega_0^{(m)} = 0.\qquad(8.101)$$

Then, one determines the scalar field $\phi(z)$ as a function of redshift using the equation [138, 320, 321].

$$\left(\frac{d\phi}{dz}\right)^2 + \frac{d^2 f}{dz^2} + \left[\frac{d(\ln H)}{dz} + \frac{2}{1+z}\right]\frac{df}{dz} - \frac{2}{1+z}\frac{d(\ln H)}{dz}f$$
$$+3(1+z)\left(\frac{H}{H_0}\right)^2 f_0\Omega_0^{(m)} = 0.\qquad(8.102)$$

Finally, with $f(z), V(z)$ and $\phi(z)$ known in parametric form, one eliminates the parameter z and finds the functions $f(\phi)$ and $V(\phi)$ characterizing the scalar-tensor theory.

5. Conclusion

This chapter is a snapshot view of a very active field of research that is rapidly changing. The early 21st century is an exciting period for cosmology, with interest spurred by the discovery of temperature anisotropies in the cosmic microwave background by *COBE* in 1992, and by the stunning 1998 discovery that the present expansion of the universe is accelerated. Such experimental findings challenge theorists to conceptualize new scenarios and provide the stimuli for new research. An exciting possibility is that the universe may not be just accelerating, but even superaccelerating: measurements of the w-parameter will confirm or reject this idea and could discriminate between simple models based on Einstein gravity and less conventional models. The reconstruction of the quintessence potential, even if model-dependent, is another worthy goal.

We should not forget the many on-going experiments that are gradually improving our knowledge of the cosmological parameters and our understanding of the universe: *Boomerang*, *MAXIMA*, *Archeops*, *WMAP*, and the future *Planck* and *SNAP* missions, among others. Studies of the Doppler peaks and of the polarization pattern in the microwave background, studies of gravitational lensing, and large scale structure surveys continue to contribute to the overall picture. While the work of astronomers is critical, there is feverish activity among theoreticians to understand the fundamental features of our universe. One thing remains certain: motivated by an essential need to understand our universe, current speculations in cosmology will turn into more definite models, or will be rejected altogether and new paradigma will emerge.

APPENDIX 8.A: Explosive solutions

In general, a first order ordinary differential equation of the form

$$\frac{dy(t)}{dt} = f(y) , \qquad (8.A.1)$$

where the function f is defined over a real interval I, admits maximally extendable solutions if the function $f(y)$ satisfies the Lipschitz condition, i.e., there exists a constant M such that

$$|f(y') - f(y'')| \leq M |y' - y''| \qquad (8.A.2)$$

for any y', y'' in I. For an example of an ordinary differential equation which does not admit a maximal extension of its solutions consider

$$\frac{dy(t)}{dt} = y^2 ; \qquad (8.A.3)$$

the solution is given by

$$y(t) = \frac{1}{t_0 - t} , \qquad (8.A.4)$$

where t_0 is an integration constant corresponding to the initial condition $y(0) = 1/t_0$. For $t_0 > 0$, consider the branch $t < t_0$: the solution explodes as $t \longrightarrow t_0$ from below and cannot be extended beyond this barrier. This qualitative behaviour of the solution is due to the fact that while y grows, its derivative grows even faster (as y^2)

and hence the solution quickly explodes. A similar situation may occur for a FLRW universe in a superacceleration regime. The coupled equations for H and ϕ in the case of a scalar field nonminimally coupled to R can be written as [451]

$$\ddot{\phi} = -6\xi H \phi \pm \frac{1}{2\kappa} \sqrt{\mathcal{F}(H,\phi)}, \qquad (8.\text{A}.5)$$

$$\dot{H} = \left[3(2\xi - 1) H^2 + 3\xi (6\xi - 1)(4\xi - 1) \kappa H^2 \phi^2 \mp \xi (6\xi - 1) \sqrt{\mathcal{F}} H \phi \right.$$
$$\left. + (1 - 2\xi) \kappa V - \kappa \xi \phi \frac{dV}{d\phi} \right] \frac{1}{1 + \kappa \xi (6\xi - 1) \phi^2}, \qquad (8.\text{A}.6)$$

where

$$\mathcal{F}(H,\phi) = 8\kappa^2 \left[\frac{3H^2}{\kappa} - V(\phi) + 3\xi (6\xi - 1) H^2 \phi^2 \right]. \qquad (8.\text{A}.7)$$

The right hand sides contain terms quadratic in H and ϕ which may assume positive sign and have the potential to make the solution explode in a big smash in a finite time, in analogy to what happens for the solution (8.A.4) of eq. (8.A.3). However, in this case the unknown functions H and ϕ are coupled and it is not easy to isolate situations in which the explosive behaviour occurs.

References

[1] Abbott, L.F. (1981), *Nucl. Phys. B* **185**, 233.

[2] Abbott, L.F. and Deser, S. (1982), *Nucl. Phys. B* **195**, 76.

[3] Abbott, L.F. and Wise, M.B. (1984), *Nucl. Phys. B* **244**, 541.

[4] Abbott, L.F., Barr, S.M. and Ellis, S.D. (1984), *Phys. Rev. D* **30**, 720.

[5] Abbott, L.F., Barr, S.M. and Ellis, S.D. (1985), *Phys. Rev. D* **31**, 673.

[6] Abramo, L.R. and Woodard, R.P. (2002), *Phys. Rev. D* **65**, 063516.

[7] Abramo, L.R., Brenig, L., Gunzig, E. and Saa, A. (2003), *Int. J. Theor. Phys.* **42**, 1145.

[8] Abramo, L.R., Brenig, L., Gunzig, E. and Saa, A. (2003), *Phys. Rev. D* **67**, 027301.

[9] Abreu, J.P., Crawford, P. and Mimoso, J.P. (1994), *Class. Quant. Grav.* **11**, 1919.

[10] Accetta, F.S. and Trester, J.J. (1989), *Phys. Rev. D* **39**, 2854.

[11] Accetta, F.S., Krauss, L.M. and Romanelli, P. (1990), *Phys. Lett. B* **248**, 146.

[12] Accetta, F.S., Zoller, D.J. and Turner, M.S. (1985), *Phys. Rev. D* **31**, 3064.

[13] Accioly, A.J., Vaidya, A.N. and Som, M.M. (1983), *Phys. Lett. A* **138**, 95.

[14] Accioly, A.J., Vaidya, A.N. and Som, M.M. (1983), *Phys. Rev. D* **27**, 2282.

[15] Adler, S.L. (1982), *Rev. Mod. Phys.* **53**, 729.

[16] Affleck, I., Dine, M. and Seiberg, N. (1983), *Phys. Rev. Lett.* **51**, 1026.

[17] Affleck, I., Dine, M. and Seiberg, N. (1984), *Nucl. Phys. B* **241**, 493.

[18] Albrecht, A. and Skordis, C. (2000), *Phys. Rev. Lett.* **84**, 2076.

[19] Alcaniz, J.S., Jain, D. and Dev, A. (2002), *Phys. Rev. D* **67**, 043514.

[20] Alcubierre, M. (1994), *Class. Quant. Grav.* **11**, L73.

[21] Allen, B. (1983), *Nucl. Phys. B* **226**, 232.

[22] Allen, B. (1989), *Phys. Rev. Lett.* **63**, 2017.

[23] Allen, B. (1990), *Gen. Rel. Grav.* **22**, 1447.

[24] Alpher, R.A. and Herman, R. (1950), *Rev. Mod. Phys.* **22**, 153.

[25] Alpher, R.A., Bethe, H.A. and Gamow, G. (1948), *Phys. Rev.* **73**, 803.

[26] Alvarez, E. and Belén Gavela, M. (1983), *Phys. Rev. Lett.* **51**, 931.

[27] Alvarez, E. and Conde, J. (2002), *Mod. Phys. Lett. A* **17**, 413.

[28] Amendola, L. (1999), *Phys. Rev. D* **60**, 043501.

[29] Amendola, L. (2000), *Mon. Not. R. Astr. Soc.* **312**, 521.

[30] Amendola, L. (2000), *Phys. Rev. D* **62**, 043511.

[31] Amendola, L. (2001), *Phys. Rev. Lett.* **86**, 196.

[32] Amendola, L. (2003), *Mon. Not. R. Astr. Soc.* **342**, 221..

[33] Amendola, L. and Quercellini, C. (2003), *Phys. Rev. D* **68**, 023514.

[34] Amendola, L. and Tocchini-Valentini, D. (2001), *Phys. Rev. D* **64**, 043509.

[35] Amendola, L. and Tocchini-Valentini, D. (2002), *Phys. Rev. D* **66**, 043528.

[36] Amendola, L., Bellisai, D. and Occhionero, F. (1993), *Phys. Rev. D* **47**, 4267.

[37] Amendola, L., Litterio, M. and Occhionero, F. (1989), *Phys. Lett. B* **231**, 43.

[38] Amendola, L., Litterio, M. and Occhionero, F. (1990), *Int. J. Mod. Phys. A* **5**, 3861.

[39] Anchordoqui, L.A., Perez-Bergliaffa, S.E., Trobo, M.L. and Birman, G.S. (1998), *Phys. Rev. D* **57**, 829.

[40] Anderson, G.W. and Carroll, S. (1997), in *Proceedings COSMO-97, 1st International Workshop on Particle Physics and the Early Universe*, Ambleside, England, 1997, Roszkowski, L., editor. Singapore: World Scientific.

[41] Antoniadis, I., Bachas, C., Ellis, J. and Nanopoulos, D.V. (1988), *Phys. Lett. B* **211**, 393.

[42] Appelquist, T. and Chodos, A. (1983), *Phys. Rev. D* **28**, 772.

[43] Appelquist, T. and Chodos, A. (1983), *Phys. Rev. Lett.* **50**, 141.

[44] Appelquist, T., Chodos, A. and Freund, P.G.O. (1987), *Modern Kaluza-Klein Theories*. Menlo Park: Addison-Wesley.

REFERENCES

[45] Arias, O. and Quiros, I. (2003), *Class. Quant. Grav.* **20**, 2563.

[46] Armendariz-Picon, C., Mukhanov, V.F. and Steinhardt, P.J. (2001), *Phys. Rev. D* **63**, 103510.

[47] Arnowitt, R., Deser, S. and Misner, C.W. (1962), in *Gravitation: An Introduction to Current Research*, Witten, L., editor. New York: Wiley.

[48] Aurich, R. and Steiner, F. (2003), *Phys. Rev. D* **67**, 123511.

[49] Baccigalupi, C. and Perrotta, F. (1998), preprint astro-ph/9811385.

[50] Baccigalupi, C., Matarrese, S. and Perrotta, F. (2000), *Phys. Rev. D* **62**, 123510.

[51] Baccigalupi, C., Matarrese, S. and Perrotta, F. (2000), in *Proceedings COSMO-99, 3rd International Conference on Particle Physics and the Early Universe*, Trieste, Italy, 1999, Cotti, U., Jeannerot, R., Sejanovic, G. and Smirnov, A., editors. Singapore: World Scientific.

[52] Bailin, D. and Love, A. (1987), *Rep. Prog. Phys.* **50**, 1087.

[53] Balakin, A.B., Pavón, D., Schwarz, D.J. and Zimdhal, W. (2003), *New J. Phys.* **5**, 85.

[54] Banday, A.J. *et al.* (1997), *Astrophys. J.* **475**, 393.

[55] Banerjee, A., Banerjee, N. and Santos, N.O. (1985), *J. Math. Phys.* **26**, 3125.

[56] Banerjee, N. and Pavón, D. (2001), *Class. Quant. Grav.* **18**, 593.

[57] Banerjee, N. and Pavón, D. (2001), *Phys. Rev. D* **63**, 043504.

[58] Banerjee, N. and Sen, S. (1997), *Phys. Rev. D* **56**, 1334.

[59] Baptista, A.B., Fabris, J.C. and Gonçalves, S.V.B. (1997), *Astr. Sp. Sci.* **246**, 315.

[60] Barber, G.A. (2003), preprint gr-qc/0302088.

[61] Barcelo, C. and Visser, M. (1999), *Phys. Lett. B* **466**, 127.

[62] Barcelo, C. and Visser, M. (2000), *Nucl. Phys. B* **584**, 415.

[63] Barcelo, C. and Visser, M. (2000), *Class. Quant. Grav.* **17**, 3843.

[64] Barcelo, C. and Visser, M. (2001), *Phys. Lett. B* **466**, 127.

[65] Bardeen, J.M. (1980), *Phys. Rev. D* **22**, 1882.

[66] Bar-kana, R. (1996), *Phys. Rev. D* **54**, 7138.

[67] Barker, B. (1978), *Astrophys. J.* **219**, 5.

[68] Barr, S.M. and Brown, L.S. (1984), *Phys. Rev. D* **29**, 2779.

[69] Barreiro, T., Copeland, E.J. and Nunes, N.J. (2000), *Phys. Rev. D* **61**, 127301.

[70] Barros, A. and Romero, C. (1998), *Phys. Lett. A* **245**, 31.

[71] Barroso, A., Casasayas, J., Crawford, P., Moniz, P. and Nunes, A. (1992), *Phys. Lett. B* **275**, 264.

[72] Barrow, J.D. (1978), *Mon. Not. R. Astr. Soc.* **184**, 677.

[73] Barrow, J.D. (1978), *Nature* **272**, 211.

[74] Barrow, J.D. (1988), *Nucl. Phys. B* **310**, 743.

[75] Barrow, J.D. (1990), *Phys. Lett. B* **249**, 406.

[76] Barrow, J.D. (1993), *Phys. Rev. D* **47**, 5329.

[77] Barrow, J.D. (1994), *Phys. Rev. D* **51**, 2729.

[78] Barrow, J.D. and Carr, B. (1996), *Phys. Rev. D* **54**, 3920.

[79] Barrow, J.D. and Cotsakis, S. (1988), *Phys. Lett. B* **214**, 515.

[80] Barrow, J.D. and Cotsakis, S. (1991), *Phys. Lett. B* **258**, 299.

[81] Barrow, J.D. and Maeda, K. (1990), *Nucl. Phys. B* **341**, 294.

[82] Barrow, J.D. and Mimoso, J.P. (1994), *Phys. Rev. D* **50**, 3746.

[83] Barrow, J.D. and Parsons, P. (1997), *Phys. Rev. D* **55**, 1906.

[84] Barrow, J.D. and Saich, P. (1990), *Phys. Lett. B* **249**, 406.

[85] Barrow, J.D. and Tipler, F.J. (1986), *The Anthropic Cosmological Principle.* Oxford: Oxford University Press.

[86] Barrow, J.D., Mimoso, J.P. and de Garcia Maia, M.R. (1993), *Phys. Rev. D* **48**, 3630.

[87] Bartolo, N. and Pietroni, M. (1999), *Phys. Rev. D* **61**, 023518.

[88] Barut, A.O. and Haugen, F.B. (1972), *Ann. Phys. (NY)* **71**, 519.

[89] Barvinsky, A.O. (1999), *Nucl. Phys. B* **561**, 159.

[90] Barvinsky, A.O., Kamenshchik, A. Yu. and Mishakov, I.V. (1977), *Nucl. Phys. B* **491**, 378.

[91] Bassett, B. and Liberati, S. (1998), *Phys. Rev. D* **58**, 021302.

[92] Bastero-Gil, M., Frampton, P.H. and Mersini, L. (2002), *Phys. Rev. D* **65**, 106002.

[93] Batakis, N.A. (1995), *Phys. Lett. B* **353**, 450.

[94] Batakis, N.A. and Kehagias, A.A. (1995), *Nucl. Phys. B* **449**, 248.

[95] Bateman, M. (1910), *Proc. Lon. Math. Soc.* **8**, 223.

REFERENCES

[96] Batista, A.B., Fabris, J.C., Gonçalves, S.V.B. and Tossa, J. (2001), *Int. J. Mod. Phys. A* **16**, 4527.

[97] Battye, R.A., Bucher, M. and Spergel, D. (1999), preprint astro-ph/9908047.

[98] Bayin, S.Ş., Cooperstock, F.I. and Faraoni, V. (1994), *Astrophys. J.* **428**, 439.

[99] Bean, R. (2001), *Phys. Rev. D* **64**, 123516.

[100] Bean, R. and Dore, O. (2003), *Phys. Rev. D* **68**, 023515.

[101] Bean, R. and Magueijo, J. (2001), *Phys. Lett. B* **517**, 177.

[102] Bechmann, O. and Lechtenfeld, O. (1995), *Class. Quant. Grav.* **12**, 1473.

[103] Bekenstein, J.D. (1974), *Ann. Phys. (NY)* **82**, 535.

[104] Bekenstein, J.D. (1975), *Phys. Rev. D* **11**, 2072.

[105] Bekenstein, J.D. (1977), *Phys. Rev. D* **15**, 1458.

[106] Bekenstein, J.D. (1979), *Comm. Astrophys.* **8**, 89.

[107] Bekenstein, J.D. and Meisels (1978), *Phys. Rev. D* **18**, 4378.

[108] Bekenstein, J.D. and Meisels (1980), *Astrophys. J.* **237**, 342.

[109] Belinskii, V.A. and Khalatnikov, I.M. (1973), *Sov. Phys. JETP*, **36**, 591.

[110] Bellucci, S. and Faraoni, V. (2002), *Nucl. Phys. B* **640**, 453.

[111] Bellucci, S., Faraoni, V. and Babusci, D. (2001), *Phys. Lett. A* **282**, 357.

[112] Benabed, K. and Bernardeau, F. (2001), *Phys. Rev. D* **64**, 083501.

[113] Benaoum, H.B. (2002), preprint hep-th/0205140.

[114] Bennett, C.L. *et al.* (1996), *Astrophys. J.* **464**, L1.

[115] Bento, M.C., Bertolami, O. and Sa, P.M. (1991), *Phys. Lett. B* **262**, 11.

[116] Bento, M.C., Bertolami, O. and Santos, N.C. (2001), preprint gr-qc/0111047.

[117] Bento, M.C., Bertolami, O. and Santos, N.C. (2002), *Phys. Rev. D* **65**, 067301.

[118] Bento, M.C., Bertolami, O. and Sen, A. (2002), *Phys. Rev. D* **66**, 043507.

[119] Bento, M.C., Bertolami, O. and Sen, A. (2002), preprint astro-ph/0210375.

[120] Berezin, V.A. (1989), *Gen. Rel. Grav.* **21**, 1177.

[121] Bergmann, P.G. (1968), *Int. J. Theor. Phys.* **1**, 25.

[122] Berkin, A.L. and Maeda, K. (1991), *Phys. Rev. D* **44**, 1691.

[123] Bertolami, O. and Martins, J. (2000), *Phys. Rev. D* **61**, 064007.

[124] Bertotti, B. (1971), in *General Relativity and Cosmology, XLII Course of the Varenna Summer School*, Sachs, R.K., editor. New York: Academic Press.

[125] Bertotti, B. and Catenacci, R. (1975), *Gen. Rel. Grav.* **6**, 329.

[126] Bhamra, K.S. (1974), *Austral. J. Phys.* **27**, 541.

[127] Bicknell, G.V. (1974), *J. Phys. A* **7**, 1061.

[128] Bicknell, G.V. and Klotz, A.H. (1976), *J. Phys. A* **9**, 1647.

[129] Bilic, N., Tupper, G.B. and Viollier, R.D. (2002), *Phys. Lett. B* **535**, 17.

[130] Billyard, A. and Coley, A.A. (1997), *Mod. Phys. Lett. A* **12**, 2121.

[131] Billyard, A., Coley, A.A. and van der Hoogen, R.J. (1998), *Phys. Rev. D* **58**, 123501.

[132] Billyard, A., Coley, A.A. and Ibanez, J. (1999), *Phys. Rev. D* **59**, 023507.

[133] Binétruy, P. (1999), *Phys. Rev. D* **60**, 063502.

[134] Binétruy, P. (2000), *Int. J. Theor. Phys.* **39**, 1859.

[135] Birrell, N.D. and Davies, P.C. (1980), *Phys. Rev. D* **22**, 322.

[136] Birrell, N.D. and Davies, P.C. (1980), *Quantum Fields in Curved Space*. Cambridge: Cambridge University Press.

[137] Blanco, S., Domenech, G., El Hasi, C. and Rosso, O.A. (1994), *Gen. Rel. Grav.* **26**, 1131.

[138] Boisseau, B., Esposito-Farèse, G., Polarski, D. and Starobinsky, A.A. (2000), *Phys. Rev. Lett.* **85**, 2236.

[139] Bombelli, L., Koul, R.K., Kunstatter, G., Lee, J. and Sorkin, R.D. (1987), *Nucl. Phys. B* **289**, 735.

[140] Bombelli, L., Lombardo, F. and Castagnino, M. (1998), *J. Math. Phys.* **39**, 6040.

[141] Bonanno, A. (1995), *Phys. Rev. D* **52**, 969.

[142] Bonanno, A. and Zappalá, D. (1997), *Phys. Rev. D* **55**, 6135.

[143] Bondi, H. (1957), *Rev. Mod. Phys.* **29**, 423.

[144] Bonnor, W.B. (1960), *J. Math. Mechanics* **9**, 439.

[145] Bose, S. and Lohiya, D. (1999), *Phys. Rev. D* **59**, 044019.

[146] Bracco, C. (1997), *Astron. Astrophys.* **321**, 985.

[147] Bracco, C. and Teyssandier, P. (1998), *Astron. Astrophys.* **339**, 921.

[148] Braginsky, V.B. *et al.* (1990), *Nuovo Cimento* **105B**, 1141.

[149] Branch, D. and Tammann, G.A. (1992), *Annu. Rev. Astron. Astrophys.* **30**, 359.

[150] Brans, C.H. (1988), *Class. Quant. Grav.* **5**, L197.

[151] Brans, C.H. (1997), preprint gr-qc/9705069.

[152] Brans, C.H. and Dicke, R.H. (1961), *Phys. Rev.* **124**, 925.

[153] Brax, P. and Martin, J. (1999), *Phys. Lett. B* **468**, 40.

[154] Brax, P. and Martin, J. (2002), preprint astro-ph/0210533.

[155] Bremer, M.S., Duff, M.J., Lu, H., Pope, C.N. and Stelle, K.S. (1999), *Nucl. Phys. B* **543**, 321.

[156] Broadhurst, T.J., Ellis, R.S., Koo, D.C. and Szalay, A.S. (1990), *Nature* **343**, 726.

[157] Bronnikov, K.A. (2001), *Phys. Rev. D* **64**, 064013.

[158] Bronnikov, K.A. (2002), *J. Math. Phys.* **43**, 6096.

[159] Bronnikov, K.A. and Shikin, G.N. (2002) *Gravit. Cosmol.* **8**, 107.

[160] Brustein, R., Gasperini, M., Giovannini, M., Mukhanov, V.F. and Veneziano, G. (1995), *Phys. Rev. D* **51**, 6744.

[161] Buchbinder, I.L. (1986), *Fortschr. Phys.* **34**, 605.

[162] Buchbinder, I.L. and Odintsov, S.D. (1983), *Sov. J. Nucl. Phys.* **40**, 848.

[163] Buchbinder, I.L. and Odintsov, S.D. (1985), *Lett. Nuovo Cimento* **42**, 379.

[164] Buchbinder, I.L., Odintsov, S.O. and Lichzier, I. (1989), *Class. Quant. Grav.* **6**, 605.

[165] Buchbinder, I.L., Odintsov, S.D. and Shapiro, I. (1986), in *Group-Theoretical Methods in Physics*, Markov, M., editor. Moscow: Nauka.

[166] Buchbinder, I.L., Odintsov, S.D. and Shapiro, I. (1992), *Effective Action in Quantum Gravity*. Bristol: IOP Publishing.

[167] Bucher, M. and Spergel, D. (1999), *Phys. Rev. D* **60**, 043505.

[168] Caldwell, R.R. (2002), *Phys. Lett. B* **545**, 23.

[169] Caldwell, R.R., Dave, R. and Steinhardt, P.J. (1998), *Phys. Rev. Lett.* **80**, 1582.

[170] Caldwell, R.R., Doran, M., Müller, C.M., Schäfer, G. and Wetterich, C. (2003), *Astrophys. J. (Lett.)* **591**, L75.

[171] Caldwell, R.R., Kamionkowski, M. and Weinberg, N.N. (2003), *Phys. Rev. Lett.* **91**, 071301.

[172] Callan, C.G. Jr., Coleman, S. and Jackiw, R. (1970) *Ann. Phys. (NY)* **59**, 42.

[173] Callan, C.G., Friedan, D., Martinec, E.J. and Perry, M.J. (1985), *Nucl. Phys.* B **262**, 593.

[174] Calzetta, E. and El Hasi, C. (1993), *Class. Quant. Grav.* **10**, 1825.

[175] Calzetta, E. and El Hasi, C. (1995), *Phys. Rev.* D **51**, 2713.

[176] Canuto, V., Adams, P.J., Hsieh, S.-H. and Tsiang, E. (1977), *Phys. Rev.* D **16**, 1643.

[177] Canuto, V., Hsieh, S.-H. and Adams, P.J. (1977), *Phys. Rev. Lett.* **39**, 429.

[178] Capozziello, S. and de Ritis, R. (1993), *Phys. Lett. A* **177**, 1.

[179] Capozziello, S. and de Ritis, R. (1994), *Class. Quant. Grav.* **11**, 107.

[180] Capozziello, S., Cardone, V.F., Carloni, S. and Troisi, A. (2003), preprint astro-ph/0307018.

[181] Capozziello, S., Carloni, S. and Troisi, A. (2003), preprint astro-ph/0303041.

[182] Capozziello, S., Demianski, M., de Ritis, R. and Rubano, C. (1995), *Phys. Rev.* D **52**, 3288.

[183] Capozziello, S., de Ritis, R. and Marino, A.A. (1997), *Class. Quant. Grav.* **14**, 3243.

[184] Capozziello, S., de Ritis, R. and Marino, A.A. (1997), *Class. Quant. Grav.* **14**, 3259.

[185] Caron, B. *et al.* (1997), *Class. Quant. Grav.* **14**, 1461.

[186] Carroll, S.M. (1998), *Phys. Rev. Lett.* **81**, 3097.

[187] Carroll, S.M. (2001), *Living Rev. Relativity* **4**, 1.

[188] Carroll, S.M. and Field, G.B. (1997), *Phys. Rev.* D **79**, 2394.

[189] Carroll, S.M., Duvvuri, V., Trodden, M. and Turner, M.S. (2003), preprint astro-ph/0306438.

[190] Carroll, S.M., Field, G.B. and Jackiw, R. (1990), *Phys. Rev.* D **41**, 1231.

[191] Carroll, S.M., Field, G.B. and Jackiw, R. (1991), *Phys. Rev.* D **43**, 3789.

[192] Carroll, S.M., Hoffman, M. and Trodden, M. (2003), *Phys. Rev.* D **68**, 023509.

[193] Casadio, R. and Gruppuso, A. (2002), *Int. J. Mod. Phys.* D **11**, 703.

[194] Casadio, R. and Harms, B. (1999), *Mod. Phys. Lett. A* **14**, 1089.

[195] Casas, J.A., Garcia-Bellido, J. and Quiros, M. (1992), *Class. Quant. Grav.* **9**, 1371.

[196] Casas, J.A., Garcia-Bellido, J. and Quiros, M. (1995), *Mod. Phys. Lett. A* **7**, 447.

[197] Castagnino, M., Giacomini, H. and Lara, L. (2000), *Phys. Rev. D* **61**, 107302.

[198] Cecotti, S. (1987), *Phys. Lett. B* **190**, 86.

[199] Cervantes-Cota, J.L. (1999), *Class. Quant. Grav.* **16**, 3903.

[200] Chamseddine, A.H. (1981), *Nucl. Phys. B* **185**, 403.

[201] Chandrasekhar, S. (1937), *Nature* **139**, 757.

[202] Chatterjee, S. and Banerjee, A. (1993), *Class. Quant. Grav.* **10**, L1.

[203] Chauvet, P. and Cervantes-Cota, J.L. (1995), *Phys. Rev. D* **52**, 3416.

[204] Chauvet, P. and Guzman, E. (1986), *Astr. Sp. Sci.* **126**, 133.

[205] Chauvet, P. and Obregon, D. (1979), *Astr. Sp. Sci.* **66**, 515.

[206] Chauvineau, B. (2003), *Class. Quant. Grav.* **20**, 2617.

[207] Chen, X. (2001), in *Proceedings COSMO 2000, 4th International Workshop on Particle Physics and the Early Universe*, Cheju Island, Korea, 2000, Kim, J.E., Ko, P. and Lee, K., editors. Singapore: World Scientific.

[208] Chen, X. and Kamionkowski, M. (1999), *Phys. Rev. D* **60**, 104036.

[209] Chen, X., Scherrer, R.J. and Steigman, G. (2001), *Phys. Rev. D* **63**, 123504.

[210] Chernikov, N.A. and Tagirov, E.A. (1968), *Ann. Inst. H. Poincarè A* **9**, 109.

[211] Chiba, T. (1999), *Phys. Rev. D* **60**, 083508.

[212] Chiba, T. (2001), *Phys. Rev. D* **64**, 103503.

[213] Chiba, T. (2003), *Phys. Lett. B* **9**, 33.

[214] Chiba, T. and Nakamura, T. (1998), *Prog. Theor. Phys.* **100**, 1077.

[215] Chiba, T., Okabe, T. and Yamaguchi, M. (2000), *Phys. Rev. D* **62**, 023511.

[216] Chiba, T., Sugiyama, N. and Yokoyama, J. (1998), *Nucl. Phys. B* **530**, 304.

[217] Chimento, L.P., Jakubi, A. and Pavón, D. (2000), *Phys. Rev. D* **62**, 063508.

[218] Chimento, L.P., Jakubi, A. and Zuccalá, N.A. (2001), *Phys. Rev. D* **63**, 103508.

[219] Chiueh, T. (2002), *Phys. Rev. D* **65**, 123502.

[220] Cho, Y.M. (1990), *Phys. Rev. D* **41**, 2462.

[221] Cho, Y.M. (1992), *Phys. Rev. Lett.* **68**, 3133.

[222] Cho, Y.M. (1994), in *Evolution of the Universe and its Observational Quest*, Yamada, Japan 1993, Sato, H., editor, p. 99. Tokyo: Universal Academy Press.

[223] Cho, Y.M. (1997), *Class. Quant. Grav.* **14**, 2963.

[224] Cho, Y.M. and Keum, Y.Y. (1998), *Mod. Phys. Lett. A* **13**, 109.

[225] Cho, Y.M. and Yoon, J.M. (1993), *Phys. Rev. D* **47**, 3465.

[226] Choi, K. (2000), *Phys. Rev. D* **62**, 043509.

[227] Choi, K. (2000), in *Proceedings, International Symposium on Supersymmetry, Supergravity and Superstring*, Seoul, Korea, 1999, Kim, J.E. and Lee, C., editors. Singapore: World Scientific.

[228] Chung, D.J.H., Everett, L.L. and Riotto, A. (2003), *Phys. Lett. B* **556**, 61.

[229] CMBFAST homepage http://physics.nyu.edu/matiasz/CMBFAST/cmbfast.html

[230] Coble, K., Dodelson, S. and Frieman, J. (1997), *Phys. Rev. D* **55**, 1851.

[231] Coley, A.A., Ibanez, J. and van der Hogen, R.J. (1997), *Int. J. Mod. Phys. D* **38**, 5256.

[232] Colistete, R. Jr., Fabris, J.C., Gonçalves, S.V.B. and de Souza, P.E. (2002), preprint gr-qc/0210079.

[233] Colistete, R. Jr., Fabris, J.C., Gonçalves, S.V.B. and de Souza, P.E. (2003), preprint astro-ph/0303338.

[234] Collins, P.D.B., Martin, A.D. and Squires, E.J. (1989), *Particle Physics and Cosmology*. New York: Wiley.

[235] Constantinidis, C.P., Fabris, J.C., Furtado, R.G. and Picco, M. (2000), *Phys. Rev. D* **61**, 043503.

[236] Contreras, C., Herrera, R. and del Campo, S. (1995), *Phys. Rev. D* **52**, 4349.

[237] Copeland, E.J., Lahiri, A. and Wands, D. (1994), *Phys. Rev. D* **50**, 4868.

[238] Copeland, E.J., Liddle, A.R. and Lidsey, J.E. (2001), *Phys. Rev. D* **64**, 023509.

[239] Copeland, E.J., Liddle, A.R. and Wands, D. (1998), *Phys. Rev. D* **57**, 4686.

[240] Copeland, E.J., Nunes, N.J. and Rosati, F. (2000), *Phys. Rev. D* **62**, 123503.

[241] Cornish, N.J. and Shellard, E.P.S. (1998), *Phys. Rev. Lett.* **81**, 3571.

[242] Cotsakis, S. (1993), *Phys. Rev. D* **47**, 1437; errata (1994), *Phys. Rev. D* **49**, 1145.

[243] Cotsakis, S. (1995), *Phys. Rev. D* **52**, 6199.

[244] Cotsakis, S. and Saich, P.J. (1994), *Class. Quant. Grav.* **11**, 383.

[245] Coule, D.H. (1992), *Class. Quant. Grav.* **9**, 2353.

[246] Coule, D.H. (1998), *Class. Quant. Grav.* **15**, 2803.

[247] Coule, D.H. (1999), *Phys. Lett. B* **450**, 48.

[248] Coule, D.H. and Maeda, K. (1990), *Class. Quant. Grav.* **7**, 955.

[249] Crittenden, R. and Steinhardt, P.J. (1992), *Phys. Lett. B* **293**, 32.

[250] Cunningham, E. (1909), *Proc. Lon. Math. Soc.* **8**, 77.

[251] Daly, R.A. (2002), preprint astro-ph/0212107.

[252] Daly, R.A. and Guerra, E.J. (2002), *Astron. J.* **124**, 1831.

[253] Daly, R.A., Mory, M.P. and Guerra, E.J. (2002), preprint astro-ph/0203113.

[254] Damour, T. and Esposito–Farèse, G. (1992), *Class. Quant. Grav.* **9**, 2093.

[255] Damour, T. and Esposito–Farèse, G. (1998), *Phys. Rev. D* **58**, 044003.

[256] Damour, T. and Nordtvedt, K. (1993), *Phys. Rev. Lett.* **70**, 2217.

[257] Damour, T. and Nordtvedt, K. (1993), *Phys. Rev. D* **48**, 3436.

[258] Damour, T. and Pichon, B. (1999), *Phys. Rev. D* **59**, 123502.

[259] Damour, T. and Polyakov, A.M. (1994), *Nucl. Phys. B* **423**, 532.

[260] Damour, T., Gibbons, G.W. and Gundlach, C. (1990), *Phys. Rev. Lett.* **64**, 123.

[261] da Silva, C. and Williams, R.M. (2000), *Class. Quant. Grav.* **17**, 1827.

[262] Dautcourt, G. (1969), *Mon. Not. Roy. Astr. Soc.* **144**, 255.

[263] Dautcourt, G. (1975), *Astron. Astrophys.* **38**, 344.

[264] Dautcourt, G. (1975), in *Confrontation of Cosmological Theories with Observation*, IAU Symposium No. 63, Longair, M.S., editor. Dordrecht: Reidel.

[265] Dautcourt, G. (1977), *Astron. Nachr.* **298**, 81.

[266] de Bernardis, P. *et al.* (2000), *Nature* **400**, 955.

[267] Deffayet, C., Dvali. G. and Gabadadze, G. (2002), *Phys. Rev. D* **65**, 044023.

[268] Dehnen, H. and Obregon, O. (1971), *Astr. Sp. Sci.* **14**, 454.

[269] Dehnen, H. and Obregon, O. (1972), *Astr. Sp. Sci.* **15**, 326.

[270] Demianski, M. *et al.* (1991), *Phys. Rev. D* **44**, 3136.

[271] de la Macorra, A. (2002), *Int. J. Mod. Phys. D* **11**, 653.

[272] de la Macorra, A. and Stephan-Otto, C. (2002), *Phys. Rev.D* **65**, 083520.

[273] de la Macorra, A. and Vucetich, H. (2002), *Int. J. Mod. Phys. D* **11**, 653.

[274] de la Macorra, A. and Vucetich, H. (2002), preprint astro-ph/0212302.

[275] Derendinger, J.P., Ibanez, L.E. and Nilles, H.P. (1985), *Phys. Lett. B* **155**, 65.

[276] de Ritis, R. and Marino, A.A. (2001), *Phys. Rev. D* **64**, 083509.

[277] de Ritis, R., Marino, A.A., Rubano, C. and Scudellaro, P. (2000), *Phys. Rev. D* **62**, 043506.

[278] Deruelle, N., Garriga, J. and Verdaguer, E. (1991), *Phys. Rev. D* **43**, 1032.

[279] Deruelle, N., Gundlach, C. and Langlois, D. (1992), *Phys. Rev. D* **46**, 5337.

[280] Deser, S. (1970), *Ann. Phys. (NY)* **59**, 248.

[281] Deser, S. (1984), *Phys. Lett. B* **134**, 419.

[282] Deser, S. and Nepomechie, R.I. (1984), *Ann. Phys. (NY)* **154**, 396.

[283] Dev, A., Alcaniz, J.S. and Jain, D. (2003), *Phys. Rev. D* **67**, 023515.

[284] De Witt, B.S. and Brehme, R.W. (1960), *Ann. Phys. (NY)* **9**, 220.

[285] De Witt, B.S. and Nicolai, H. (1986), *Nucl. Phys. B* **274**, 363.

[286] Diaz-Rivera, L.M. and Pimentel, L.O. (2003), *Int. J. Mod. Phys. A* **18**, 651.

[287] Dick, R. (1988), *Gen. Rel. Grav.* **30**, 435.

[288] Dick, R. (2001), *Class. Quant. Grav.* **18**, R1.

[289] Dick, R. (2003), preprint gr-qc/0307052.

[290] Dicke, R.H. (1962), *Phys. Rev.* **125**, 2163.

[291] Dicke, R.H. (1968), *Astrophys. J.* **152**, 1.

[292] Dine, M., Rohm, R., Seiberg, N. and Witten, E. (1985), *Phys. Lett. B* **156**, 55.

[293] Dirac, P.A.M. (1937), *Nature* **139**, 323.

[294] Dirac, P.A.M. (1937), *Nature* **139**, 1001.

[295] Dirac, P.A.M. (1938), *Proc. R. Soc. Lon. A* **165**, 199.

[296] Dirac, P.A.M. (1973), *Proc. R. Soc. Lond. A* **333**, 403.

[297] Di Pietro, E. and Demaret, J. (1999), *Int. J. Mod. Phys. D* **10**, 231.

[298] Dolgov, A.D. (1983), in *The Very Early Universe*, Gibbons, G.W., Hawking, S.W. and Siklos, T.C., editors. Cambridge: Cambridge University Press.

[299] Dolgov, A.D. (1985), *Sov. Phys. JETP (Lett.)*, **41**, 345.

[300] Dominguez-Tenreiro, R. and Yepes, G (1987), *Astron. Astrophys.* **177**, 5.

[301] Doran, M., Schwindt, J.M. and Wetterich, C. (2001), *Phys. Rev. D* **64**, 123520.

[302] Drinkwater, M.J. et al. (1998), *Mon. Not. R. Astr. Soc.* **295**, 457.

[303] Duff, M.J., Khuri, R.R. and Lu, J.X. (1995), *Phys. Rep.* **259**, 213.

[304] Durrer, R. (1994), *Phys. Rev. Lett.* **72**, 3301.

[305] Dyson, F.J. (1972), in *Aspects of Quantum Theory*, Salam, A. and Wigner, E.P., editors, p. 212. Cambridge: Cambridge University Press.

REFERENCES

[306] Dyson, F.J. and Damour, T. (1996), *Nucl. Phys. B* **480**, 37.

[307] Eardley, D.M., Lee, D.L. and Lightman, A.P. (1973), *Phys. Rev. D* **10**, 3308.

[308] Eardley, D.M., Lee, D.L., Lightman, A.P., Wagoner, R.V. and Will, C.M. (1973), *Phys. Rev. Lett.* **30**, 884.

[309] Easther, R. (1994), preprint NZ-CAN-RE-94/1, astro-ph/9405034.

[310] Easther, R. (1996), *Class. Quant. Grav.* **13**, 1775.

[311] Efstathiou, G., Bridle, S.L., Lasenby, A.N., Hobson, M.P. and Ellis, R.S. (1998), preprint astro-ph/9812226.

[312] Einstein, A. (1987), *The Collected Papers of Albert Einstein*, vol. 4, doc. 17. Princeton: Princeton University Press.

[313] Eisenstein, D.J. and Bunn, E.F. (1997), *Phys. Rev. Lett.* **79**, 1957.

[314] Elizalde, E. and Odintsov, S.D. (1994), *Phys. Lett. B* **333**, 331.

[315] Ellis, G.F.R. and Bruni, M. (1989), *Phys. Rev. D* **40**, 1804.

[316] Ellis, G.F.R., Hwang, J.-C. and Bruni, M. (1989), *Phys. Rev. D* **40**, 1819.

[317] Ellis, G.F.R., Bruni, M. and Hwang, J.-C. (1990), *Phys. Rev. D* **42**, 1035.

[318] Ellis, J. et al. (1984), *Phys. Lett. B* **134**, 429.

[319] Epstein, H., Glaser, V. and Jaffe, A. (1965), *Nuovo Cimento* **36**, 1016.

[320] Esposito-Farèse, G. (2000), preprint gr-qc/0011115.

[321] Esposito-Farèse, G. and Polarski (2001), *Phys. Rev. D* **63**, 063504.

[322] Fabris, J.C. and Tossa, J. (1997), *Gravit. Cosmol.* **3**, 165.

[323] Fabris, J.C., Batista, A.B. and Baptista, J.P. (1989), *C. Rend. Acad. Sci. Paris* **309**, 791.

[324] Fabris, J.C., Furtado, R.G., Peter, P. and Pinto-Neto, N. (2002), *Phys. Rev. D* **67**, 124003.

[325] Fabris, J.C., Gonçalves S.V.B. and de Souza, P.E. (2002), *Gen. Rel. Grav.* **34**, 53.

[326] Fabris, J.C., Kerner, R. and Tossa, J. (2000), *Int. J. Mod. Phys. D* **9**, 111.

[327] Fakir, R. (1990), *Phys. Rev. D* **41**, 3012.

[328] Fakir, R. (1993), *Astrophys. J.* **418**, 202.

[329] Fakir, R. (1994), *Astrophys. J.* **426**, 74.

[330] Fakir, R. (1994), *Phys. Rev. D* **50**, 3795.

[331] Fakir, R. (1995), preprint astro-ph/9507112.

[332] Fakir, R. (1997), *Int. J. Mod. Phys. D* **6**, 49.

[333] Fakir, R. and Habib, S. (1993), *Mod. Phys. Lett. A* **8**, 2827.

[334] Fakir, R. and Unruh, W.G. (1990), *Phys. Rev. D* **41**, 1783.

[335] Fakir, R. and Unruh, W.G. (1990), *Phys. Rev. D* **41**, 1792.

[336] Fakir, R. and Unruh, W.G. (1990), *Phys. Rev. D* **41**, 3012.

[337] Fakir, R., Habib, S. and Unruh, W.G. (1992), *Astrophys. J.* **394**, 396.

[338] Faraoni, V. (1992), *Astrophys. J.* **398**, 425.

[339] Faraoni, V. (1992), in *Gravitational Lenses*, Proceedings, Hamburg, Germany 1991, Kayser, R., Schramm, T. and Nieser, L., editors. Berlin: Springer-Verlag.

[340] Faraoni, V. (1996), *Astrophys. Lett. Comm.* **35**, 305.

[341] Faraoni, V. (1996), *Phys. Rev. D* **53**, 6813.

[342] Faraoni, V. (1997), *Gen. Rel. Grav.* **29**, 251.

[343] Faraoni, V. (1997), *Gravit. Cosmol.* **4**, 321.

[344] Faraoni, V. (1997), preprint gr-qc/9807066.

[345] Faraoni, V. (1998), *Int. J. Mod. Phys. D* **7**, 429.

[346] Faraoni, V. (1998), *Phys. Lett. A* **245**, 26.

[347] Faraoni, V. (1998), in *Proceedings of the Asia-Pacific Conference on Gravitation and Cosmology*, Seoul 1996, Cho, Y.M., Lee, C.H. and Kim, S.-W., editors. Singapore: World Scientific.

[348] Faraoni, V. (1999), *Am. J. Phys.* **67**, 732.

[349] Faraoni, V. (1999), *Phys. Rev. D* **59**, 084021.

[350] Faraoni, V. (2000), *Phys. Lett. A* **269**, 209.

[351] Faraoni, V. (2000), *Phys. Rev. D* **62**, 023504.

[352] Faraoni, V. (2001), *Am. J. Phys.* **69**, 372.

[353] Faraoni, V. (2001), *Int. J. Theor. Phys.* **40**, 2259.

[354] Faraoni, V. (2002), *Int. J. Mod. Phys. D* **11**, 471.

[355] Faraoni, V. (2003), *Phys. Rev. D* **68**, 063508.

[356] Faraoni, V. and Cooperstock, F.I. (1998), *Eur. J. Phys.* **19**, 419.

[357] Faraoni, V. and Cooperstock, F.I. (2003), *Astrophys. J.* **587**, 483.

[358] Faraoni, V. and Gunzig, E. (1998), *Astron. Astrophys.* **332**, 1154.

[359] Faraoni, V. and Gunzig, E. (1998), *Int. J. Mod. Phys. D* **8**, 177.

[360] Faraoni, V. and Gunzig, E. (1999), *Int. J. Theor. Phys.* **38**, 199.

[361] Faraoni, V. and Gunzig, E. (1999), *Int. J. Theor. Phys.* **38**, 217.

[362] Faraoni, V. and Gunzig, E. (1999), *Int. J. Theor. Phys.* **38**, 2931.

[363] Faraoni, V., Gunzig, E. and Nardone, P. (1998), *Fund. Cosm. Phys.* **20**, 121.

[364] Faraoni, V. and Sonego, S. (1994), in *Proceedings of the 5th Canadian Conference on General Relativity and Relativistic Astrophysics*, Waterloo, Canada, 1993, Mann, R.B. and McLenaghan, R.G., editors, p. 386. Singapore: World Scientific.

[365] Faraoni, V., Cooperstock, F.I. and Overduin, J.M. (1995), *Int. J. Mod. Phys. A* **4**, 387.

[366] Faraoni, V., Cooperstock, F.I. and Overduin, J.M. (1997), *Fields Institute Comm.* **15**, 237.

[367] Fay, S. (2000), *Class. Quant. Grav.* **17**, 891.

[368] Fay, S. (2001), *Class. Quant. Grav.* **18**, 45.

[369] Feinstein, A. (2002), *Phys. Rev. D* **66**, 063511.

[370] Felder, G., Frolov, A., Kofman, L. and Linde, A. (2002), *Phys. Rev. D* **66**, 023507.

[371] Felder, G., Kofman, L. and Linde, A. (1999), *Phys. Rev. D* **60**, 103505.

[372] Ferraris, M. (1986), in *Atti del 6° Convegno Nazionale di Relatività Generale e Fisica della Gravitazione*, Firenze 1984, Modugno, M., editor, p. 127. Bologna: Tecnoprint.

[373] Ferreira, P.G. and Joyce, M. (1998), *Phys. Rev. D* **58**, 023503.

[374] Ferreira, P.G. and Joyce, M. (1998), *Phys. Rev. Lett.* **79**, 4740.

[375] Fierz, M. (1956), *Helv. Phys. Acta* **29**, 128.

[376] Filippenko, A.V. and Riess, A.G. (1998), *Phys. Rep.* **307**, 31.

[377] Fishbach, E. and Talmadge, C. (1992), *Nature* **356**, 207.

[378] Fishbach, E. *et al.* (1986), *Phys. Rev. Lett.* **56**, 3.

[379] Fishbach, E. *et al.* (1988), *Ann. Phys. (NY)* **182**, 1.

[380] Flanagan, E.E. (2003), preprint astro-ph/0308111.

[381] Flanagan, E.E. and Wald, R.M. (1996), *Phys. Rev. D* **54**, 6233.

[382] Ford, L.H. (1987), *Phys. Rev. D* **32**, 1409.

[383] Ford, L.H. (1987), *Phys. Rev. D* **35**, 2339.

[384] Ford, L.H. and Roman, T.A. (2001), *Phys. Rev. D* **64**, 024023.

[385] Ford, L.H. and Toms, D.J. (1982), *Phys. Rev. D* **25**, 1510.

[386] Foster, S. (1998), *Class. Quant. Grav.* **15**, 3485.

[387] Fradkin, E.S. and Tseytlin, A.A. (1985), *Nucl. Phys. B* **261**, 1.

[388] Frampton, P.H. (2002), in *Proceedings, 30th Conference on High Energy Physics and Cosmology*, Coral Gables, Florida, 2001, Kursunoglu, B.N., Mintz, S.L. and Perlmutter, A., editors. Melville: AIP.

[389] Frampton, P.H. (2003), *Phys. Lett. B* **555**, 139.

[390] Frampton, P.H. (2003), preprint hep-th/0302007.

[391] Frampton, P.H. and Takahashi, T. (2003), *Phys. Lett. B* **557**, 135.

[392] Freund, P.G.O. (1982), *Nucl. Phys. B* **209**, 146.

[393] Frewin, R.A. and Lidsey, J.E. (1993), *Int. J. Mod. Phys. D* **2**, 3323.

[394] Friedlander, F.G. (1975), *The Wave Equation on a Curved Spacetime*. Cambridge: Cambridge University Press.

[395] Frieman, J.A., Harari, D.D. and Surpi, G.C. (1994), *Phys. Rev. D* **50**, 4895.

[396] Frieman, J.A., Hill, C., Stebbins, A. and Waga, J. (1995), *Phys. Rev. D* **50**, 4895.

[397] Frolov, V., Kofman, L. and Starobinsky, A.A. (2002), *Phys. Lett. B* **545**, 8.

[398] Froyland, J. (1982), *Phys. Rev. D* **25**, 1470.

[399] Fujii, Y. (1991), *Int. J. Mod. Phys. A* **6**, 3505.

[400] Fujii, Y. (2000), *Gravit. Cosmol.* **6**, 107.

[401] Fujii, Y. (2000), *Phys. Rev. D* **62**, 044011.

[402] Fujii, Y. and Nishioka, T. (1990), *Phys. Rev. D* **42**, 361.

[403] Fukuyama, T. *et al.* (1997), *Int. J. Mod. Phys. D* **6**, 69.

[404] Futamase, T. and Maeda, K. (1989), *Phys. Rev. D* **39**, 399.

[405] Futamase, T. and Tanaka, M. (1999), *Phys. Rev. D* **60**, 063511.

[406] Futamase, T., Rothman, T. and Matzner, R. (1989), *Phys. Rev. D* **39**, 405.

[407] Garcia-Bellido, J. (1993), *Int. J. Mod. Phys. D* **2**, 85.

[408] Garcia-Bellido, J. (1994), *Nucl. Phys. B* **423**, 221.

[409] Garcia-Bellido, J. and Linde, A. (1995), *Phys. Rev. D* **52**, 6730.

[410] Garcia-Bellido, J., Linde, A. and Linde, D. (1994), *Phys. Rev. D* **50**, 730.

[411] Garcia-Bellido, J. and Quirós, M. (1990), *Phys. Lett. B* **243**, 45.

REFERENCES

[412] Garcia–Bellido, J. and Quiròs, M. (1992), *Nucl. Phys. B* **368**, 463.

[413] Garcia-Bellido, J. and Wands, D. (1995), *Phys. Rev. D* **52**, 5636.

[414] Garcia-Bellido, J. and Wands, D. (1995), *Phys. Rev. D* **52**, 6739.

[415] Garnavich, P.M. *et al.* (1998), *Astrophys. J.* **509**, 74.

[416] Garriga, J. and Tanaka, T. (2000), *Phys. Rev. Lett.* **84**, 2778.

[417] Gasperini, M. (1999), *Phys. Lett. B* **470**, 67.

[418] Gasperini, M. (2001), *Phys. Rev. D* **64**, 043510.

[419] Gasperini, M. and Ricci, R. (1993), *Class. Quant. Grav.* **12**, 677.

[420] Gasperini, M. and Veneziano, G. (1992), *Phys. Lett. B* **277**, 256.

[421] Gasperini, M. and Veneziano, G. (1993), *Astropart. Phys.* **1**, 317.

[422] Gasperini, M., Maharana, J. and Veneziano, G. (1991), *Phys. Lett. B* **272**, 277.

[423] Gasperini, M., Piazza, F. and Veneziano, G. (2002), *Phys. Rev. D* **65**, 023508.

[424] Gasperini, M., Ricci, R. and Veneziano, G. (1993), *Phys. Lett. B* **319**, 438.

[425] Gaztanaga, E. and Lobo, J.A. (2001), *Astrophys. J.* **548**, 47.

[426] Gerard, J.-M. and Mahara, I. (1995), *Phys. Lett. B* **346**, 35.

[427] Geroch, R. (1969), *Comm. Math. Phys.* **13**, 180.

[428] Gibbons, G.W. (2002), *Phys. Lett. B* **537**, 1.

[429] Gibbons, G.W. (2003), *Class. Quant. Grav.* **20**, S231.

[430] Gibbons, G.W. and Whiting, B.F. (1981), *Nature* **291**, 636.

[431] Gillies, G.T. (1997), *Rep. Prog. Phys.* **60**, 151.

[432] Gilliland, R.L., Nugent, P.E. and Phillips, M.M. (1999), *Astrophys. J.* **521**, 30.

[433] Giovannini, M. (1999), *Class. Quant. Grav.* **16**, 2905.

[434] Giovannini, M. (1999), *Phys. Rev. D* **60**, 123511.

[435] Giveon, A., Porrati, M. and Rabinovici, E. (1994), *Phys. Rep.* **244**, 443.

[436] Gonzalez, J.A., Quevedo, H., Salgado, M. and Sudarsky, D. (2000), *Astron. Astrophys.* **362**, 835.

[437] Gonzalez, J.A., Quevedo, H., Salgado, M. and Sudarsky, D. (2000), *Phys. Rev. D* **64**, 047504.

[438] Goobar, A. and Perlmutter, S. (1995), *Astrophys. J.* **450**, 14.

[439] Gorini, V., Kamenshchik, A. Yu. and Moschella, U. (2003), *Phys. Rev. D* **67**, 063509.

[440] Górski, K.M. et al. (1996), *Astrophys. J.* **464**, L11.

[441] Gott, S., Schmidt, H.-J. and Starobinsky, A.A. (1990), *Class. Quant. Grav.* **7**, 893.

[442] Green, A.M. and Liddle, A.R. (1996), *Phys. Rev. D* **54**, 2557.

[443] Green, B., Schwarz, J.M. and Witten, E. (1987), *Superstring Theory*. Cambridge: Cambridge University Press.

[444] Greenstein, G.S. (1968), *Astrophys. Lett.* **1**, 139.

[445] Greenstein, G.S. (1968), *Astr. Sp. Sci.* **2**, 155.

[446] Grib, A.A. and Poberii, E.A. (1995), *Helv. Phys. Acta* **68**, 380.

[447] Grib, A.A. and Rodrigues, W.A. (1995), *Gravit. Cosmol.* **1**, 273.

[448] Gross, D.J. and Perry, M.J. (1983), *Nucl. Phys. B* **226**, 29.

[449] Guckenheimer, J. and Holmes, P (1983), *Nonlinear Oscillations, Dynamical Systems and Bifurcation of Vector Fields*. New York: Springer-Verlag.

[450] Guenther, D.B., Krauss, L.M. and Demarque, P. (1998), *Astrophys. J.* **498**, 871.

[451] Gunzig, E. et al. (2000), *Class. Quant. Grav.* **17**, 1783.

[452] Gunzig, E. et al. (2000), *Int. J. Theor. Phys.* **39**, 1901.

[453] Gunzig, E. et al. (2001), *Phys. Rev. D* **63**, 067301.

[454] Gurevich, L.E., Finkelstein, A.M. and Ruban, V.A. (1973), *Astr. Sp. Sci.* **22**, 231.

[455] Guth, A.H. (1981), *Phys. Rev. D* **23**, 347.

[456] Guth, A.H. (1997), *The Inflationary Universe*. Reading, Massachusetts: Addison-Wesley.

[457] Guth, A.H. and Jain, B. (1992), *Phys. Rev. D* **45**, 426.

[458] Gwinn, C.R. et al. (1997), *Astrophys. J.* **485**, 87.

[459] Hadamard, J. (1952), *Lectures on Cauchy's Problem in Linear Partial Differential Equations*. New York: Dover.

[460] Halford, W.D. (1970), *Austral. J. Phys.* **23**, 863.

[461] Halliwell, J.J. (1987), *Phys. Lett. B* **185**, 341.

[462] Halliwell, J.J. and Laflamme, R. (1989), *Class. Quant. Grav.* **6**, 1839.

[463] Hanany, S. et al (2000), *Astrophys. J. (Lett.)* **545**, L5.

[464] Hannestad, S. and Mortsell, E. (2002), *Phys. Rev. D* **66**, 063508.

REFERENCES

[465] Harrison, E.R. (1967), *Mon. Not. R. Astr. Soc.* **137**, 69.

[466] Harrison, E.R. (1972), *Phys. Rev. D* **6**, 2077.

[467] Hawking, S.W. (1972), *Comm. Math. Phys.* **25**, 167.

[468] Hawking, S.W. and Ellis, G.F.R. (1973), *The Large Scale Structure of Spacetime*. Cambridge: Cambridge University Press.

[469] Hawking, S.W. and Penrose, R. (1969), *Proc. R. Soc. Lon. A* **314**, 529.

[470] Hawking, S.W. and Reall, H.S. (1999), *Phys. Rev. D* **59**, 023502.

[471] Hebecker, A. and Wetterich, C. (2000), *Phys. Rev. Lett.* **85**, 3339.

[472] Heckmann, O. (1956), *Z. Astrophys.* **40**, 278.

[473] Hellermann, S., Kaloper, N. and Susskind, L. (2001), *JHEP* **0106**, 003.

[474] Helmi, A. and Vucetich, H. (1997), *Phys. Lett. A* **230**, 153.

[475] Herrera, R., Contreras, C. and del Campo, S. (1995), *Class. Quant. Grav.* **12**, 1937.

[476] Higgs, P.W. (1959), *Nuovo Cimento* **11**, 816.

[477] Hill, C.T. and Leibovitch, A.K. (2002), *Phys. Rev. D* **66**, 075010.

[478] Hill, C.T. and Salopek, D.S. (1992), *Ann. Phys. (NY)* **213**, 21.

[479] Hinshaw, G. *et al.* (1996), *Astrophys. J.* **464**, L17.

[480] Hirsch, M.W. and Smale, S. (1974), *Differential Equations, Dynamical Systems, and Linear Algebra*. New York: Academic Press.

[481] Hiscock, W.A. (1990), *Class. Quant. Grav.* **7**, L35.

[482] Holden, D.J. and Wands, D. (1998), *Class. Quant. Grav.* **15**, 3271.

[483] Holden, D.J. and Wands, D. (2000), *Phys. Rev. D* **61**, 043506.

[484] Holman, R., Kolb, E.W. and Wang, Y. (1990), *Phys. Rev. Lett.* **65**, 17.

[485] Holman, R., Kolb, E.W., Vadas, S. and Wang, Y. (1991), *Phys. Lett. B* **269**, 252.

[486] Hosotani, Y. (1985), *Phys. Rev. D* **32**, 1949.

[487] Hoyle, F. and Tayler, R.J. (1964), *Nature* **203**, 1108.

[488] Hoyle, F., Burbidge, G. and Narlikar, J. (1993), *Astrophys. J.* **410**, 437.

[489] Hoyle, F., Burbidge, G. and Narlikar, J. (1994), *Astron. Astrophys.* **289**, 729.

[490] Hoyle, F., Burbidge, G. and Narlikar, J. (1994), *Mon. Not. R. Astr. Soc.* **267**, 1007.

[491] Hoyle, F., Burbidge, G. and Narlikar, J. (1995), *Proc. R. Soc. Lon. A* **448**, 191.

[492] Hoyle, F., Burbidge, G. and Narlikar, J. (2000), *A Different Approach to Cosmology*. Cambridge: Cambridge University Press.

[493] Hu, B.-L. (1982), *Phys. Lett. A* **90**, 375.

[494] Hu, B.-L. and Parker, L. (1978), *Phys. Rev. D* **17**, 933.

[495] Hu, Y., Turner, M.S. and Weinberg, E.J. (1994), *Phys. Rev. D* **49**, 3830.

[496] Huey, G. and Lidsey, J. (2001), *Phys. Lett. B* **514**, 217.

[497] Huterer, D. and Turner, M.S. (1999), *Phys. Rev. D* **60**, 081301.

[498] Hwang, J.-C. (1990), *Class. Quant. Grav.* **7**, 1613.

[499] Hwang, J.-C. (1990), *Phys. Rev. D* **42**, 2601;

[500] Hwang, J.-C. (1991), *Class. Quant. Grav.* **8**, 195.

[501] Hwang, J.-C. (1991), *Class. Quant. Grav.* **8**, 1047.

[502] Hwang, J.-C. (1996), *Phys. Rev. D* **53**, 762.

[503] Hwang, J.-C. (1997), *Class. Quant. Grav.* **14**, 1981.

[504] Hwang, J.-C. (1997), *Class. Quant. Grav.* **14**, 3327.

[505] Hwang, J.-C. (1998), *Class. Quant. Grav.* **15**, 1387.

[506] Hwang, J.-C. (1998), *Class. Quant. Grav.* **15**, 1401.

[507] Hwang, J.-C. and Noh, H. (1996), *Phys. Rev. D* **54**, 1460.

[508] Hwang, J.-C. and Noh, H. (1998), *Phys. Rev. Lett.* **81**, 5274.

[509] Hwang, J.-C. and Vishniac, E.T. (1990), *Astrophys. J.* **353**, 1.

[510] Iorio, A., O'Raifeartaigh, L., Sachs, I. and Wiesendanger, C. (1997), *Nucl. Phys. B* **495**, 433.

[511] Isaacson, R.A. (1968), *Phys. Rev.* **166**, 1263.

[512] Ishikawa, J. (1983), *Phys. Rev. D* **28**, 2445.

[513] Israel, W. and Stewart, J.M. (1979), *Ann. Phys. (NY)* **118**, 341.

[514] Jafry, Y.R., Cornelisse, J. and Reinhard, R. (1994), *ESA Bull.* **18**, 219.

[515] Jeavons, J.S., McIntosh, C.B.G. and Sen, D.K. (1975), *J. Math. Phys.* **16**, 320.

[516] Jensen, L.G. and Stein-Schabes, J. (1987), *Phys. Rev. D* **35**, 1146.

[517] Jetzer, P. (1993), *Phys. Rep.* **220**, 163.

[518] Johri, V.B. (2002), *Class. Quant. Grav.* **19**, 5959.

REFERENCES

[519] Jordan, P. (1938), *Naturwiss.* **26**, 417.

[520] Jordan, P. (1948), *Astr. Nachr.* **276**, 193.

[521] Jordan, P. (1952), *Schwerkraft und Weltfall, Grundlagen der Theoretische Kosmologie.* Braunschweig: Vieweg und Sohns.

[522] Jordan, P. (1959), *Z. Phys.* **157**, 112.

[523] Joyce, M. and Prokopec, T. (1998), *Phys. Rev. D* **7**, 6022.

[524] Kaiser, D.I. (1995), *Phys. Rev. D* **52**, 4295.

[525] Kaiser, D.I. (1995), preprint astro-ph/9507048.

[526] Kaiser, N. and Jaffe, A. (1997), *Astrophys. J.* **484**, 545.

[527] Kallosh, R. and Linde, A.D., (2003), *J. Cosm. Astrop. Phys.* **02**, 1.

[528] Kallosh, R., Linde, A.D., Prokushkin, S. and Shmakova, M. (2002), *Phys. Rev. D* **65**, 105016.

[529] Kaloper, N. and Olive, K.A. (1998), *Phys. Rev. D* **57**, 811.

[530] Kaluza, Th. (1921), *Sitz. Preuss. Akad. Wiss. Berlin, Math. Phys. K* **1**, 966.

[531] Kalyana Rama, S. (1997), *Phys. Rev. D* **56**, 6130.

[532] Kalyana Rama, S. (1997), *Phys. Rev. Lett.* **78**, 1620.

[533] Kamenshchik, A. Yu., Khalatnikov, I.M. and Toporensky, A.V. (1995), *Phys. Lett. B* **357**, 35.

[534] Kamenshchik, A. Yu., Moschella, U. and Pasquier, V. (2001), *Phys. Lett. B* **511**, 265.

[535] Kawasaki, M., Moroi, T. and Takahashi, T. (2001), *Phys. Rev. D* **2001**, 083009.

[536] Kim, J.E. (1999), *JHEP* **9905**, 022.

[537] Klein, O. (1926), *Z. Phys.* **37**, 895.

[538] Kofman, L., Linde, A.A. and Starobinsky, A.A. (1996), *Phys. Rev. Lett.* **76**, 1011.

[539] Kofman, L., Linde, A.A. and Starobinsky, A.A. (1997), *Phys. Rev. D* **56**, 3258.

[540] Kolb, E.W. (1991), *Physica Scripta* **T36**, 376.

[541] Kolb, E.W. and Turner, M.S. (1994), *The Early Universe.* Reading, Massachusetts: Addison-Wesley.

[542] Kolb, E.W., Lidley, D. and Seckel, D. (1984), *Phys. Rev. D* **30**, 1205.

[543] Kolb, E.W., Salopek, D. and Turner, M.S. (1990), *Phys. Rev. D* **42**, 3925.

[544] Kolitch, S.J. (1996), *Ann. Phys. (NY)* **246**, 121.

[545] Kolitch, S.J. and Eardley, D.M. (1995), *Ann. Phys. (NY)* **241**, 128.

[546] Komatsu, E. and Futamase, T. (1998), *Phys. Rev. D* **58**, 023004.

[547] Komatsu, E. and Futamase, T. (1999), *Phys. Rev. D* **59**, 064029.

[548] Kopeikin, S.M., Schäfer, G., Gwinn, C.R. and Marshall Eubanks, T. (1999), *Phys. Rev. D* **59**, 084023.

[549] Kothari, D.S. (1938), *Nature* **142**, 354.

[550] Kovner, I. (1990), *Astrophys. J.* **351**, 114.

[551] Kubyshin, Yu. and Martin, J. (1995), preprint UB–ECM–PF 95/13, LGCR 95/06/05, DAMPT R95, gr–qc/9507010.

[552] La, D. and Steinhardt, P.J. (1989), *Phys. Rev. Lett.* **62**, 376.

[553] La, D., Steinhardt, P.J. and Bertschinger, E. (1989), *Phys. Lett. B* **231**, 231.

[554] Labeyrie, A. (1993), *Astron. Astrophys.* **268**, 823.

[555] Landau, L.D. and Lifschitz, E.M. (1989), *The Classical Theory of Fields*. Oxford: Pergamon Press.

[556] Lange, A.E. *et al.* (2001), *Phys. Rev. D* **63**, 042001.

[557] Laycock, A.M. and Liddle, A.R. (1994), *Phys. Rev. D*, **49**, 1827.

[558] Leahy, J.P. (1997), preprint astro-ph/9704285.

[559] Lee, J. *et al.* (1999), *Phys. Rev. D* **61**, 027301.

[560] Lemos, J.P.S. and Sá, P.M. (1994), *Class. Quant. Grav.* **11**, L11.

[561] Lessner, G. (1974), *Astr. Sp. Sci.* **30**, L5.

[562] Levin, J.J. and Freese, K. (1994), *Nucl. Phys. B* **421**, 635.

[563] Li, X.-Z. and Hao, J.-G. (2003), preprint hep-th/0303093.

[564] Li, X.-Z, Hao, J.-G. and Liu, D.-J. (2002), *Class. Quant. Grav.* **19**, 6049.

[565] Liddle, A.R. (1995), *Phys. Rev. D* **51**, 5347.

[566] Liddle, A.R. (1999), in *Proceedings of the 1998 Summer School in High Energy Physics and Cosmology*, ICTP Series in Theoretical Physics, Vol. 15, Masiero, A., Sejanovic, G. and Smirnov, A., editors. Singapore: World Scientific.

[567] Liddle, A.R. and Lyth, D.H. (1992), *Phys. Lett. B* **291**, 391.

[568] Liddle, A.R. and Lyth, D.H. (1993), *Phys. Rep.* **231**, 1.

[569] Liddle, A.R. and Lyth, D.H. (2000), *Cosmological Inflation and Large Scale Structure*. Cambridge: Cambridge University Press.

[570] Liddle, A.R. and Madsen, M.S. (1992), *Int. J. Mod. Phys. D* **1**, 101.

REFERENCES

[571] Liddle, A.A. and Scherrer, R.J. (1999), *Phys. Rev. D* **59**, 023509.

[572] Liddle, A.R. and Wands, D. (1991), *Mon. Not. R. Astr. Soc.* **253**, 637.

[573] Liddle, A.R. and Wands, D. (1992), *Phys. Rev. D* **45**, 2665.

[574] Liddle, A.A., Parsons, P. and Barrow, J.D. (1994), *Phys. Rev. D* **50**, 7222.

[575] Lidsey, J.E. (1992), *Class. Quant. Grav.* **9**, 149.

[576] Lidsey, J.E. (1995), *Phys. Rev. D* **52**, 5407.

[577] Lidsey, J.E. (1996), *Class. Quant. Grav.* **13**, 2449.

[578] Lidsey, J.E. *et al.* (1997), *Rev. Mod. Phys.* **69**, 373.

[579] Lidsey, J.E., Wands, D. and Copeland, E.J. (2000), *Phys. Rep.* **337**, 343.

[580] Linde, A.D. (1980), *Sov. Phys. JETP (Lett.)* **30**, 447.

[581] Linde, A.D. (1990), *Particle Physics and Inflationary Cosmology*. Chur, Switzerland: Hardwood Academic.

[582] Linde, A.D. (1990), *Phys. Lett., B* **249**, 18.

[583] Linde, A.D. (2001), *JHEP* **0111**, 052.

[584] Linder, E.V. (1986), *Phys. Rev. D* **34**, 1759.

[585] Linder, E.V. (1988), *Astrophys. J.* **328**, 77.

[586] Loredo, T.J., Flanagan, E.A. and Wasserman, I.M. (1997), *Phys. Rev. D* **56**, 7507.

[587] Lorentz, M.A. (1937), *Collected Papers*, vol. 5, p. 363. The Hague: Nijhoff.

[588] Lorentz-Petzold, D. (1984), *Astr. Sp. Sci.* **98**, 101.

[589] Lorentz-Petzold, D. (1984), *Astr. Sp. Sci.* **98**, 249.

[590] Lorentz-Petzold, D. (1984), *Astr. Sp. Sci.* **106**, 419.

[591] Lorentz-Petzold, D. (1984), in *Lecture Notes in Physics*, Vol. 105, Hoenselaers, C. and Dietz, W., editors. Springer-Verlag: Berlin.

[592] Lorentz-Petzold, D. (1984), *Math. Proc. Camb. Phil. Soc.* **96**, 183.

[593] Lovelace, C. (1985), *Nucl. Phys. B* **273**, 413.

[594] Lucchin, F. and Matarrese, S. (1985), *Phys. Rev. D* **32**, 1316.

[595] Lucchin, F., Matarrese, S. and Pollock, M.D. (1986), *Phys. Lett. B* **167**, 163.

[596] Lukash, V.N. (1980), *Sov. Phys. JETP (Lett.)* **31**, 596.

[597] Lukash, V.N. (1980), *Sov. Phys. JETP* **52**, 807.

[598] Lyra, G. (1951), *Math. Z.* **54**, 2.

[599] Lyth, D.H. and Stewart, E.D. (1992), *Phys. Lett. B* **274**, 168.

[600] Maartens, R., Wands, D., Bassett, B.A. and Heard, I.P.C. (2000), *Phys. Rev. D* **62**, 041301.

[601] Madsen, M.S. (1988), *Class. Quant. Grav.* **5**, 627.

[602] Madsen, M.S. (1993), *Gen. Rel. Grav.* **25**, 855.

[603] Maeda, K. (1986), *Class. Quant. Grav.* **3**, 233.

[604] Maeda, K. (1986), *Class. Quant. Grav.* **3**, 651.

[605] Maeda, K. (1986), *Phys. Lett. B* **166**, 59.

[606] Maeda, K. (1987), in *Modern Kaluza-Klein Theories*, Appelquist, T., Chodos, A. and Freund, P.G.O., editors. Reading, Massachusetts: Addison-Wesley.

[607] Maeda, K. (1987), *Phys. Lett. B* **186**, 33.

[608] Maeda, K. (1988), *Phys. Rev. D* **37**, 858.

[609] Maeda, K. (1989), *Phys. Rev. D* **39**, 3159.

[610] Maeda, K. et al. (1982), *Phys. Lett. B* **108**, 98.

[611] Maggiore, M. and Nicolis, A. (2000), *Phys. Rev. D* **62**, 024004.

[612] Magnano, G. (1996), in *General Relativity and Gravitational Physics*, Proceedings, Trieste, Italy 1994, M. Carfora et al., editors. Singapore: World Scientific.

[613] Magnano, G. and Sokolowski, L.M. (1994), *Phys. Rev. D* **50**, 5039.

[614] Mak, M.K. and Harko, T. (2002), *Europhys. Lett.* **60**, 155.

[615] Mak, M.K. and Harko, T. (2002), *Int. J. Mod. Phys. D* **11**, 1389.

[616] Makino, N. and Sasaki, M. (1991), *Prog. Theor. Phys.* **86**, 103.

[617] Makler, M., Quinet de Oliveira, S. and Waga, J. (2003), *Phys. Lett. B* **555**, 1.

[618] Malaney, R.A. and Mathews, G. (1993), *Phys. Rep.* **229**, 147.

[619] Mallik, S. and Rai Chaudhuri, D. (1997), *Phys. Rev. D* **56**, 625.

[620] Maor, I., Brustein, R., McMahon, J. and Steinhardt, P.J. (2002), *Phys. Rev. D* **65**, 123003.

[621] Marleau, F.R. and Starkman, G.D. (1996), preprint astro-ph/9605066.

[622] Marshall Eubanks, T. et al. (1997), *Bull. Am. Phys. Soc.* Abstract K11.05.

[623] Martin, J. and Schwarz, D.J. (2001), *Phys. Lett. B* **500**, 1.

[624] Masiero, A., Pietroni, M. and Rosati, F. (2000), *Phys. Rev. D* **61**, 023504.

[625] Mathiazhagan, C. and Johri, V.B. (1984), *Class. Quant. Grav.* **1**, L29.

REFERENCES

[626] Matsuda, T. (1962), *Prog. Theor. Phys.* **47**, 738.

[627] McBreen, B. and Metcalfe, L. (1988), *Nature* **332**, 234.

[628] McDonald, J. (1993), *Phys. Rev. D* **48**, 2462.

[629] McInnes, B. (2002), *JHEP* **0208**, 029.

[630] McInnes, B. (2002), preprint astro-ph/0210321.

[631] Meisels, A. (1982), *Astrophys. J.* **252**, 403.

[632] Melchiorri, A. et al. (2000), *Astrophys. J. (Lett.)* **536**, L63.

[633] Melchiorri, A., Mersini, L., Ödman, C.J. and Trodden, M. (2003), *Phys. Rev. D* **68**, 043509.

[634] Meng, X. and Wang, P. (2003), preprint astro-ph/0308031.

[635] Mersini, L., Bastero-Gil, M. and Kanti, P. (2001), *Phys. Rev. D* **64**, 043508.

[636] Mignemi, S. and Schmidt, H.-J. (1995), *Class. Quant. Grav.* **12**, 849.

[637] Mihic, M.B. and Stein-Schabes, J. (1988), *Phys. Lett. B* **203**, 353.

[638] Miller, A.D. et al. (1999), *Astrophys. J. (Lett.)* **524**, L1.

[639] Mimoso, J.P. and Wands, D. (1995), *Phys. Rev. D* **51**, 477.

[640] Misner, C.W., Thorne, K.S. and Wheeler, J.A. (1973), *Gravitation*. San Francisco: Freeman.

[641] Miyazaki, A. (1982), *Nuovo Cimento* **68B**, 126.

[642] Miyazaki, A. (2000), preprint gr-qc/0012104.

[643] Miyazaki, A. (2001), preprint gr-qc/0102003.

[644] Morganstern, R.E. (1971), *Phys. Rev. D* **3**, 2946.

[645] Morganstern, R.E. (1971), *Phys. Rev. D* **4**, 282.

[646] Morganstern, R.E. (1971), *Phys. Rev. D* **4**, 946.

[647] Morikawa, M. (1990), *Astrophys. J. (Lett.)* **362**, L37.

[648] Morikawa, M. (1991), *Astrophys. J.* **369**, 20.

[649] Morris, J.R. (2001), *Class. Quant. Grav.* **18**, 2997.

[650] Moss, I. and Sahni, V. (1986), *Phys. Lett.* **178**, 2118.

[651] Motter, A.E. and Letelier, P.S. (2002), *Phys. Rev. D* **65**, 068502.

[652] Mukhanov, V.F. (1988), *Sov. Phys. JETP* **68**, 1297.

[653] Mukhanov, V.F., Feldman, H.A. and Brandenberger, R.H. (1992), *Phys. Rep.* **215**, 203.

[654] Mukohyama, S. (2002), *Phys. Rev. D* **66**, 024009.

[655] Mukohyama, S. (2003), preprint hep-th/0306208.

[656] Mukohyama, S. and Randall, L. (2003), preprint hep-th/0306108.

[657] Muller, V., Schmidt, H.-J. and Starobinsky, A.A. (1990), *Class. Quant. Grav.* **7**, 1163.

[658] Muslimov, A.G. (1990), *Class. Quant. Grav.* **7**, 231.

[659] Muta, T.S. and Odintsov, S.D. (1991), *Mod. Phys. Lett. A* **6**, 3641.

[660] Myers, R.C. (1987), *Phys. Lett. B* **199**, 371.

[661] Nakamura, T. and Chiba, T. (1999), *Mon. Not. R. Astr. Soc.* **306**, 696.

[662] Nariai, H. (1968), *Prog. Theor. Phys.* **40**, 49.

[663] Nariai, H. (1971), *Prog. Theor. Phys.* **47**, 1824.

[664] Nariai, H. (1972), *Prog. Theor. Phys.* **48**, 703.

[665] Nelson, B. and Panangaden, P. (1982), *Phys. Rev. D* **25**, 1019.

[666] Ng, S.C.C. (2000), *Phys. Lett. B* **485**, 1.

[667] Nishino, H. and Szegin, E. (1984), *Phys. Lett. B* **144**, 187.

[668] Nishioka, T. and Fujii, Y. (1992), *Phys. Rev. D* **45**, 2140.

[669] Nodland, B. and Ralston, J.P. (1997), *Phys. Rev. Lett.* **78**, 3043.

[670] Noh, H. and Hwang, J.-C. (2001), *Phys. Lett. B* **515**, 231.

[671] Nojiri, S. and Odintsov, S.D. (2003), preprint hep-th/0307288.

[672] Nojiri, S. *et al.* (2001), *Phys. Rev. D* **64**, 043505.

[673] Nordström, G. (1912), *Phys. Zeitschr.* **13**, 1126.

[674] Nordström, G. (1913), *Ann. Phys. (Leipzig)* **40**, 856.

[675] Nordström, G. (1913), *Ann. Phys. (Leipzig)* **42**, 533.

[676] Nordström, G. (1914), *Phys. Zeitschr.* **15**, 504.

[677] Nordtvedt, K. (1968), *Phys. Rev. D* **169**, 1017.

[678] Nordtvedt, K. (1970), *Astrophys. J.* **161**, 1059.

[679] Obregon, O. and Chauvet, P. (1978), *Astr. Sp. Sci.* **56**, 335.

[680] Odintsov, S.D. (1991), *Fortschr. Phys.* **39**, 621.

[681] O'Hanlon, J. and Tupper, B. (1972), *Nuovo Cimento B* **7**, 305.

[682] Okamura, T. (1998), in *Proceedings of the Asia-Pacific Conference on Gravitation and Cosmology*, Seoul, Korea 1996, Cho, Y.M., Lee, C.H. and Kim, S.-W., editors. Singapore: World Scientific.

[683] Olive, K.A. *et al.* (1990), *Phys. Lett. B* **36**, 454.

[684] Onemli, V.K. and Woodard, R.P. (2002), *Class. Quant. Grav.* **19**, 4607.

[685] Oukuiss, A. (1997), *Nucl. Phys. B* **486**, 413.

[686] Overduin, J.M. and Cooperstock, F.I. (1998), *Phys. Rev. D* **58**, 043506.

[687] Overduin, J.M. and Wesson, P.S. (1997), *Phys. Rep.* **283**, 303.

[688] Ozsváth, I. and Schücking, E. (1962), in *Recent Developments in General Relativity*, p. 339. New York: Pergamon Press.

[689] Padmanabhan, T. (2002), *Phys. Rev. D* **66**, 021301.

[690] Pagel, B.E.J. (1997), *Nucleosynthesis and Chemical Evolution of Galaxies*. Cambridge: Cambridge University Press.

[691] Paiva, F.M. and Romero, C. (1993), *Gen. Rel. Grav.* **25**, 1305.

[692] Paiva, F.M., Reboucas, M. and MacCallum, M. (1993), *Class. Quant. Grav.* **10**, 1165.

[693] Parker, L. and Raval, A. (2000), *Phys. Rev. D* **62**, 083503.

[694] Parker, L. and Toms, D.J. (1985), *Phys. Rev. D* **29**, 1584.

[695] Parker, L. and Toms, D.J. (1985), *Phys. Rev. D* **32**, 1409.

[696] Peebles, P.J.E. (1996), *Astrophys. J.* **146**, 542.

[697] Peebles, P.J.E. and Ratra, B. (1988), *Astrophys. J. (Lett.)* **325**, L17.

[698] Peebles, P.J.E. and Vilenkin, A. (1999), *Phys. Rev. D* **59**, 063505.

[699] Peebles, P.J.E. and Yu, J.T. (1970), *Astrophys. J.* **162**, 815.

[700] Peloso, M. and Rosati, F. (1999), *JHEP* **9912**, 026.

[701] Penrose, R. (1964), in *Relativity, Groups and Topology*, Les Houches Lectures, Morette, C. and De Witt, B.S., editors. New York: Gordon and Breach.

[702] Penrose, R. (1965), *Rev. Mod. Phys.* **37**, 1.

[703] Perlmutter, S. *et al.* (1998), *Nature* **391**, 51.

[704] Perlmutter, S. *et al.* (1999), *Astrophys. J.* **517**, 565.

[705] Perlmutter, S., Turner, M.S. and White, M. (1999), *Phys. Rev. Lett.* **83**, 670.

[706] Perrotta, F. and Baccigalupi, C. (2002), *Phys. Rev. D* **65** 123505.

[707] Perrotta, F. and Baccigalupi, C. (2002), preprint astro-ph/0205217.

[708] Perrotta, F. and Baccigalupi, C. (2002), preprint astro-ph/0205245.

[709] Perrotta, F., Baccigalupi, C. and Matarrese, S. (2000), *Phys. Rev. D* **61**, 023507.

[710] Peters, P.C. (1969), *J. Math. Phys.* **10**, 1029.

[711] Pimentel, L.O. (1989), *Class. Quant. Grav.* **6**, L263.

[712] Pimentel, L.O. and Stein-Schabes, J. (1989), *Phys. Lett. B* **216**, 27.

[713] *PLANCK* homepage http://astro.estec.esa.nl/SA-general/Projects/Planck

[714] Pogrebenko, S. *et al.* (1994), in *Abstracts XXIInd GA IAU Meeting*, p. 105. The Netherlands: Twin Press.

[715] Pogrebenko, S. *et al.* (1994), in *Proceedings of the 2nd EVN/JIVE Symposium*, Torun, Poland 1994, Kus, A.J., Schilizzi, R.T., Borkowski, K.M. and Gurvits, L.I., editors, p. 33. Torun, Poland: Radio Astronomy Observatory.

[716] Pogrebenko, S. *et al.* (1996), in *Compact Stars in Binaries*, Proc. IAU Symp. 165, The Hague, Netherlands 1994, Van Paradijs, J., Van den Heuvel, J. and Kuulkers, E.P.J., editors, p. 546. Dordrecht: Kluwer.

[717] Polchinski, J. (1998), *String Theory*. Cambridge: Cambridge University Press.

[718] Pollock, M.D. (1982), *Phys. Lett. B* **108**, 386.

[719] Pollock, M.D. (1985), *Phys. Lett. B* **156**, 301.

[720] Pollock, M.D. (1988), *Phys. Lett. B* **215**, 635.

[721] Pollock, M.D. and Sadhev, D. (1989), *Phys. Lett. B* **222**, 1.

[722] Pradhan, A., Iotemshi, I. and Singh, G.P. (2002), preprint gr-qc/0207024.

[723] Prestage, J.D., Tjoelker, R.L. and Maleki, L. (1995), *Phys. Rev. Lett.* **74**, 3511.

[724] Quevedo, H., Salgado, M. and Sudarsky, D. (1997), *Astrophys. J.* **488**, 14.

[725] Quiros, I. (2000), *Phys. Rev. D* **61**, 124026.

[726] Ragazzoni, R., Valente, G. and Marchetti, E. (2003), *Mon. Not. R. Astr. Soc.* **345**, 100.

[727] Rainer, M. and Zuhk, A. (1996), *Phys. Rev. D* **54**, 6186.

[728] Randall, L. and Sundrum, R. (1999), *Phys. Rev. Lett.* **83**, 3370.

[729] Randall, L. and Sundrum, R. (1999), *Phys. Rev. Lett.* **83**, 4690.

[730] Ratra, B. and Peebles, P.J. (1988), *Phys. Rev. D* **37**, 3406.

[731] Reasenberg, R.D. *et al.* (1979), *Astrophys. J. (Lett.)* **234**, L219.

[732] Rees, M.J. (1971), *Mon. Not. R. Astron. Soc.* **154**, 187.

REFERENCES

[733] Rees, M.J. and Sciama, D.W. (1968), *Nature* **217**, 511.

[734] Reeves, H. (1994), *Rev. Mod. Phys.* **66**, 193.

[735] Reuter, M. (1994), *Phys. Rev. D* **49**, 6379.

[736] Riazuelo, A. and Uzan, J.-P. (2000), *Phys. Rev. D* **62**, 083506.

[737] Riazuelo, A. and Uzan, J.-P. (2002), *Phys. Rev. D* **66**, 023535.

[738] Riess, A.G. et al. (1998), *Astron. J.* **116**, 1009.

[739] Riess, A.G. et al. (1999), *Astron. J.* **118**, 2668.

[740] Riess, A.G. et al. (2000), *Astrophys. J.* **536**, 62.

[741] Riess, A.G. et al. (2001), *Astrophys. J.* **560**, 49.

[742] Rocha Filho, T.M. et al. (2000), *Int. J. Theor. Phys.* **39**, 1933.

[743] Romero, C. and Barros, A. (1992), *Astr. Sp. Sci.* **192**, 263.

[744] Romero, C. and Barros, A. (1993), *Gen. Rel. Grav.* **23**, 491.

[745] Romero, C. and Barros, A. (1993), *Gen. Rel. Grav.* **25**, 1305.

[746] Romero, C. and Barros, A. (1993), *Phys. Lett. A* **173**, 243.

[747] Rosati, F. (2000), in *Proceedings COSMO-99, 3rd International Conference on Particle Physics and the Early Universe*, Trieste, Italy, 1999, Cotti, U., Jeannerot, R., Sejanovic, G. and Smirnov, A., editors, p. 126. Singapore: World Scientific.

[748] Rosati, F. (2001), *Nucl. Phys. Proc. Suppl.* **95**, 74.

[749] Ross, D.K. (1980), *J. Phys. A* **13**, 557.

[750] Rubakov, V.A. (2000), *Phys. Rev. D* **61**, 061501.

[751] Ruban, V.A. (1977), *Sov. Phys. JETP* **45**, 269.

[752] Ruban, V.A. and Finkelstein, A.M. (1976), *Astrofizika* **12**, 371.

[753] Saa, A. et al. (2001), *Int. J. Theor. Phys.* **40**, 2295.

[754] Sachs, R.K. (1961), *Proc. R. Soc. Lon. A* **264**, 309.

[755] Sahni, V. (2002), *Class. Quant. Grav.* **19**, 3435.

[756] Sahni, V. and Shtanov, Yu. (2002), preprint astro-ph/0202346.

[757] Sahni, V. and Starobinsky, A.A. (2000), *Int. J. Mod. Phys. D* **9**, 373.

[758] Sahni, V. and Wang, L. (2000), *Phys. Rev. D* **62**, 103517.

[759] Sahni, V., Sami, M. and Souradeep, T. (2002), *Phys. Rev. D* **65**, 023518.

[760] Saini, T.D., Raychaudhuri, S., Sahni, V. and Starobinsky, A.A. (2000), *Phys. Rev. Lett.* **85**, 1162.

[761] Sakharov, A.D. (1968), *Sov. Phys. Dokl.* **12**, 1040.

[762] Salgado, M. (2002), preprint gr-qc/0202082.

[763] Salgado, M., Sudarsky, D. and Quevedo, H. (1996), *Phys. Rev. D* **53**, 6771.

[764] Salgado, M., Sudarsky, D. and Quevedo, H. (1997), *Phys. Lett. B* **408**, 69.

[765] Salopek, D.S. (1992), *Phys. Rev. Lett.* **69**, 3602.

[766] Salopek, D.S. and Bond, J.R. (1990), *Phys. Rev. D* **42**, 3936.

[767] Salopek, D.S., Bond, J.R. and Bardeen, J.M. (1989), *Phys. Rev. D* **40**, 1753.

[768] Santiago, D.I. and Silbergleit, A.S. (2001), *Gen. Rel. Grav.* **32**, 565.

[769] Santiago, D.I., Kalligas, D. and Wagoner, R.V. (1997), *Phys. Rev. D* **56**, 7627.

[770] Santiago, D.I., Kalligas, D. and Wagoner, R.V. (1999), *Phys. Rev. D* **58**, 124005.

[771] Santos, C. and Gregory, R. (1997), *Ann. Phys. (NY)* **258**, 111.

[772] Sanyal, A.K. (2002), *Phys. Lett. B* **524**, 177.

[773] Sanyal, A.K. and Modak, B. (2001), *Class. Quant. Grav.* **18**, 3767.

[774] Sarkar, R.P. and Banerjee, S. (1993), *Gen. Rel. Grav.* **25**, 1107.

[775] Scheel, M.A., Shapiro, S.L. and Teukolsky, S.A. (1995), *Phys. Rev. D* **51**, 4236.

[776] Scherk, J. and Schwarz, J.H. (1979), *Nucl. Phys. B* **153**, 61.

[777] Schmidt, B.R. et al. (1998), *Astrophys. J.* **507**, 46.

[778] Schmidt, H.-J. (1987), *Astr. Nachr.* **308**, 183.

[779] Schmidt, H.-J. (1990), *Class. Quant. Grav.* **7**, 1023.

[780] Schmidt, H.-J. (1995), *Phys. Rev. D* **52**, 6198.

[781] Schneider, P., Ehlers, J. and Falco, E.E. (1992), *Gravitational Lenses*. Berlin: Springer Verlag.

[782] Schramm, D.N. and Turner, M.S. (1998), *Rev. Mod. Phys.* **70**, 303.

[783] Schramm, D.N. and Wagoner, R.V. (1977), *Ann. Rev. Nucl. Sci.* **27**, 37.

[784] Schuecker, P. et al. (2003), *Astron. Astrophys.* **402**, 53.

[785] Schulz, A.E. and White, M. (2001), *Phys. Rev. D* **64**, 043514.

[786] Schwinger, J. (1970), *Particles, Sources and Fields*. Reading, Massachusetts: Addison-Wesley.

[787] Sen, A.A. (2002), *JHEP* **0207**, 065.

[788] Sen, A.A. (2002), *Mod. Phys. Lett. A* **17**, 1797.

[789] Sen, A.A. and Sen, S. (2001), *Mod. Phys. Lett. A* **16**, 1303.

[790] Sen, A.A. and Seshadri, T.R. (2000), *Int. J. Mod. Phys. D* **12**, 445.

[791] Sen, A.A. and Sethi, S. (2002), *Phys. Lett. B* **532**, 159.

[792] Sen, A.A., Sen, S. and Sethi, S. (2001), *Phys. Rev. D* **63**, 107501.

[793] Sen, D.K. (1957), *Z. Phys.* **149**, 311.

[794] Sen, D.K. (1960), *Can. Math. Bull.* **3**, 255.

[795] Sen, D.K. and Dunn, K.A. (1971), *J. Math. Phys.* **12**, 219.

[796] Sen, D.K. and Vanstone, J.R. (1972), *J. Math. Phys.* **13**, 990.

[797] Serna, A. and Alimi, J.-M. (1996), *Phys. Rev. D* **50**, 7304.

[798] Serna, A., Alimi, J.-M. and Navarro, A. (2002), *Class. Quant. Grav.* **19**, 857.

[799] Serna, A., Dominguez-Tenreiro, R. and Yepes, G. (1992), *Astrophys. J.* **391**, 433.

[800] Shapiro, I.L. and Takata, H. (1995), *Phys. Lett. B* **361**, 31.

[801] Shibata, M., Nakao, K. and Nakamura, T. (1994), *Phys. Rev. D* **50**, 7304.

[802] Shlyakhter, A.I. (1976), *Nature* **264**, 340.

[803] Silva, P.T. and Bertolami, O. (2003), preprint astro-ph/0303353.

[804] Singh, P., Sami, M. and Dadhich, N. (2003), *Phys. Rev. D* **68**, 023522.

[805] Singh, T. (1975), *J. Math. Phys.* **16**, 219.

[806] Singh, T. and Rai, L.N. (1983), *Gen. Rel. Grav.* **15**, 875.

[807] Sisterna, P. and Vucetich, H. (1990), *Phys. Rev. D* **41**, 1034.

[808] Skordis, C. and Albrecht, A. (2000), *Phys. Rev. D* **66**, 043523.

[809] Smoot, G.F. *et al.* (1992), *Astrophys. J. (Lett.)* **396**, L1.

[810] Sokolowski, L.M. (1989), *Class. Quant. Grav.* **6**, 59.

[811] Sokolowski, L.M. (1989), *Class. Quant. Grav.* **6**, 2045.

[812] Sokolowski, L.M. (1997), in *Proceedings of the 14th International Conference on General Relativity and Gravitation*, Firenze, Italy 1995, Francaviglia, M., Longhi, G., Lusanna, L. and Sorace, E., editors, p. 337. Singapore: World Scientific.

[813] Sokolowski, L.M. and Carr, B. (1986), *Phys. Lett. B* **176**, 334.

[814] Sokolowski, L.M. and Golda, Z.A. (1987), *Phys. Lett. B* **195**, 349.

[815] Soleng, H.H. (1988), *Class. Quant. Grav.* **5**, 1489.

[816] Soleng, H.H. (1988), *Class. Quant. Grav.* **5**, 1501.

[817] Sonego, S. and Faraoni, V. (1993), *Class. Quant. Grav.* **10**, 1185.

[818] Soussa, M.E. and Woodard, R.P. (2003), preprint astro-ph/0308114.

[819] Spergel, D. and Pen, U.-L. (1997), *Astrophys. J. (Lett.)* **491**, L67.

[820] Spokoiny, B.J. (1984), *Phys. Lett. B* **147**, 39.

[821] Stabell, R. and Refsdal, S. (1966), *Mon. Not. R. Astr. Soc* **132**, 379.

[822] Starobinsky, A.A. (1980), *Phys. Lett. B* **91**, 99.

[823] Starobinsky, A.A. (1980), *Sov. Astron. (Lett.)* **10**, 135.

[824] Starobinsky, A.A. (1981), *Sov. Astron. (Lett.)* **7**, 36.

[825] Starobinsky, A.A. (1983), *Sov. Phys. JETP (Lett.)* **37**, 66.

[826] Starobinsky, A.A. (1987), in *Proceedings of the 4th Seminar on Quantum Gravity*, Markov, M.A. and Frolov, V.P., editors. Singapore: World Scientific.

[827] Starobinsky, A.A. (1998), *Sov. Phys. JETP (Lett.)* **68**, 757.

[828] Starobinsky, A.A. (2000), *Gravit. Cosmol.* **6**, 157.

[829] Starobinsky, A.A. and Yokoyama, J. (1995), in *Proceedings of the 4th Workshop on General Relativity and Gravitation*, Nakao, K. et al., editors, p. 381. Tokyo: Yukawa Institute.

[830] Steinhardt, P.J. and Accetta, F.S. (1990), *Phys. Rev. Lett.* **64**, 2740.

[831] Steinhardt, P.J., Wang, L. and Zlatev, I. (1999), *Phys. Rev. D* **59**, 123504.

[832] Stelle, K.S. (1977), in *Evolution of the Universe and Its Observational Quest*, Proceedings, Yamada, Japan, Sato, H. et al., editors. Tokyo: Universal Academy Press.

[833] Streater, R. and Wightman, A. (1964), *PCT, Spin and Statistics, and All That*. New York: Benjamin.

[834] Suen, W.-M. and Will, C.M. (1988), *Phys. Lett. B* **205**, 447.

[835] Susperregi, M. (1997), *Phys. Rev. D* **55**, 560.

[836] Susperregi, M. (1997), preprint gr-qc/9712031.

[837] Susperregi, M. (1998), *Phys. Lett. B* **440**, 257.

[838] Susperregi, M. (1998), *Phys. Rev. D* **58**, 083512.

[839] Synge, J.L. (1955), *Relativity: the General Theory*. Amsterdam: North Holland.

REFERENCES

[840] Tauber, E.G. (1967), *J. Math. Phys.* **8**, 118.

[841] Taylor, T.R. and Veneziano, G. (1988), *Phys. Lett. B* **213**, 450.

[842] Taylor, T.R., Veneziano, G. and Yankielowicz, S. (1983), *Nucl. Phys. B* **218**, 493.

[843] Teller, E. (1948), *Phys. Rev.* **73**, 801.

[844] Teyssandier, P. (1995), *Phys. Rev. D* **52**, 6195.

[845] Teyssandier, P. and Tourrenc, P. (1983), *J. Math. Phys.* **24**, 2793.

[846] Thirry, Y.R. (1948), *Compt. Rend. Acad. Sci. Paris* **226**, 216.

[847] Thorsett, S.E. (1996), *Phys. Rev. Lett.* **77**, 1432.

[848] Tocchini-Valentini, D. and Amendola, L. (2002), *Phys. Rev. D* **65**, 063508.

[849] Tomita, K. and Ishihara, H. (1985), *Phys. Rev. D* **32**, 1935.

[850] Tonry, J.L. *et al.* (2001), preprint astro-ph/0105413.

[851] Toporensky, A.V. (1999), *Int. J. Mod. Phys. D* **8**, 739.

[852] Torres, D.F. (1995), *Phys. Lett. B* **359**, 249.

[853] Torres, D.F. (1996), preprint gr-qc/9612048.

[854] Torres, D.F. (1997), *Phys. Lett. A* **225**, 13.

[855] Torres, D.F. (2002), *Phys. Rev. D* **66**, 043522.

[856] Torres, D.F. and Vucetich, H. (1996), *Phys. Rev. D* **54**, 7373.

[857] Torres, D.F., Schunk, F.E. and Liddle, A.R. (1998), *Class. Quant. Grav.* **15**, 3701.

[858] Townsend, P.K. (2001), *JHEP* **0111**, 042.

[859] Tseytlin, A.A. and Vafa, C. (1992) *Nucl. Phys. B* **372**, 443.

[860] Tsujikawa, S. and Bassett, B. (2002), *Phys. Lett. B* **536**, 9.

[861] Tsujikawa, S., Maeda, K. and Torii, T. (2000), *Phys. Rev. D* **61**, 103501.

[862] Turner, M.S. (1993), in *Recent Directions in Particle Theory – Superstrings and Black Holes to the Standard Model*, Proceedings of the Theoretical Advanced Study Institute in Elementary Particle Physics, Boulder, Colorado 1992, Harvey, J. and Polchinski, J., editors. Singapore: World Scientific.

[863] Turner, M.S. and Riess. A. (2001), preprint astro-ph/0106051.

[864] Turner, M.S. and Widrow, E.W. (1988), *Phys. Rev. D* **37**, 3428.

[865] Turner, M.S., Salopek, D.S. and Kolb, E.W. (1990), *Phys. Rev. D* **42**, 3925.

[866] Urena-Lopez, L.A. and Matos, T. (2000), *Phys. Rev. D* **62**, 081302.

[867] Uzan, J.-P. (1999), *Phys. Rev. D* **59**, 123510.

[868] Vajk, J.P. (1969), *J. Math. Phys.* **10**, 1145.

[869] Van den Bergh, N. (1980), *Gen. Rel. Grav.* **12**, 863

[870] Van den Bergh, N. (1982), *Gen. Rel. Grav.* **14**, 17.

[871] Van den Bergh, N. (1983), *Gen. Rel. Grav.* **15**, 441.

[872] Van den Bergh, N. (1983), *Gen. Rel. Grav.* **15**, 449.

[873] Van den Bergh, N. (1983), *Gen. Rel. Grav.* **15**, 1043.

[874] Van den Bergh, N. (1983), *Gen. Rel. Grav.* **16**, 2191.

[875] Van den Bergh, N. (1986), *Lett. Math. Phys.* **12**, 43.

[876] Van der Bij, J.J. and Gleiser, M. (1987), *Phys. Lett. B* **194**, 482.

[877] Van der Bij, J.J. and Radu, E. (2000), *Nucl. Phys. B* **585**, 637.

[878] Veneziano, G. (1991), *Phys. Lett. B* **265**, 287.

[879] Viana, P.T.P. and Liddle, A.R. (1998), *Phys. Rev. D* **57**, 674.

[880] Vilenkin, A. (1984), *Phys. Rev. Lett.* **53**, 1016.

[881] Vilenkin, A. (2001), preprint hep-th/0106083.

[882] Visser, M. (1995), *Lorentzian Wormholes: from Einstein to Hawking*. Woodbury, New York: American Institute of Physics Press.

[883] Visser, M. (2002), *Mod. Phys. Lett. A* **17**, 977.

[884] Visser, M. and Barcelo, C. (2000), in *Proceedings COSMO-99, 3rd International Conference on Particle Physics and the Early Universe*, Trieste, Italy, 1999, Cotti, U., Jeannerot, R., Sejanovic, G. and Smirnov, A., editors. Singapore: World Scientific.

[885] Vollick, D.N. (2003), *Phys. Rev. D* **68**, 063510.

[886] Voloshin, M.B. and Dolgov, A.D. (1982), *Sov. J. Nucl. Phys.* **35**, 120.

[887] Wagoner, R.V. (1970), *Phys. Rev. D* **1**, 3209.

[888] Wagoner, R.V. (1973), *Astrophys. J.* **179**, 343.

[889] Wagoner, R.V., Fowler, W.A. and Hoyle, F. (1967), *Astrophys. J.* **148**, 3.

[890] Wainwright, J. and Ellis, G.F.R. (1997), *Dynamical Systems in Cosmology*. Cambridge: Cambridge University Press.

[891] Wald, R.M. (1983), *Phys. Rev. D* **28**, 2118.

REFERENCES

[892]   Wald, R.M. (1984), *General Relativity*. Chicago: Chicago University Press.

[893]   Walker, T.P. et al. (1991), *Astrophys. J.* **376**, 51.

[894]   Wands, D. (1994), *Class. Quant. Grav.* **11**, 269.

[895]   Wands, D. and Mimoso, J.P. (1995), *Phys. Rev. D* **52**, 5612.

[896]   Wands, D. Copeland, E.J. and Liddle, A.R. (1993), *Ann. N.Y. Acad. Sci.* **688**, 647.

[897]   Weinberg, E.J. (1989), *Phys. Rev. D* **40**, 3950.

[898]   Weinberg, S. (1972), *Gravitation and Cosmology*. New York: Wiley.

[899]   Weinberg, S. (1989), *Rev. Mod. Phys.* **61**, 1.

[900]   Weinberg, S. (1996), in *Proceedings, Critical Dialogues in Cosmology*, Princeton, NJ, 1996, Turok, N., editor, p. 195. Singapore: World Scientific.

[901]   Weinstein, S. (1996), *Phil. Sci.* **63**, S63.

[902]   Wesson, P.S. (1985), *Astron. Astrophys.* **151**, 276.

[903]   Wetterich, C. (1988), *Nucl. Phys. B* **302**, 668.

[904]   Wetterich, C. (1995), *Astron. Astrophys.* **301**, 321.

[905]   Weyl, H. (1919), *Ann. Phys. (Leipzig)* **59**, 101.

[906]   Weyl, H. (1922), *Space, Time, Matter*. London: Methuen.

[907]   Wheeler, J.A. (1961), in *Rendiconti della Società Italiana di Fisica, 11th Course of the Varenna Summer School*, p. 67. New York: Academic Press.

[908]   Whitt, B. (1984), *Phys. Lett. B* **145**, 176.

[909]   Will, C.M. (1993), *Theory and Experiment in Gravitational Physics*, revised edition. Cambridge: Cambridge University Press.

[910]   Will, C.M. (2001), in *Proceedings, SLAC Summer Institute on Particle Physics – From the Hubble Length to the Planck Length*, Stanford, CA, 1998, Burke, D., Dixon, L. and Prescott, C., editors, p. 15. Stanford: SLAC.

[911]   Winterberg, F. (1968), *Nuovo Cimento* **53B**, 1096.

[912]   Witten, E. (1985), *Phys. Lett. B* **155**, 151.

[913]   *WMAP* homepage   http://map.gsfc.nasa.gov/

[914]   Yoon, J.M. and Brill, D.R. (1990), *Class. Quant, Grav.* **7**, 1253.

[915]   Zee, F. (1979), *Phys. Rev. Lett.* **42**, 417.

[916]   Zel'dovich, Ya. B. (1970), *Sov. Phys. JETP (Lett.)* **12**, 307.

[917]   Zel'dovich, Ya. B. and Starobinsky, A.A. (1971), *Sov. Phys. JETP* **34**, 1159.

[918] Ziaeepour, H. (2000), preprint astro-ph/0002400.

[919] Ziaeepour, H. (2003), preprint astro-ph/0301640.

[920] Zimdhal, W., Schwarz, D.J., Balakin, A.B. and Pavón, D. (2001), *Phys. Rev. D* **64**, 063501.

[921] Zipoy, D.M. (1966), *Phys. Rev.* **142**, 825.

[922] Zipoy, D.M. and Bertotti, B. (1968), *Nuovo Cimento* **56B**, 195.

[923] Zlatev, I., Wang, L. and Steinhardt, P.J. (1999), *Phys. Rev. Lett.* **82**, 896.

[924] Zuhk, A. (1996), *Gravit. Cosmol.* **2**, 319.

[925] Zuhk, A. (1997), *Gravit. Cosmol.* **3**, 24.

Index

ADM decomposition, 34
ADM mass, 34, 61
Annihilation operator, 131, 138, 185
Anomalous coupling, 55, 69–70, 81
Antigravity, 49, 156
Attractor mechanism, 26, 124, 174
Barker's theory, 77
BD field equations, 11, 65, 74, 76, 99, 104
BD parameter, 13–16, 22, 27, 31, 42, 104, 106, 118, 222–224, 226
BD quintessence, 223–224
Bianchi model, 20, 29, 50, 109–110, 112–113, 192
Big bang, 5, 73, 93–95, 97, 107, 115
Big bubble problem, 118–121
Big smash, 219–221, 228
Blue spectrum, 121, 140–141, 192
Boomerang, 128, 197, 227
Bosonic string theory, 2, 10, 29, 56, 217
Boson star, 69, 147
Brane-world, 31, 200, 205
Brans-Dicke theory, 1–3, 8, 38, 95, 215, 220
Canonical kinetic energy, 23
Cauchy problem, 49–50, 66
Chaotic inflation, 145–146, 148, 161, 164, 168
Chaplygin gas, 200
Coasting universe, 44, 106–107, 109, 223
COBE, 44, 119, 127, 146, 173, 183, 227
Comoving wavelength, 127
Compactification, 3–4, 10, 19, 30–31, 42
Conformal coupling, 143, 151, 167–168, 183
Conformal factor, 34, 36, 38, 40, 47, 51
Conformal invariance, 15, 29, 36, 56, 59, 149
Conformal time, 60, 109, 182, 185, 195
Conformal transformation, 14, 19, 33–34, 37–42, 46–48, 52, 55–56, 58, 61, 63, 67–70, 72, 84, 149, 160–163, 171–173
Conservation equation, 1, 5, 7, 12, 15, 24, 36–41, 76, 78, 81, 154, 216

Consistency relation, 141
Cosmic coincidence problem, 199, 201–202
Cosmic microwave background, 43, 45, 115, 118–120, 127, 134, 141, 145–146, 183, 189, 193, 206, 210, 212–213, 225, 227
Cosmic no-hair theorems, 7, 20, 108, 192–193
Cosmological birifringence, 207
Cosmological constant, 5, 7–8, 20, 36, 43, 46, 72, 94, 100, 102, 106–108, 113, 116–117, 119, 121, 165–166, 168, 176, 192, 198–199, 202, 216, 221
Cosmological constant problem, 192, 221
Coupled quintessence, 205
Coupling function, 26–27, 77–78, 120, 224–225
Creation operator, 131, 138, 185
Dark energy, 3, 70, 198–199, 206, 221
Dark matter, 198, 206, 210–211
Density perturbations, 8, 43–44, 61, 83, 119–121, 184, 206, 221, 223, 225–226
De Sitter space, 3, 7, 35, 95, 97, 103–104, 106, 108–109, 117, 133, 140, 146, 151, 163–164, 173–174, 176–178, 180–183, 193, 195, 215, 218–219
Dilaton, 2, 17, 29–30, 41–42, 87, 205, 217, 220
Doppler peaks, 193, 212–213, 225, 227
Duality, 13, 29, 97, 204
Effective stress-energy tensor, 55, 63–66, 74–76, 129, 151, 157, 159
Einstein-Cartan theory, 45
Einstein frame, 20, 30, 33, 38–42, 45–50, 52–53, 55–58, 60–64, 66–70, 73–74, 78–79, 81, 83–85, 90, 92–93, 113, 123, 160–164, 172, 206
Energy conditions, 61, 71–73, 151, 157, 219
Equivalence principle, 30, 39, 41, 46, 69, 85, 149–151, 200, 217

265

Euler-Lagrange equations, 28
Exponential potential, 22, 43–44, 169, 203
Extended inflation, 31, 100, 106, 116, 119–120, 221
Extended quintessence, 208, 221
Extra spatial dimensions, 3, 10, 17–20, 42
False vacuum, 116, 118
Fifth force, 41, 61, 69–70, 206
Freeze-out, 122–123
Friedmann equations, 6, 157, 165, 220
Friedmann-Lemaitre-Robertson-Walker metric, 5
Gauge-dependence, 128, 176
Gauge function, 20–21
Gauge-invariance, 129
Generalized inflation, 145–146, 164–165, 174–175, 190
Generalized Nariai solution, 102
Generalized scalar-tensor theories, 32–33, 47, 77
General relativity, 5, 8–10, 13, 18–19, 21, 24, 27, 39, 49, 65, 67, 83, 87, 90, 105, 109, 116–117, 119, 121, 123–124, 147, 149, 151, 157, 164, 199, 208, 221, 223, 225
Geodesic deviation, 39–40, 85–86
Geodesic deviation equation, 30, 85
Geodesic equation, 30, 37, 40–41, 86, 154
Geometric reheating, 148
Graceful exit, 3, 118
Gravitational coupling, 1–2, 9, 14, 16–18, 26, 31, 48, 57, 75–78, 95, 119, 153, 155, 171, 177, 192, 210, 220–224, 226
Gravitational dragging, 211
Gravitational lensing, 87, 89, 208, 225
Gravitational wave background, 91–92
Gravitational waves, 8, 35, 40, 62–65, 75, 83–84, 87–89, 91, 118, 140, 159, 183, 191–192, 205, 210
Graviton, 22, 29, 49–50, 66, 68
Green function, 68, 150
GUTs, 148
Hamiltonian constraint, 6, 24, 28, 49–50, 159, 171, 175, 177, 194
Harrison-Zeldovich spectrum, 119, 121, 136, 191
Heisenberg picture, 130, 185
Hesitating universe, 107
Horizon, 127, 134, 164, 188
Horizon crossing, 127, 136, 172, 188
Huygens' principle, 149
Hyperextended inflation, 26, 31, 120
Imperfect fluid, 75, 129
Improved energy momentum tensor, 76
Induced gravity, 2, 22, 31, 50, 213, 215
Inflation, 3, 6–8, 19–20, 43, 50, 72, 83, 108, 116–122, 127, 129, 132, 134–136, 141, 145, 148, 156, 163–165, 168, 172–174, 190, 205, 214, 220
Initial data, 12–13, 28
Intermediate inflation, 121, 169
Isotropization, 20, 110, 113
Jordan frame, 11, 19, 22, 33, 35, 38–39, 42, 45–46, 49–50, 55–70, 72–74, 76, 78–79, 84–85, 87, 89–90, 94, 121, 160–164
Kaluza-Klein theory, 17, 19–20, 22, 42
K-essence, 199, 216
Klein-Gordon equation, 7, 35, 143–144, 146, 160, 162, 167–169, 175, 194, 210–211
Large number hypothesis, 1
Laser interferometers, 87, 141, 205
Late time mild inflation, 168
Limit to general relativity, 13
Lyra's geometry, 20
Mach's principle, 2, 9, 16, 22, 94
MAXIMA, 128, 197, 227
Minimal coupling, 141, 143, 163, 166–168, 170, 176, 192
Nariai solution, 98–100, 102, 105–106, 117, 222
Noether symmetries, 27, 113
Nonminimal coupling, 72, 143, 145–147, 156, 162, 164–171, 174–175, 178, 190, 192–193, 203, 208–210, 212–213, 217–218, 220, 222
Nonminimally coupled quintessence, 211, 219
O'Hanlon-Tupper solution, 96–97, 102, 107
Optical scalars, 88
Parametrized Post-Newtonian (PPN) formalism, 3, 77
Perfect fluid, 5, 11–12, 24, 30, 37, 66, 93–94, 98, 110, 112, 151, 214
Perturbations, 127
Phantom energy, 71, 208, 214
Physical wavelength, 127
PLANCK, 146, 183
Planck length, 121, 220
Planck mass, 4, 39
Planck scale, 172, 198
Plausible double inflation, 119
Pole-like inflation, 121–122, 140, 190, 215, 217
Post-Newtonian parameters, 125
Power-law attractor, 3, 224
Power-law inflation, 44, 106, 117, 122, 146, 169, 173
Power spectrum, 133–134, 138–139, 188, 191
Pre-big bang cosmology, 99, 121–122, 215, 220
Primordial nucleosynthesis, 115, 122, 124, 147, 199
Quantum gravity, 32, 116
Quintessence, 19, 72, 76–77, 95, 156, 164–165, 169, 171–172, 193, 197–202,

INDEX

204–205, 207–211, 213–217, 219–225, 227
Quintessential inflation, 205, 210
Ratra-Peebles potential, 169, 202, 204, 209–210, 213, 225
Rees-Sciama effect, 210, 213
Reheating, 8, 117, 207, 213
Scalar gravitational waves, 83, 159
Scalar perturbations, 127–128, 130, 135–136, 141
Scalar-tensor theory, 1, 5, 22–23, 26–27, 29, 45, 57, 61, 83–84, 87, 93, 112, 213, 221, 225–226
Singularities, 60–61, 73–74
Slow-roll approximation, 7, 117, 120, 132, 135, 146, 162–163, 173–174, 178, 182, 190
Slow-roll inflation, 135
Slow-roll parameters, 8, 135, 140, 178, 182, 191, 195
Spectral index, 44, 119, 135–136, 140, 189–192
Speedup factor, 123–124, 199
Spherical harmonics, 128, 141
Stiff fluid, 94, 113, 170
Stochastic inflation, 119
String frame, 29
String theory, 3, 13, 15, 29–30, 41–42, 56, 99, 103, 116

Strong coupling, 166, 172, 224
Superacceleration, 103, 214–217, 219–222, 228
Supergravity, 3, 29, 43, 47, 68–69, 165, 202, 205, 212, 216
Superinflation, 214
Superquintessence, 71, 208, 213, 216, 221
Temperature fluctuations, 43, 127, 134–135, 141, 146, 183, 189, 210, 227
Tensor perturbations, 127, 136, 139–141
Tensor spherical harmonics, 128
Topological defects, 199
Torsion, 4, 20–21, 45
Tracking solution, 201–204, 209, 222, 224
True vacuum bubbles, 116–118, 120
Tunneling, 116, 118, 120
Type Ia supernovae, 165, 197, 207
Vacillating universe, 103, 107
Vacuum energy, 101, 118, 120, 199, 201
Vacuum state, 61, 116, 130, 133, 140, 185, 188
Vector spherical harmonics, 128
Viscosity, 151, 199, 216
Weak coupling, 172, 209
Weak field limit, 3, 10, 14, 48, 154
Weyl tensor, 35, 88
Weyl theory, 47, 57
WMAP, 146, 183, 227
Wormholes, 71, 147, 151, 153, 192, 194

Fundamental Theories of Physics

Series Editor: Alwyn van der Merwe, University of Denver, USA

1. M. Sachs: *General Relativity and Matter.* A Spinor Field Theory from Fermis to Light-Years. With a Foreword by C. Kilmister. 1982   ISBN 90-277-1381-2
2. G.H. Duffey: *A Development of Quantum Mechanics.* Based on Symmetry Considerations. 1985   ISBN 90-277-1587-4
3. S. Diner, D. Fargue, G. Lochak and F. Selleri (eds.): *The Wave-Particle Dualism.* A Tribute to Louis de Broglie on his 90th Birthday. 1984   ISBN 90-277-1664-1
4. E. Prugovečki: *Stochastic Quantum Mechanics and Quantum Spacetime.* A Consistent Unification of Relativity and Quantum Theory based on Stochastic Spaces. 1984; 2nd printing 1986   ISBN 90-277-1617-X
5. D. Hestenes and G. Sobczyk: *Clifford Algebra to Geometric Calculus.* A Unified Language for Mathematics and Physics. 1984   ISBN 90-277-1673-0; Pb (1987) 90-277-2561-6
6. P. Exner: *Open Quantum Systems and Feynman Integrals.* 1985   ISBN 90-277-1678-1
7. L. Mayants: *The Enigma of Probability and Physics.* 1984   ISBN 90-277-1674-9
8. E. Tocaci: *Relativistic Mechanics, Time and Inertia.* Translated from Romanian. Edited and with a Foreword by C.W. Kilmister. 1985   ISBN 90-277-1769-9
9. B. Bertotti, F. de Felice and A. Pascolini (eds.): *General Relativity and Gravitation.* Proceedings of the 10th International Conference (Padova, Italy, 1983). 1984   ISBN 90-277-1819-9
10. G. Tarozzi and A. van der Merwe (eds.): *Open Questions in Quantum Physics.* 1985   ISBN 90-277-1853-9
11. J.V. Narlikar and T. Padmanabhan: *Gravity, Gauge Theories and Quantum Cosmology.* 1986   ISBN 90-277-1948-9
12. G.S. Asanov: *Finsler Geometry, Relativity and Gauge Theories.* 1985   ISBN 90-277-1960-8
13. K. Namsrai: *Nonlocal Quantum Field Theory and Stochastic Quantum Mechanics.* 1986   ISBN 90-277-2001-0
14. C. Ray Smith and W.T. Grandy, Jr. (eds.): *Maximum-Entropy and Bayesian Methods in Inverse Problems.* Proceedings of the 1st and 2nd International Workshop (Laramie, Wyoming, USA). 1985   ISBN 90-277-2074-6
15. D. Hestenes: *New Foundations for Classical Mechanics.* 1986   ISBN 90-277-2090-8; Pb (1987) 90-277-2526-8
16. S.J. Prokhovnik: *Light in Einstein's Universe.* The Role of Energy in Cosmology and Relativity. 1985   ISBN 90-277-2093-2
17. Y.S. Kim and M.E. Noz: *Theory and Applications of the Poincaré Group.* 1986   ISBN 90-277-2141-6
18. M. Sachs: *Quantum Mechanics from General Relativity.* An Approximation for a Theory of Inertia. 1986   ISBN 90-277-2247-1
19. W.T. Grandy, Jr.: *Foundations of Statistical Mechanics.* Vol. I: *Equilibrium Theory.* 1987   ISBN 90-277-2489-X
20. H.-H von Borzeszkowski and H.-J. Treder: *The Meaning of Quantum Gravity.* 1988   ISBN 90-277-2518-7
21. C. Ray Smith and G.J. Erickson (eds.): *Maximum-Entropy and Bayesian Spectral Analysis and Estimation Problems.* Proceedings of the 3rd International Workshop (Laramie, Wyoming, USA, 1983). 1987   ISBN 90-277-2579-9
22. A.O. Barut and A. van der Merwe (eds.): *Selected Scientific Papers of Alfred Landé.* [1888-1975]. 1988   ISBN 90-277-2594-2

Fundamental Theories of Physics

23. W.T. Grandy, Jr.: *Foundations of Statistical Mechanics.* Vol. II: *Nonequilibrium Phenomena.* 1988　ISBN 90-277-2649-3
24. E.I. Bitsakis and C.A. Nicolaides (eds.): *The Concept of Probability.* Proceedings of the Delphi Conference (Delphi, Greece, 1987). 1989　ISBN 90-277-2679-5
25. A. van der Merwe, F. Selleri and G. Tarozzi (eds.): *Microphysical Reality and Quantum Formalism, Vol. 1.* Proceedings of the International Conference (Urbino, Italy, 1985). 1988　ISBN 90-277-2683-3
26. A. van der Merwe, F. Selleri and G. Tarozzi (eds.): *Microphysical Reality and Quantum Formalism, Vol. 2.* Proceedings of the International Conference (Urbino, Italy, 1985). 1988　ISBN 90-277-2684-1
27. I.D. Novikov and V.P. Frolov: *Physics of Black Holes.* 1989　ISBN 90-277-2685-X
28. G. Tarozzi and A. van der Merwe (eds.): *The Nature of Quantum Paradoxes.* Italian Studies in the Foundations and Philosophy of Modern Physics. 1988　ISBN 90-277-2703-1
29. B.R. Iyer, N. Mukunda and C.V. Vishveshwara (eds.): *Gravitation, Gauge Theories and the Early Universe.* 1989　ISBN 90-277-2710-4
30. H. Mark and L. Wood (eds.): *Energy in Physics, War and Peace.* A Festschrift celebrating Edward Teller's 80th Birthday. 1988　ISBN 90-277-2775-9
31. G.J. Erickson and C.R. Smith (eds.): *Maximum-Entropy and Bayesian Methods in Science and Engineering.* Vol. I: *Foundations.* 1988　ISBN 90-277-2793-7
32. G.J. Erickson and C.R. Smith (eds.): *Maximum-Entropy and Bayesian Methods in Science and Engineering.* Vol. II: *Applications.* 1988　ISBN 90-277-2794-5
33. M.E. Noz and Y.S. Kim (eds.): *Special Relativity and Quantum Theory.* A Collection of Papers on the Poincaré Group. 1988　ISBN 90-277-2799-6
34. I.Yu. Kobzarev and Yu.I. Manin: *Elementary Particles. Mathematics, Physics and Philosophy.* 1989　ISBN 0-7923-0098-X
35. F. Selleri: *Quantum Paradoxes and Physical Reality.* 1990　ISBN 0-7923-0253-2
36. J. Skilling (ed.): *Maximum-Entropy and Bayesian Methods.* Proceedings of the 8th International Workshop (Cambridge, UK, 1988). 1989　ISBN 0-7923-0224-9
37. M. Kafatos (ed.): *Bell's Theorem, Quantum Theory and Conceptions of the Universe.* 1989　ISBN 0-7923-0496-9
38. Yu.A. Izyumov and V.N. Syromyatnikov: *Phase Transitions and Crystal Symmetry.* 1990　ISBN 0-7923-0542-6
39. P.F. Fougère (ed.): *Maximum-Entropy and Bayesian Methods.* Proceedings of the 9th International Workshop (Dartmouth, Massachusetts, USA, 1989). 1990　ISBN 0-7923-0928-6
40. L. de Broglie: *Heisenberg's Uncertainties and the Probabilistic Interpretation of Wave Mechanics.* With Critical Notes of the Author. 1990　ISBN 0-7923-0929-4
41. W.T. Grandy, Jr.: *Relativistic Quantum Mechanics of Leptons and Fields.* 1991　ISBN 0-7923-1049-7
42. Yu.L. Klimontovich: *Turbulent Motion and the Structure of Chaos.* A New Approach to the Statistical Theory of Open Systems. 1991　ISBN 0-7923-1114-0
43. W.T. Grandy, Jr. and L.H. Schick (eds.): *Maximum-Entropy and Bayesian Methods.* Proceedings of the 10th International Workshop (Laramie, Wyoming, USA, 1990). 1991　ISBN 0-7923-1140-X
44. P. Pták and S. Pulmannová: *Orthomodular Structures as Quantum Logics.* Intrinsic Properties, State Space and Probabilistic Topics. 1991　ISBN 0-7923-1207-4
45. D. Hestenes and A. Weingartshofer (eds.): *The Electron.* New Theory and Experiment. 1991　ISBN 0-7923-1356-9

Fundamental Theories of Physics

46. P.P.J.M. Schram: *Kinetic Theory of Gases and Plasmas.* 1991 ISBN 0-7923-1392-5
47. A. Micali, R. Boudet and J. Helmstetter (eds.): *Clifford Algebras and their Applications in Mathematical Physics.* 1992 ISBN 0-7923-1623-1
48. E. Prugovečki: *Quantum Geometry.* A Framework for Quantum General Relativity. 1992 ISBN 0-7923-1640-1
49. M.H. Mac Gregor: *The Enigmatic Electron.* 1992 ISBN 0-7923-1982-6
50. C.R. Smith, G.J. Erickson and P.O. Neudorfer (eds.): *Maximum Entropy and Bayesian Methods.* Proceedings of the 11th International Workshop (Seattle, 1991). 1993 ISBN 0-7923-2031-X
51. D.J. Hoekzema: *The Quantum Labyrinth.* 1993 ISBN 0-7923-2066-2
52. Z. Oziewicz, B. Jancewicz and A. Borowiec (eds.): *Spinors, Twistors, Clifford Algebras and Quantum Deformations.* Proceedings of the Second Max Born Symposium (Wrocław, Poland, 1992). 1993 ISBN 0-7923-2251-7
53. A. Mohammad-Djafari and G. Demoment (eds.): *Maximum Entropy and Bayesian Methods.* Proceedings of the 12th International Workshop (Paris, France, 1992). 1993 ISBN 0-7923-2280-0
54. M. Riesz: *Clifford Numbers and Spinors* with Riesz' Private Lectures to E. Folke Bolinder and a Historical Review by Pertti Lounesto. E.F. Bolinder and P. Lounesto (eds.). 1993 ISBN 0-7923-2299-1
55. F. Brackx, R. Delanghe and H. Serras (eds.): *Clifford Algebras and their Applications in Mathematical Physics.* Proceedings of the Third Conference (Deinze, 1993) 1993 ISBN 0-7923-2347-5
56. J.R. Fanchi: *Parametrized Relativistic Quantum Theory.* 1993 ISBN 0-7923-2376-9
57. A. Peres: *Quantum Theory: Concepts and Methods.* 1993 ISBN 0-7923-2549-4
58. P.L. Antonelli, R.S. Ingarden and M. Matsumoto: *The Theory of Sprays and Finsler Spaces with Applications in Physics and Biology.* 1993 ISBN 0-7923-2577-X
59. R. Miron and M. Anastasiei: *The Geometry of Lagrange Spaces: Theory and Applications.* 1994 ISBN 0-7923-2591-5
60. G. Adomian: *Solving Frontier Problems of Physics: The Decomposition Method.* 1994 ISBN 0-7923-2644-X
61. B.S. Kerner and V.V. Osipov: *Autosolitons.* A New Approach to Problems of Self-Organization and Turbulence. 1994 ISBN 0-7923-2816-7
62. G.R. Heidbreder (ed.): *Maximum Entropy and Bayesian Methods.* Proceedings of the 13th International Workshop (Santa Barbara, USA, 1993) 1996 ISBN 0-7923-2851-5
63. J. Peřina, Z. Hradil and B. Jurčo: *Quantum Optics and Fundamentals of Physics.* 1994 ISBN 0-7923-3000-5
64. M. Evans and J.-P. Vigier: *The Enigmatic Photon.* Volume 1: The Field $\boldsymbol{B}^{(3)}$. 1994 ISBN 0-7923-3049-8
65. C.K. Raju: *Time: Towards a Constistent Theory.* 1994 ISBN 0-7923-3103-6
66. A.K.T. Assis: *Weber's Electrodynamics.* 1994 ISBN 0-7923-3137-0
67. Yu. L. Klimontovich: *Statistical Theory of Open Systems.* Volume 1: A Unified Approach to Kinetic Description of Processes in Active Systems. 1995 ISBN 0-7923-3199-0; Pb: ISBN 0-7923-3242-3
68. M. Evans and J.-P. Vigier: *The Enigmatic Photon.* Volume 2: Non-Abelian Electrodynamics. 1995 ISBN 0-7923-3288-1
69. G. Esposito: *Complex General Relativity.* 1995 ISBN 0-7923-3340-3

Fundamental Theories of Physics

70. J. Skilling and S. Sibisi (eds.): *Maximum Entropy and Bayesian Methods.* Proceedings of the Fourteenth International Workshop on Maximum Entropy and Bayesian Methods. 1996
ISBN 0-7923-3452-3
71. C. Garola and A. Rossi (eds.): *The Foundations of Quantum Mechanics Historical Analysis and Open Questions.* 1995 ISBN 0-7923-3480-9
72. A. Peres: *Quantum Theory: Concepts and Methods.* 1995 (see for hardback edition, Vol. 57)
ISBN Pb 0-7923-3632-1
73. M. Ferrero and A. van der Merwe (eds.): *Fundamental Problems in Quantum Physics.* 1995
ISBN 0-7923-3670-4
74. F.E. Schroeck, Jr.: *Quantum Mechanics on Phase Space.* 1996 ISBN 0-7923-3794-8
75. L. de la Peña and A.M. Cetto: *The Quantum Dice.* An Introduction to Stochastic Electrodynamics. 1996 ISBN 0-7923-3818-9
76. P.L. Antonelli and R. Miron (eds.): *Lagrange and Finsler Geometry.* Applications to Physics and Biology. 1996 ISBN 0-7923-3873-1
77. M.W. Evans, J.-P. Vigier, S. Roy and S. Jeffers: *The Enigmatic Photon.* Volume 3: Theory and Practice of the $B^{(3)}$ Field. 1996 ISBN 0-7923-4044-2
78. W.G.V. Rosser: *Interpretation of Classical Electromagnetism.* 1996 ISBN 0-7923-4187-2
79. K.M. Hanson and R.N. Silver (eds.): *Maximum Entropy and Bayesian Methods.* 1996
ISBN 0-7923-4311-5
80. S. Jeffers, S. Roy, J.-P. Vigier and G. Hunter (eds.): *The Present Status of the Quantum Theory of Light.* Proceedings of a Symposium in Honour of Jean-Pierre Vigier. 1997
ISBN 0-7923-4337-9
81. M. Ferrero and A. van der Merwe (eds.): *New Developments on Fundamental Problems in Quantum Physics.* 1997 ISBN 0-7923-4374-3
82. R. Miron: *The Geometry of Higher-Order Lagrange Spaces.* Applications to Mechanics and Physics. 1997 ISBN 0-7923-4393-X
83. T. Hakioğlu and A.S. Shumovsky (eds.): *Quantum Optics and the Spectroscopy of Solids.* Concepts and Advances. 1997 ISBN 0-7923-4414-6
84. A. Sitenko and V. Tartakovskii: *Theory of Nucleus.* Nuclear Structure and Nuclear Interaction. 1997 ISBN 0-7923-4423-5
85. G. Esposito, A.Yu. Kamenshchik and G. Pollifrone: *Euclidean Quantum Gravity on Manifolds with Boundary.* 1997 ISBN 0-7923-4472-3
86. R.S. Ingarden, A. Kossakowski and M. Ohya: *Information Dynamics and Open Systems.* Classical and Quantum Approach. 1997 ISBN 0-7923-4473-1
87. K. Nakamura: *Quantum versus Chaos.* Questions Emerging from Mesoscopic Cosmos. 1997
ISBN 0-7923-4557-6
88. B.R. Iyer and C.V. Vishveshwara (eds.): *Geometry, Fields and Cosmology.* Techniques and Applications. 1997 ISBN 0-7923-4725-0
89. G.A. Martynov: *Classical Statistical Mechanics.* 1997 ISBN 0-7923-4774-9
90. M.W. Evans, J.-P. Vigier, S. Roy and G. Hunter (eds.): *The Enigmatic Photon.* Volume 4: New Directions. 1998 ISBN 0-7923-4826-5
91. M. Rédei: *Quantum Logic in Algebraic Approach.* 1998 ISBN 0-7923-4903-2
92. S. Roy: *Statistical Geometry and Applications to Microphysics and Cosmology.* 1998
ISBN 0-7923-4907-5
93. B.C. Eu: *Nonequilibrium Statistical Mechanics.* Ensembled Method. 1998
ISBN 0-7923-4980-6

Fundamental Theories of Physics

94. V. Dietrich, K. Habetha and G. Jank (eds.): *Clifford Algebras and Their Application in Mathematical Physics.* Aachen 1996. 1998   ISBN 0-7923-5037-5
95. J.P. Blaizot, X. Campi and M. Ploszajczak (eds.): *Nuclear Matter in Different Phases and Transitions.* 1999   ISBN 0-7923-5660-8
96. V.P. Frolov and I.D. Novikov: *Black Hole Physics.* Basic Concepts and New Developments. 1998   ISBN 0-7923-5145-2; Pb 0-7923-5146
97. G. Hunter, S. Jeffers and J-P. Vigier (eds.): *Causality and Locality in Modern Physics.* 1998   ISBN 0-7923-5227-0
98. G.J. Erickson, J.T. Rychert and C.R. Smith (eds.): *Maximum Entropy and Bayesian Methods.* 1998   ISBN 0-7923-5047-2
99. D. Hestenes: *New Foundations for Classical Mechanics (Second Edition).* 1999   ISBN 0-7923-5302-1; Pb ISBN 0-7923-5514-8
100. B.R. Iyer and B. Bhawal (eds.): *Black Holes, Gravitational Radiation and the Universe.* Essays in Honor of C. V. Vishveshwara. 1999   ISBN 0-7923-5308-0
101. P.L. Antonelli and T.J. Zastawniak: *Fundamentals of Finslerian Diffusion with Applications.* 1998   ISBN 0-7923-5511-3
102. H. Atmanspacher, A. Amann and U. Müller-Herold: *On Quanta, Mind and Matter Hans Primas in Context.* 1999   ISBN 0-7923-5696-9
103. M.A. Trump and W.C. Schieve: *Classical Relativistic Many-Body Dynamics.* 1999   ISBN 0-7923-5737-X
104. A.I. Maimistov and A.M. Basharov: *Nonlinear Optical Waves.* 1999   ISBN 0-7923-5752-3
105. W. von der Linden, V. Dose, R. Fischer and R. Preuss (eds.): *Maximum Entropy and Bayesian Methods Garching, Germany 1998.* 1999   ISBN 0-7923-5766-3
106. M.W. Evans: *The Enigmatic Photon Volume 5: O(3) Electrodynamics.* 1999   ISBN 0-7923-5792-2
107. G.N. Afanasiev: *Topological Effects in Quantum Mecvhanics.* 1999   ISBN 0-7923-5800-7
108. V. Devanathan: *Angular Momentum Techniques in Quantum Mechanics.* 1999   ISBN 0-7923-5866-X
109. P.L. Antonelli (ed.): *Finslerian Geometries A Meeting of Minds.* 1999   ISBN 0-7923-6115-6
110. M.B. Mensky: *Quantum Measurements and Decoherence Models and Phenomenology.* 2000   ISBN 0-7923-6227-6
111. B. Coecke, D. Moore and A. Wilce (eds.): *Current Research in Operation Quantum Logic.* Algebras, Categories, Languages. 2000   ISBN 0-7923-6258-6
112. G. Jumarie: *Maximum Entropy, Information Without Probability and Complex Fractals.* Classical and Quantum Approach. 2000   ISBN 0-7923-6330-2
113. B. Fain: *Irreversibilities in Quantum Mechanics.* 2000   ISBN 0-7923-6581-X
114. T. Borne, G. Lochak and H. Stumpf: *Nonperturbative Quantum Field Theory and the Structure of Matter.* 2001   ISBN 0-7923-6803-7
115. J. Keller: *Theory of the Electron.* A Theory of Matter from START. 2001   ISBN 0-7923-6819-3
116. M. Rivas: *Kinematical Theory of Spinning Particles.* Classical and Quantum Mechanical Formalism of Elementary Particles. 2001   ISBN 0-7923-6824-X
117. A.A. Ungar: *Beyond the Einstein Addition Law and its Gyroscopic Thomas Precession.* The Theory of Gyrogroups and Gyrovector Spaces. 2001   ISBN 0-7923-6909-2
118. R. Miron, D. Hrimiuc, H. Shimada and S.V. Sabau: *The Geometry of Hamilton and Lagrange Spaces.* 2001   ISBN 0-7923-6926-2

Fundamental Theories of Physics

119. M. Pavšič: *The Landscape of Theoretical Physics: A Global View.* From Point Particles to the Brane World and Beyond in Search of a Unifying Principle. 2001   ISBN 0-7923-7006-6
120. R.M. Santilli: *Foundations of Hadronic Chemistry.* With Applications to New Clean Energies and Fuels. 2001   ISBN 1-4020-0087-1
121. S. Fujita and S. Godoy: *Theory of High Temperature Superconductivity.* 2001
    ISBN 1-4020-0149-5
122. R. Luzzi, A.R. Vasconcellos and J. Galvão Ramos: *Predictive Statitical Mechanics.* A Nonequilibrium Ensemble Formalism. 2002   ISBN 1-4020-0482-6
123. V.V. Kulish: *Hierarchical Methods.* Hierarchy and Hierarchical Asymptotic Methods in Electrodynamics, Volume 1. 2002   ISBN 1-4020-0757-4; Set: 1-4020-0758-2
124. B.C. Eu: *Generalized Thermodynamics.* Thermodynamics of Irreversible Processes and Generalized Hydrodynamics. 2002   ISBN 1-4020-0788-4
125. A. Mourachkine: *High-Temperature Superconductivity in Cuprates.* The Nonlinear Mechanism and Tunneling Measurements. 2002   ISBN 1-4020-0810-4
126. R.L. Amoroso, G. Hunter, M. Kafatos and J.-P. Vigier (eds.): *Gravitation and Cosmology: From the Hubble Radius to the Planck Scale.* Proceedings of a Symposium in Honour of the 80th Birthday of Jean-Pierre Vigier. 2002   ISBN 1-4020-0885-6
127. W.M. de Muynck: *Foundations of Quantum Mechanics, an Empiricist Approach.* 2002
    ISBN 1-4020-0932-1
128. V.V. Kulish: *Hierarchical Methods.* Undulative Electrodynamical Systems, Volume 2. 2002
    ISBN 1-4020-0968-2; Set: 1-4020-0758-2
129. M. Mugur-Schächter and A. van der Merwe (eds.): *Quantum Mechanics, Mathematics, Cognition and Action.* Proposals for a Formalized Epistemology. 2002   ISBN 1-4020-1120-2
130. P. Bandyopadhyay: *Geometry, Topology and Quantum Field Theory.* 2003
    ISBN 1-4020-1414-7
131. V. Garzó and A. Santos: *Kinetic Theory of Gases in Shear Flows.* Nonlinear Transport. 2003
    ISBN 1-4020-1436-8
132. R. Miron: *The Geometry of Higher-Order Hamilton Spaces.* Applications to Hamiltonian Mechanics. 2003   ISBN 1-4020-1574-7
133. S. Esposito, E. Majorana Jr., A. van der Merwe and E. Recami (eds.): *Ettore Majorana: Notes on Theoretical Physics.* 2003   ISBN 1-4020-1649-2
134. J. Hamhalter. *Quantum Measure Theory.* 2003   ISBN 1-4020-1714-6
135. G. Rizzi and M.L. Ruggiero: *Relativity in Rotating Frames.* Relativistic Physics in Rotating Reference Frames. 2004   ISBN 1-4020-1805-3
136. L. Kantorovich: *Quantum Theory of the Solid State: an Introduction.* 2004
    ISBN 1-4020-1821-5
137. A. Ghatak and S. Lokanathan: *Quantum Mechanics: Theory and Applications.* 2004
    ISBN 1-4020-1850-9
138. A. Khrennikov: *Information Dynamics in Cognitive, Psychological, Social, and Anomalous Phenomena.* 2004   ISBN 1-4020-1868-1
139. V. Faraoni: *Cosmology in Scalar-Tensor Gravity.* 2004   ISBN 1-4020-1988-2
140. P.P. Teodorescu and N.-A. P. Nicorovici: *Applications of the Theory of Groups in Mechanics and Physics.* 2004   ISBN 1-4020-2046-5